Ecological Restoration of Aquatic and Semi-Aquatic Ecosystems in the Netherlands (NW Europe)

Developments in Hydrobiology 166

Series editor
H. J. Dumont

Ecological Restoration of Aquatic and Semi-Aquatic Ecosystems in the Netherlands (NW Europe)

Edited by

P.H. Nienhuis[1,2] & R.D. Gulati[2]

[1] *Department of Environmental Studies, University of Nijmegen, The Netherlands*
[2] *Centre for Limnology, Netherlands Institute of Ecology, Nieuwersluis, The Netherlands*

Reprinted from Hydrobiologia, volume 478 (2002)

Kluwer Academic Publishers
Dordrecht / Boston / London

Library of Congress Cataloging-in-Publication Data

A C.I.P. Catalogue record for this book is available from the Library of Congress.

ISBN 1-4020-1023-0

Published by Kluwer Academic Publishers,
P.O. Box 17, 3300 AA Dordrecht, The Netherlands

Sold and distributed in North, Central and South America
by Kluwer Academic Publishers,
101 Philip Drive, Norwell, MA 02061, U.S.A.

In all other countries, sold and distributed
by Kluwer Academic Publishers Group,
P.O. Box 322, 3300 AH Dordrecht, The Netherlands

Printed on acid-free paper

Printed in the Netherlands

TABLE OF CONTENTS

Hydrobiologia **478:** vii, 2002.
P.H. Nienhuis & R.D. Gulati (eds), Ecological Restoration of Aquatic and Semi-Aquatic Ecosystems in the Netherlands (NW Europe).

Preface

'The Netherlands Water Land' is no commonplace cliché. This small, low-lying delta country, mainly a sedimentary deposit by a number of large European rivers, and situated on the North Sea, contains a large diversity of water bodies. These include both stagnant water and running water; waters in the climax stage of growing solid by peat formation, and shallow, exposed open water bodies; wide, slowly running, turbid rivers and narrow, overshadowed, clear rivulets. In 1989 a symposium was held, under the title 'Netherlands-Wetlands', aiming at presenting an overview of our present knowledge of the structure and functioning of the typical wetlands in the country. The proceedings were published in *Hydrobiologia* **265** (1993), as well as in *Developments in Hydrobiology* **88**. Major threats to the Dutch wetlands were analysed, and among them the annihilation of original ecosystems, the fragmentation of landscapes, and the acidification and eutrophication of waters and sediments. Emerging perspectives of recovery and restoration, together with a plea for a better insight into the governing ecological processes, were postulated in 1989.

The present multi-author volume, with its twelve main chapters, can be seen as a self-evident successor to the results of the 'Netherlands-Wetlands' symposium. The focus is now fully on the achievements and failures of ecological rehabilitation measures over the past 25 years. Many of these are, indeed, rather impressive. Restoration ecology appears to be a matter of trial and error: in general, progressive restoration endeavours are greatly benefiting from experiences in the past. Many of the completed studies now exemplify an integration of hydrological, geomorphological and ecological knowledge. Brian Moss, a well-known aquatic restoration ecologist stated in *Environmental Conservation* **27** (2000): "The unit of a freshwater system is not defined by the wetted perimeter of the lake or river but by the limits of the catchment area from which the water is drawn. Conservation of functional values is thus not possible without control of the way the catchment is managed". This general awareness is more and more becoming the dominant way of thinking among the Dutch water policymakers. The stakeholders' attitude is changing towards retreating and abandoning economic activities from vulnerable catchment areas. The trend is now to strive after economic 'win–win' situations, and to reduce human-induced flood hazards in favour of the rehabilitation of more 'natural' wetlands. Sustainable management of aquatic ecosystems means that user functions of the catchments (e.g. navigation, agriculture, urbanisation, recreation) should be accommodated to the dynamics of the natural wetland system, and not the other way around.

The contributions to this book are from invited Dutch experts, all having long standing experience in their respective fields. The authors are affiliated to the state and university research institutions in the Netherlands where restoration ecology and ecological restoration are important disciplines of research, the University of Nijmegen, the State University of Groningen, the University of Utrecht, the Netherlands Institute of Ecology, Alterra and the Netherlands Institute for Fisheries Research. The manuscripts were peer-reviewed by anonymous experts, following the procedures set out for *Hydrobiologia*.

PIET H. NIENHUIS, *Nijmegen*
RAMESH D. GULATI, *Nieuwersluis*

Hydrobiologia **478:** 1–6, 2002.
P.H. Nienhuis & R.D. Gulati (eds), Ecological Restoration of Aquatic and Semi-Aquatic Ecosystems in the Netherlands (NW Europe).

Ecological restoration of aquatic and semi-aquatic ecosystems in the Netherlands: an introduction*

P.H. Nienhuis[1,2] & R.D. Gulati[2]
[1]*Department of Environmental Studies, Faculty of Science, Mathematics and Informatics, University of Nijmegen, P.O. Box 9010, 6500 GL Nijmegen, The Netherlands*
[2]*Centre for Limnology, Netherlands Institute of Ecology, Rijksstraatweg 6, 3631 AC Nieuwersluis, The Netherlands*

Key words: ecological restoration, rehabilitation, aquatic and semi-aquatic ecosystems, The Netherlands

Abstract

In 1989, a symposium was held under the title 'Netherlands-Wetlands', aiming at the presentation of the state of the art of the existing knowledge of structure and functioning of wetlands, characteristic for the Netherlands. The present multi-author volume can be seen as a logical successor of the results of this symposium. The focus is now fully on the results of ecological restoration measures over the past 25 years. The main question to be answered is what makes a project a success in terms of ecological performance, and when are we speaking about a failure? The present volume contains 10 case studies, roughly covering all aquatic and semi-aquatic ecosystems in the Netherlands.

History

The pristine coastal environment of the Netherlands, i.e. before noticeable human impact, was dominated by a continuous sea-level rise since the last glacial period, roughly 10 000 to 15 000 years ago. The reconstruction of Zagwijn (1986; Fig. 1), depicting the geographic situation some 2000 years ago, shows an extensive area of raised bogs behind a solid dune ridge, which separated the North Sea from the dry land. Cultivation of these raised bogs by man, from the tenth century onwards, drastically changed that picture. For the purpose of the cultivation of their agricultural products, and locally for the purpose of excavating peat for fuel, the colonists were digging extensive networks of ditches and trenches in order to drain the raised bogs. Consequently the remaining peat layers oxidised, leading to compacting of the soil. We might say that the early inhabitants of the Netherlands, unaware of the consequences, made the raised bog areas accessible for the rising sea. By lowering the surface level of the ground, and the groundwater table, storm floods from the sea could have an increasing impact on the remaining land. Until the 15th century the sea took more than half of the land that was present in the tenth century. From the 17th century onwards, the reverse process took place: man successfully gained land from the sea and started with the draining and the reclamation of lakes (Edelman, 1974). In the course of the centuries, the artificial Netherlands originated, of which more than half would not exist without levees, dikes and pumping engines. The present map of the Netherlands (Fig. 2; Schultz, 1982) shows that 60% of the country have to be protected against flooding, either from the sea (50%) or from the rivers (10%).

There are three major factors to be mentioned that altered the original, natural situation of aquatic ecosystems in the Netherlands. (1) The cold temperate climate governing the seasonal changes in the Netherlands has led to a hydrological regime of low water levels in summer and high water levels during winter. These natural water level changes have been disturbed by management actions of man over the past 1000 years or so. Especially during winter, water tables are artificially kept as low as possible, in order to prevent inaccessibility of low-lying agricultural grounds early in spring. Another aspect is the general lowering of the water table by water distraction for all kinds of purposes, such as drinking water and water for industrial purposes, facilitated by a complicated artificial

* NIOO Contribution no. 3053.

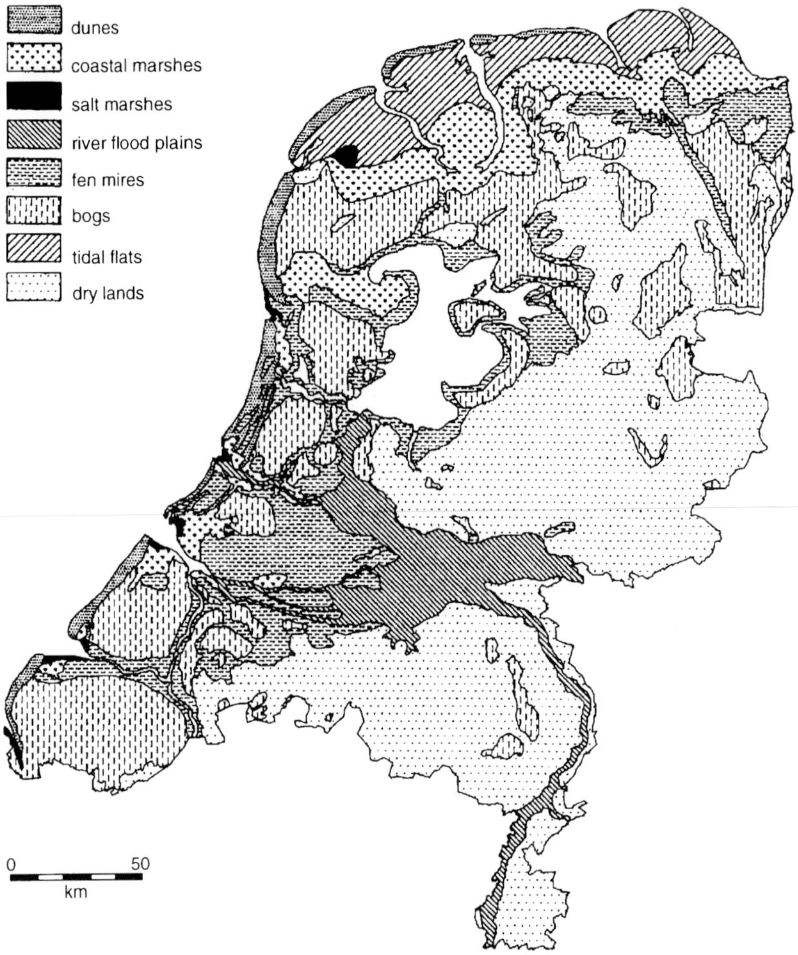

Figure 1. Reconstruction of the distribution of major aquatic ecosystems in the Netherlands, approximately 2000 BP (derived from Zagwijn, 1986).

drainage system, covering the entire country. (2) The conversion of natural aquatic ecosystems into drained agricultural land was a major cultural operation over the past centuries. Inaccessible wetlands and marshes were considered as 'wasteland', and were cultivated. Of the total area of land in the Netherlands 60% was agricultural land in 1996, but the tendency is now to set aside agricultural land in favour of nature development (CBS/RIVM, 1999). (3) Compared with the geographic situation of roughly a thousand years ago, massive habitat loss took place, owing to numerous causes. Obvious causes are the regulation of rivers and water courses, the reclamation of wetlands for infrastructure and urban sprawl, pollution with persistent and toxic chemicals, acidification by atmospheric deposits, eutrophication by enriched river water, desiccation (see argument 1), and total loss or fragmentation of habitats.

In 1989, a symposium was held, under the title 'Netherlands-Wetlands', aiming at the presentation of the state of the art of the existing knowledge of structure and function of wetland types characteristic for the Netherlands (Fig. 3). The proceedings of the symposium (Best & Bakker, 1993) offered much information about the status of the aquatic ecosystems, among which many were liable to severe environmental deterioration. Only little progress about rehabilitation measures proper could be given in 1989. The present volume (Nienhuis & Gulati, 2002) can be seen as a logical successor of the results of the symposium in 1989. Now, 12 years later, a lot of additional ecological research has been done, and the focus is more on the results of ecological restoration measures.

Table 1. Geographic position of the ecosystems covered by the various authors of this volume (Nienhuis & Gulati, 2002), using the legends of Figures 2 and 3

Author	Geographic area
De Jonge & de Jong (2002)	Tidal flats, estuaries, saline lakes (Fig. 3)
Bakker et al. (2002)	Salt marshes (Fig. 3)
Nienhuis et al. (2002)	Large rivers (Figs 2 & 3)
De Groot (2002)	Large rivers (Figs 2 & 3)
Gulati & van Donk (2002)	Freshwater lakes, fen mires (Fig. 3)
Lamers et al. (2002)	Fen mires (Fig. 3)
Verdonschot & Nijboer (2002)	Streams originating on Pleistocene sandy grounds (white, Fig. 2)
Grootjans et al. (2002a)	Brooks originating on Pleistocene sandy grounds (white, Fig. 2)
Roelofs et al. (2002)	Raised bogs, soft-water bodies on Pleistocene sandy grounds (white, Fig. 2; Fig. 3)
Grootjans et al. (2002b)	Coastal dune slacks (Fig. 3)

Table 2. Typology of freshwater bodies (LNV, 2001) and saline water bodies in the Netherlands, listed together with the authors from the present volume, covering the ecological restoration of the water bodies distinguished

Typology of water bodies	Covered by
1. Sources of streams and brooks	Verdonschot & Nijboer (2002); Grootjans et al. (2002a)
2. Streams and brooks	Verdonschot & Nijboer (2002); Grootjans et al. (2002a)
3. Rivers	Nienhuis et al. (2002; de Groot (2002)
4. Brackish inland waters	Lamers et al. (2002); de Jonge & de Jong (2002)
5. Small pools	—
6. Ditches	—
7. Fens	Lamers et al. (2002)
8. Sand- and gravel-mining pits	—
9. Lakes	Gulati & van Donk (2002)
10. Canals	—
11. Dune slacks	Grootjans et al. (2002b)
12. Pleistocene soft-water bodies	Roelofs et al. (2002)
13. Estuaries and tidal waters	De Jonge & de Jong (2002); Bakker et al. (2002)
14. Large brackish lakes	De Jonge & de Jong (2002)

Ecological restoration, theory and practice

It is only very recently, arbitrarily some 25 years ago, that the tide of ecological deterioration has been turned in the Netherlands, and ecological rehabilitation of disturbed ecosystems is in full swing now. In this volume, the status of Dutch aquatic restoration projects will be presented. All rehabilitation measures will be placed in a historical and ecological perspective. Ecological restoration often seems a process of trial and error, and practitioners proclaim why restoration ecology should bother about having a conceptual framework. Restoration scientists, however, usually stress the necessity to define and agree upon common targets in restoration projects (Hobbs & Norton, 1996; Pfadenhauer & Grootjans, 1999; Bakker et al., 2000). They advocate the use of clearly defined target communities and/or target species to measure success. A definition of targets, however, depends to a large degree on the level of ambition of particular reconstruction projects. In all cases, this comprises partly or completely structural and functional return of the

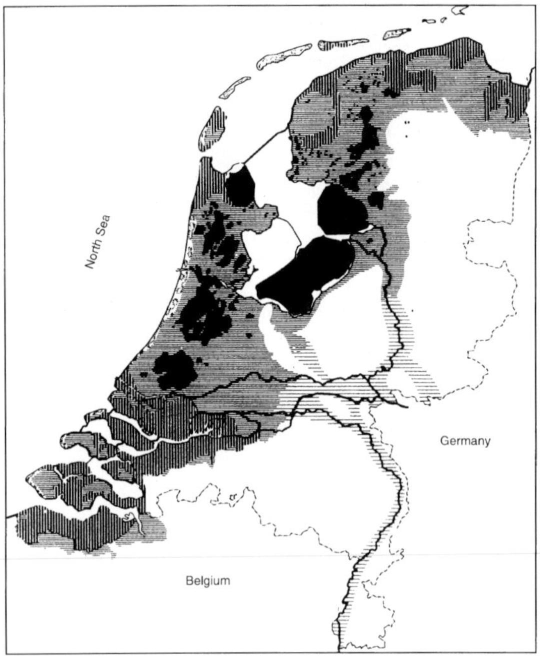

Figure 2. Areas in the Netherlands that have to be protected against flooding (Schultz, 1982). The white parts in the eastern Netherlands are elevated diluvial (Pleistocene) sandy soils, deposited between 1 000 000 y and 10 000 y BP. The black and shaded parts in the central and western Netherlands are alluvial (Holocene) mineral-rich grounds, deposited from 10 000 y BP onwards.

pre-disturbance state of the ecosystem. Van Diggelen et al. (2001) suggest three levels of ecological restoration: (1) reclamation – sometimes called enhancement – consists of attempts to increase biodiversity; (2) rehabilitation, consists of the reintroduction of certain ecosystem functions, and (3) restoration, consists of a reconstruction of a prior ecosystem; this is the most ambitious attempt. Restoration ecology is a branch of ecology and the Society for Ecological Restoration suggests the following definition: ecological restoration is the process of assisting the recovery and management of ecological integrity. Ecological integrity includes a critical range of variability in biodiversity, ecological processes and structures, regional and historical contexts and sustainable cultural practices (Van Diggelen et al., 2001). In the context of this book we implicitly use this definition. The main question that all authors have tried to answer concerns the role of ecological tools and knowledge in ecological restoration projects. Which ecological concepts have been applied during restoration processes in the Netherlands? What makes a project a success in terms of ecology, and when are we speaking about a failure?

In classical ecological textbooks, aquatic ecosystems and terrestrial ecosystems are mainly dealt with separately. One of the main features of the many shallow water bodies in the Netherlands, however, is the fact that over the seasons these ecosystems show their aquatic status mainly during winter, and their terrestrial status mainly during summer. A large part of the year the plant and animal communities compose wet-land ecosystems proper, either partly and temporarily dry or wet. This wet-land change in hydrological status demands many adaptations of terrestrial and aquatic animals and plants. The inundation of land results in either water-logging of the soil or partial or complete submergence of the shoots of plants, which asks for 'amphibious' behaviour of plants and animals. Besides the aquatic ecosystems proper, the contributions to this volume cover these semi-terrestrial (or semi-aquatic) features, characteristic for shallow aquatic ecosystems in the Netherlands. In Table 1 the geographic position of the ecosystems covered by the various authors of this volume is indicated, using the legends of Figures 2 and 3.

The present volume contains ten case studies, covering roughly the aquatic and semi-aquatic ecosystems in the Netherlands. Table 2 shows the typology of freshwater bodies (LNV, 2001) and saline water bodies in the Netherlands, together with the coverage from the chapters of this book. As can be noticed, some ecosystems have been omitted (small-drinking-pools, sand- and gravel-mining pits, ditches and canals). Particularly ditches are a characteristic part of the Netherlands landscape, altogether having a length of 350 000 km (Nijboer, 2001). Ditches are artificial ribbon-like, narrow (on average some metres) and shallow bodies of water, used for the drainage of agricultural fields and meadows. The restoration measures suggested by Lamers et al. (2002) for fen and peat-bog systems, and Verdonschot & Nijboer (2002) and Grootjans et al. (2002a) for streams and brooks, are also broadly applicable to the network of ditches and small pools in the Netherlands.

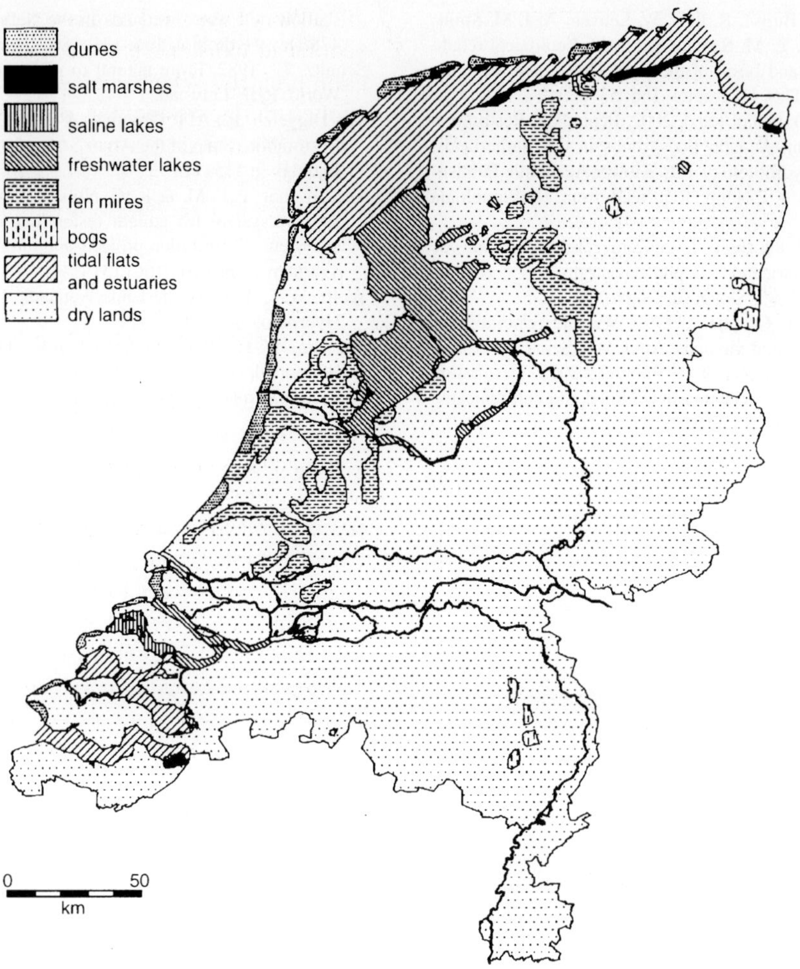

Figure 3. Present distribution of major aquatic ecosystems in the Netherlands (Wolff, 1993).

References

Bakker, J. P., P. Esselink, K. S. Dijkema, W. E. van Duin & D. J. de Jong, 2002. Restoration of salt marshes in the Netherlands. Hydrobiologia 478/Dev. Hydrobiol. 166: 29–51.

Bakker, J. P., A. P. Grootjans, M. Hermy & P. Poschlod, 2000. How to define targets for eclogical restoration? Introduction. Appl.Veg. Sci. 3: 3–6.

Best, E. P. H. & J. P. Bakker (eds), 1993. Netherlands-Wetlands. Hydrobiologia 265/Dev. Hydrobiol. 88: 328 pp.

CBS/RIVM, 1999. Milieucompendium 1999, Het Milieu in Cijfers. Centraal Bureau voor de Statistiek, Voorburg en Rijksinstituut voor Volksgezondheid en Milieu, Bilthoven: 1–341 (in Dutch).

De Groot, S. J., 2002. A review of the past and present status of anadromous fish species in the Netherlands: is restocking the Rhine feasible? Hydrobiologia 478/Dev. Hydrobiol. 166: 205–218.

De Jonge, V. N. & D. J. de Jong, 2002. Ecological restoration in coastal areas in the Netherlands: concepts, dilemmas and some examples. Hydrobiologia 478/Dev. Hydrobiol. 166: 7–28.

Edelman, T., 1974. Bijdrage tot de historische geografie van de Nederlandse kuststrook. Rijkswaterstaat Publicatie 14: 1–82 (in Dutch).

Grootjans, A. P., J. P. Bakker, A. J. M. Jansen & R. H. Kemmers, 2002a. Restoration of brook valley meadows in the Netherlands. Hydrobiologia 478/Dev. Hydrobiol. 166: 149–170.

Grootjans, A. P., H. W. T. Geelen, A. J. M. Jansen & E. J. Lammerts, 2002b. Restoration of coastal dune slacks in the Netherlands. Hydrobiologia 478/Dev. Hydrobiol. 166: 181–203.

Gulati, R. D. & E. van Donk, 2002. Lakes in the Netherlands, their origin, eutrophication and restoration: state-of-the-art review. Hydrobiologia 478/Dev. Hydrobiol. 166: 73–106.

Hobbs, R. J. & D. A. Norton, 1996. Towards a conceptual framework for restoration ecology. Restor. Ecol. 4: 93–110.

Lamers, L. P. M., A. J. P. Smolders & J. G. M. Roelofs, 2002. The restoration of fens in the Netherlands. Hydrobiologia 478/Dev. Hydrobiol. 166: 107–130.

LNV, 2001. Natuurlijke levensgemeenschappen van de Nederlandse binnenwateren. Expertisecentrum LNV & ALTERRA, Wageningen. Rapport AS-01 t/m/AS-13 (in Dutch).

Nienhuis, P. H., A. D. Buijse, R. S. E. W. Leuven, A. J. M. Smits, R. J. W. de Nooij & E. M. Samborska, 2002. Ecological rehabilitation of the lowland basin of the river Rhine (NW Europe). Hydrobiologia 478/Dev. Hydrobiol. 166: 53–72.

Nienhuis, P. H. & R. D. Gulati (eds), 2002. Ecological Restoration of Aquatic and Semi-Aquatic Ecosystems in the Netherlands. Developments in Hydrobiology 166. Kluwer Academic Publishers, Dordrecht: vii + 234 pp. Reprinted from Hydrobiologia 478.

Nijboer, R., 2001. Natuurlijke levensgemeenschappen van de Nederlandse binnenwateren. Expertisecentrum LNV & ALTERRA. Rapport AS-06, Sloten (in Dutch).

Pfadenhauer, J. & A. P. Grootjans, 1999. Wetland restoration in Central Europe: aims and methods. Appl. Veg. Sci. 2: 95–106.

Roelofs, J. G. M., E. Brouwer & R. Bobbink, 2002. Restoration of aquatic macrophyte vegetation in acidified and eutrophicated shallow soft water wetlands in the Netherlands. Hydrobiologia 478/Dev. Hydrobiol. 166: 171–180.

Schultz, E., 1982. From natural to reclaimed land. Polders of the World. RIJP, Lelystad: 17–42.

Van Diggelen, R., A. P. Grootjans & J. A. Harris, 2001. Ecological Restoration: State of the Art or State of the Science? Restoration Ecology 9: 115–118.

Verdonschot, P. F. M. & R. C. Nijboer, 2002. Towards a decision support system for stream restoration in the Netherlands: an overview of restoration projects and future needs. Hydrobiologia 478/Dev. Hydrobiol. 166: 131–148.

Wolff, W. J., 1993. Netherlands-Wetlands. Hydrobiologia 265/Dev. Hydrobiol. 88: 1–14.

Zagwijn, W. H., 1986. Nederland in het Holoceen. Geologie van Nederland, deel 1, Rijksgeologische Dienst, Haarlem / Staatsuitgeverij Den Haag: 46 pp. (in Dutch).

Hydrobiologia **478:** 7–28, 2002.
P.H. Nienhuis & R.D. Gulati (eds), Ecological Restoration of Aquatic and Semi-Aquatic Ecosystems in the Netherlands (NW Europe).

Ecological restoration in coastal areas in the Netherlands: concepts, dilemmas and some examples

V.N. de Jonge[1] & D.J. de Jong[2]
[1]*Department of Marine Biology, University of Groningen, P.O. Box 14, 9750 AA Haren, The Netherlands*
[2]*National Institute for Coastal and Marine Management/RIKZ, P.O. Box 8039, 4330 EA Middelburg, The Netherlands*

Key words: ecological restoration, coastal ecosystems, (former) estuaries, Wadden Sea, brackish inland ecosystems, process orientation, sustainable development

Abstract

This chapter gives an overview of attempts in the Netherlands to restore coastal ecosystems and habitats, and explains how scientific and non-scientific information has been used to meet the goals. Indications for successes and failures of management measures taken so far, as well as dilemmas to cope with, are given. Up to now only small scale restoration projects have been executed, while large scale projects generally are not further then the thinking or planning phase. A special type of 'restoration projects' are the large civil engineering works, particularly in the south-west of the Netherlands. Although these works were not planned and executed as restoration projects, but designed for safety against flooding from the sea, they have led to significant changes in the boundary conditions of the systems concerned. For restoration projects yet to be executed one can learn very much from these developments, particularly regarding the sensitivity of coastal systems for changes in boundary conditions and about the (im)possibilities to 'steer' ecological developments. Physical, chemical and biological processes form the basis of restoration measures of coastal habitats, and this means that a thorough knowledge of these processes is essential. Coastal ecosystems are the result of complex interactions of large-scale and small-scale processes, implying a holistic approach in scientific investigations. Consequently, restoration of these systems primarily has to be realised by influencing the basic processes. This is the only way to preserve or regain in a sustainable way ecological values, such as species composition. Focusing only at one particular species (e.g. breeding terns) or a specific habitat (e.g. a salt marsh) may easily ignore the underlying processes. In general, coastal restoration should focus on the redirection of processes towards a desired status by stimulating certain process parameters. Monitoring of the results and, if necessary, gradual readjustment of the governing factors, is an essential part of this approach.

Problem identification

Conservation and restoration of shallow coastal areas and connected wetlands is an important issue today. Apart from the fact that most coastal areas are by nature the most productive areas on earth (Holligan & de Boois, 1993), they are also the areas most heavily populated by humans. Historically seen it is evident that humans have not been able to practice their respect to the integrity of the natural environment. As a consequence, the natural areas are heavily under stress due to the occupation of land by man and the subsequent changes in land-use, the land reclamation, the production and discharge of waste and the withdrawal of fresh water from the nearby river basins.

The perspective about population developments in the coastal zone is that, as quoted in Kay & Alder (1999), within 30 years these areas may occupy as much or more people than there are now living on the entire planet. This means a number of circa 6 billion people in the coastal zone on a total human population which will have been increased by 30% to approximately 8 billion individuals. Compared to the present status, this future population growth and the accompanying increase in activities will lead to even more conflicting situations between nature on the one hand and the desired economic developments within the coastal areas on the other hand. These expectations highly stress the importance of safeguarding the last natural remnants in the coastal zone for further deterioration, and stress the importance of investment in restoration of natural systems where and if possible. If mankind is able to successfully follow this line, we may meet the requirements for 'sustainable development' as stated by the World Commission on Environment and Development chaired by G.H. Brundtland and, therefore, better known as the 'Brundtland Commission' (WCED, 1987). One of its most important messages to society was that it must be possible to achieve a route of further developing the global economy "which meets the needs of the present generation without compromising the chances of future generations to meet their own needs".

This chapter gives an overview of attempts in the Netherlands to restore coastal systems and habitats, and explains how scientific and non-scientific information has been used to meet the goals. Indications for successes and failures of management measures taken so far, as well as dilemmas to cope with are given. Up to now, only small scale restoration projects have been executed, while large scale projects generally are not further then the thinking or planning phase, therefore emphasise will be put on the small scale projects. A special type of 'restoration projects' are the large civil engineering works, particularly in the south-west of the Netherlands. Although these works were not planned and executed as restoration projects, but designed for safety or shipping purposes, they have led to significant changes in the boundary conditions of the systems concerned and by that to significant changes in the ecosystem. For restoration projects yet to be executed, one can learn very much from the developments in these systems about the sensitivity of coastal systems for changes in boundary conditions and about the (im)possibilities to 'steer' developments. Effects due to xenobiotic compounds like contaminants are not treated.

Natural and human induced changes in Dutch coastal waters

The identification and interpretation of changes in the coastal systems are quite complex as they have to be seen in the more general context of the long term and short term variations in large scale natural processes. Some examples of long term variations are the long term cyclic variations in the tidal range (cf. Führböter, 1989) and its effects to estuaries and coasts (cf. de Jonge, 1983, in press) and the occurrence of the North Atlantic Oscillation (Garrod & Colebrook, 1978, Selten et al., 1999) that also may impact the North Sea and its fringes. Short term variations like changes in meteorological conditions, resulting in a varying wind climate and varying river discharges, are responsible for significant natural fluctuations as well. Also natural variations in the life-cycle of a species, e.g. breeding circumstances in quite different areas, may influence the occurrence of species in a significant way. Altogether these natural variations can be considered as the background noise hampering the detection of human induced effects.

The present developments have also to be considered within a historical context before they can be judged on a fair and objective basis. Especially in the Netherlands, a number of major changes with significant impact to the coastal area have already been taken place over a very long period of time (de Jonge et al., 1993, 2000). Some examples are shown below.

Shortening of coastline and land reclamation

A significant shortening of the coastline since 1200 resulted in the disappearance of shallow and sheltered areas with eelgrass beds and the disappearance of most of the natural salt marshes (Dijkema, 1987). Recently (Fig. 1), the dramatic decline in shallow sheltered coastal areas was completed by closing off the Zuiderzee (in 1932), the Lauwerszee (in 1969) and the realisation of the delta scheme in the south west of the Netherlands (period 1960–1986).

Extension of intertidal areas by the occurrence of the so-called 'Voordelta' due to the recent creation of the 'Maasvlakte' (Fig. 1) and the closure of the tidal inlets in Zeeland can be considered as a sort of naturally created compensation for the above mentioned losses.

The shortening of the coastline was accompanied by a rationalisation of the water management in that the number of freshwater discharge points was drastically reduced. This had unforeseen consequences to temporal and spatial variations in salinity and thus organisms.

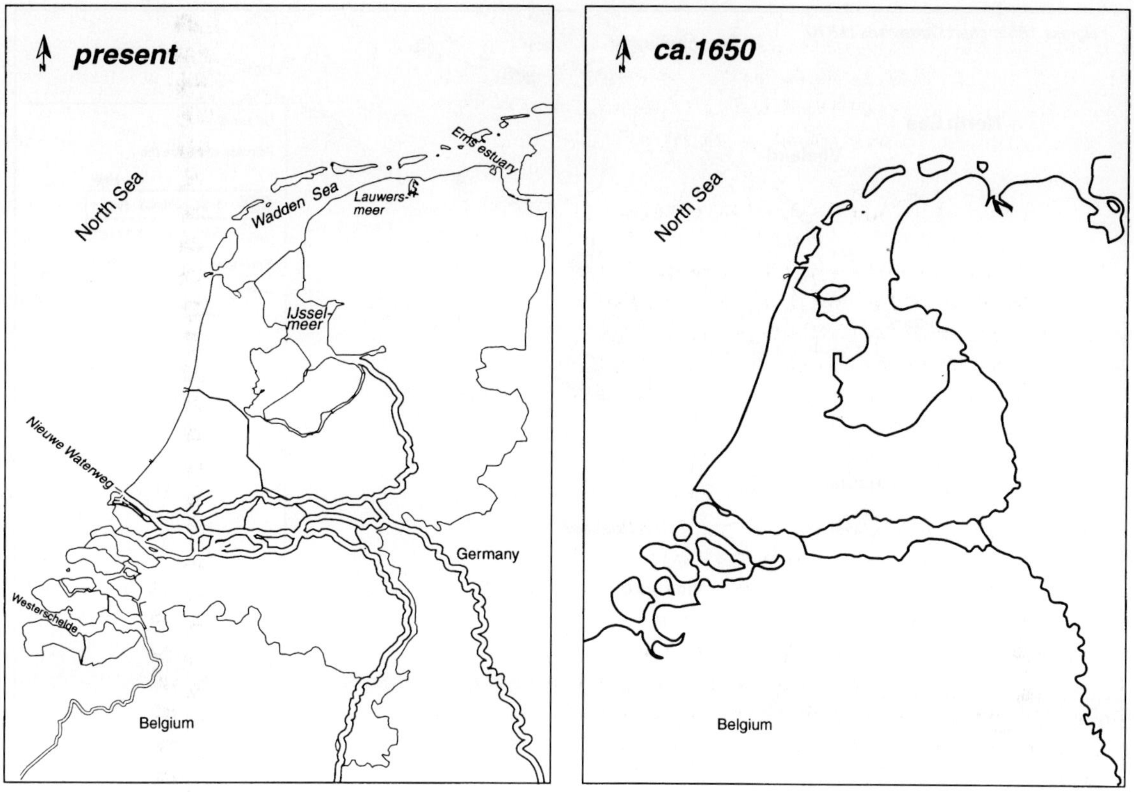

Figure 1. Changes of the Dutch coastline since circa 1650.

Disappearance of indigenous species and introduction of new species

The centuries-old intensive fishery on the flat oyster (*Ostrea edulis*) in the Wadden Sea led to the (almost) extinction of this oyster from around 1845 (van Ginkel, 1995). The strong decline of the flat oyster in the south western part of The Netherlands was due to a parasite (*Bonamia ostreae*), introduced by import of oysters from France.

The complete disappearance of sublittoral eelgrass (*Zostera marina*) and the strong decline in littoral eelgrass stands (Fig. 2) in the Wadden Sea, was probably caused by the 'wasting disease' and the engineering works in the late 1920s and early 1930s (Giesen et al., 1990a,b; de Jonge & de Jong, 1992). The (littoral) eelgrass beds in the south-western part of the Netherlands suffered severely. The beds in the Oosterschelde slowly but steadily decreased in surface area since the start of the 'delta works' in this basin. In the Grevelingenmeer, eelgrass first increased strongly due to more favourable conditions (clear and less dynamic water) after the closure of the lake, but since the late 1980s eelgrass has disappeared quite rapidly to almost extinction. There are strong indications that the reduction of the stands in both areas is related to decreased seasonal influence of freshwater input, leading to an increase in the salinity of the water mass (Wijgergangs & de Jong, 1999; Kamermans et al., 1999).

The common whelk (*Buccinum undatum*) and the common dog whelk (*Nucella lapillus*) almost became extinct in the Dutch coastal waters in the early 1970s. The most important cause of this disappearance has been the introduction of TBT as antifouling (Cadée et al., 1995 and references therein, see also Rilov et al., 2000).

Since the 1980s, the intertidal mussel beds in the Wadden Sea completely disappeared due to a combination of intense fishery and storms. The extinction of these beds resulted in a decrease in abundance of the bladder wrack (*Fucus vesiculosus*) and accompanying species like *Elachista fucicola, Littorina littorea* and *Jaera albifrons* (Reise et al., 1994). The mussel beds in the Oosterschelde disappeared already before the outbreak of World War II due to over-fishing for the seedlings of the blue mussel (*Mytilus edulis*).

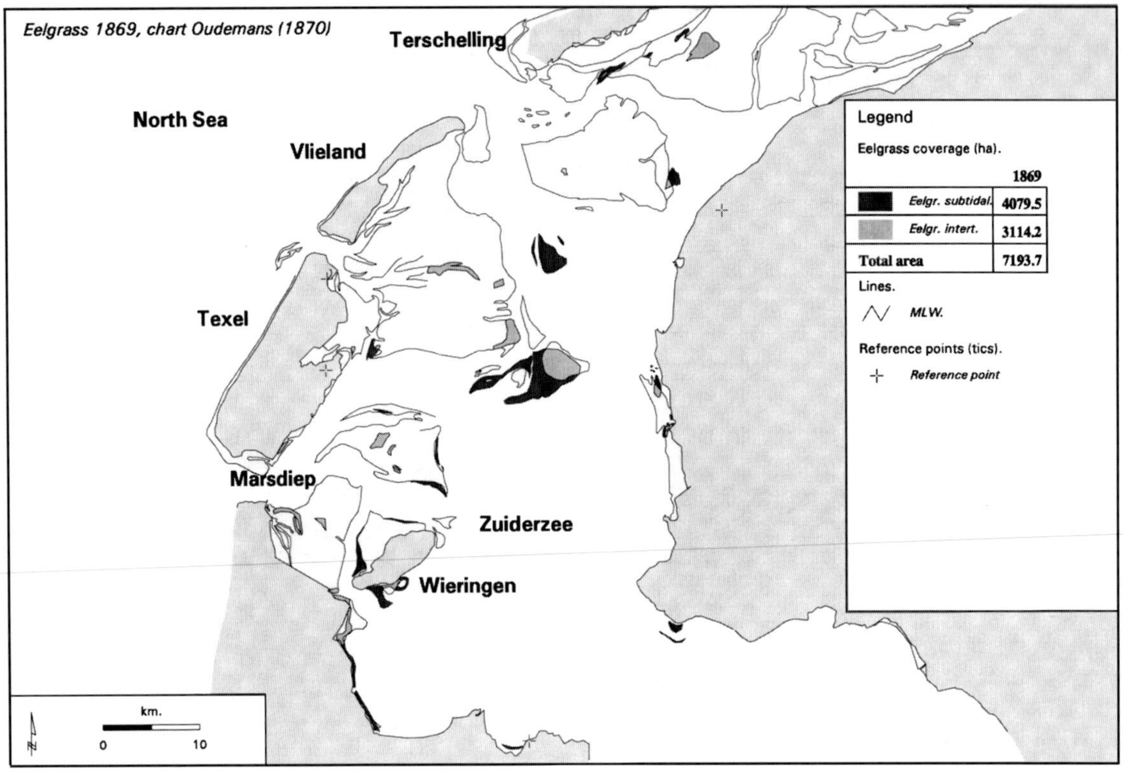

Figure 2. Eelgrass stands as mapped in 1869 (Oudemans et al., 1870) subdivided in littoral and sublittoral eelgrass stands according to de Jonge & Ruiter (1996).

In the Westerschelde, the mussel beds disappeared due to bad water quality, geomorphological changes, as a consequence of channel dredging (maintenance dredging and deepening of the navigation route), and land reclamation works. During the last decades, the fishery for the edible cockle (*Cerastoderma edule*) changed from a conventional fishery by either hand- or trawl-net into an extremely modern and efficient fishery which is based on a combination of trawling and suction dredging. This fishery today is so efficient that it strongly competes with the wader birds and ducks that are dependent on these bivalve species as a food source.

The introduction of especially new oyster species, is causing much trouble. In the Oosterschelde, the Japanese oyster (*Crassostrea gigas*) has expanded tremendously over the revetments of the seawalls, while in the sandy littoral areas the oysters even have formed natural reefs. At the moment, these dense populations are considered as a threat to the natural system because of their competition for food with other filter feeders, like blue mussel and edible cockle. In spring 2000, measures have been proposed to reduce these 'wild' oyster stands and reefs. During recent years, the Japanese oyster also expands rapidly into the Westerschelde and in the Dutch Wadden Sea. In the latter area, several natural oyster beds have been developed , but monitoring of their occurrence has not been carried out yet. For the Wadden Sea there even has been an application in 1999 for a license to exploit these natural beds.

Eutrophication

Increased eutrophication of our coastal waters presumably caused the decrease in abundance of brown algae and red algae in the Wadden Sea (Reise et al., 1994). There is sufficient evidence that eutrophication can certainly be blamed for the decline in brown macroalgae (Reise et al., 1994). Experimental evidence has shown (e.g. Sommer 1996) that macroalgae can easily be outcompeted by opportunistic species, such as representatives of the genera *Enteromorpha* and *Ulva*.

Turbidity

The light conditions in the Dutch coastal waters are not stable. Studies carried out by de Jonge (1986) and

Anonymous (1991) have demonstrated that there is a long term (over fourfold; de Jonge & de Jong, 1992, in press) natural variation in turbidity in the Wadden Sea. Although the causes are not entirely clear, it seems reasonable to assume that the disposal of dredged material from the Rotterdam harbour in the coastal zone contributes to this increase (de Kok et al., 1982) The effect of this variation in turbidity may have contributed to the deterioration of the populations of brown algae along the coast of the Netherlands (de Jong & de Jonge, 1989). Also maintenance dredging of the navigation route in estuaries has major effects on the concentrations of suspended matter and consequently the development of the populations of algae (de Jonge, 2000).

Development of motivation and goals in society and politics

Concerning the developments in appreciation of coastal systems by society and politics, it can be stated that gradually, during the last century, more needs are to be met, varying from safety via development of infrastructure and water quality to environmental concern.

Protection against flooding and salt intrusion

Numerous engineering works, drastically changing the aquatic environment, have been carried out under responsibility of the Ministry of Transport, Public Works and Water Management. As reported above, the large scale changes started already in the late 1920s in the north (Afsluitdijk) but were continued after World War II, and especially after the flooding in 1953 in the south-western part of the country. The main goal of the execution of these works was protection of the country against flooding from the sea, and controlled run off and sluicing out of freshwater to reduce salt intrusion as a measure to improve the quality of agriculture.

More recently, improvement of the waterways infrastructure became an important management item, because of the ever increasing need for high quality freshwater for domestic use, due to increased welfare and population growth.

Water quality and integral water management

Since the early 1970s, new visions have been developed, certainly also due to the activities of fishery organisations, non-governmental organisations (NGO's) and action groups that questioned the need of closing all the former estuaries. This movement resulted in a reconsideration of the closure of the Oosterschelde and the consequent construction of the storm-surge barrier in the tidal basin around 1978, instead of the envisaged construction of a massive dam. Also the development of water quality objectives and standards, together with their monitoring and evaluation, covering all territorial waters including the Exclusive Economic Zone in the North Sea, were started in this period. All these items were established in the Second Report on Water Management in 1984.

From the mid 1980s onwards, the 'environment' issue got an increasing influence on politics and management. Necessary scientific research to explore the environmental effects of policy options was started in many places. Two relevant examples are the Biological Research Ems Dollard Estuary (BOEDE, 1972–1985) to investigate the effects of the discharge into the Ems estuary of untreated waste water from the potato flour industry and the straw board industry in the north-east of the country (e.g. Anonymous, 1985; Baretta & Ruardij, 1988), and the environmental research carried out within the scope of the Deltaproject (Deltadienst, 1971–1986; e.g. Saeijs, 1982, Nienhuis et al., 1994). Among many others, a practical result of these projects was the development of Policy Analysis as a tool to take many different aspects into account in the decision-making process (e.g. Anonymous, 1976; Saeijs, 1982), developing further into the Environmental Impact Assessment (EIA) to assess and evaluate the environmental impacts of proposed human actions.

Restoration

The above mentioned way of thinking developed further and one of the most important recent documents published by the Ministry of Transport, Public Works and Water Management was the 'Derde Nota Waterhuishouding' (Third Report on Water Management; Anonymous, 1989). In this document, the improvement of water quality by reducing emissions was aimed as starting point, and 'integral water management' considering both socio-economic aspects as well as ecological targets, was launched as an prime instrument. Based on this policy ecological restoration projects were stimulated in order to improve the overall quality of the environment. The term 'sustainability' was introduced (WCED, 1987) and came into use as an operational target. It is within this array of

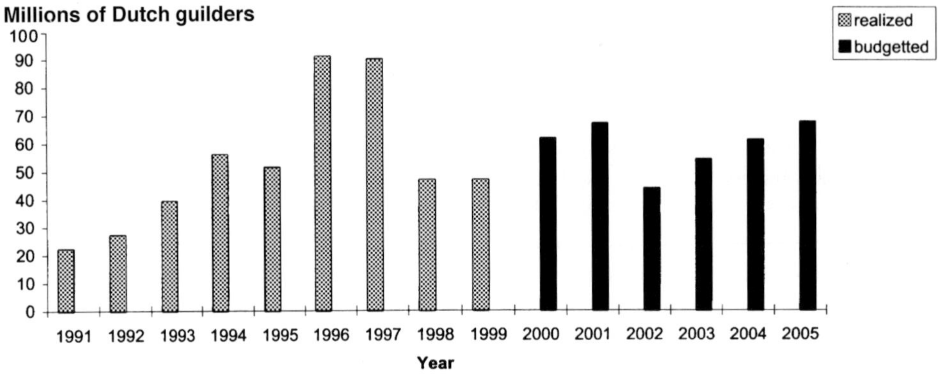

Figure 3. Invested and budgeted finances (excluding personnel) for integrated water management and ecological restoration purposes. For explanation see text.

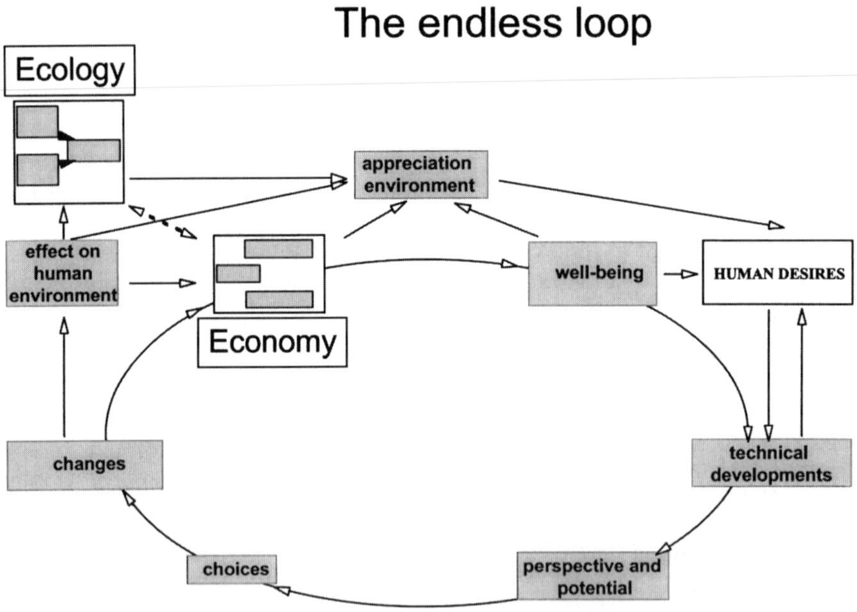

Figure 4. The 'endless loop' presenting the continuation of impulses to technological development due to human curiosity and its basic relation to economy and ecology.

changing policy plans and action plans that improvement of the environment became really an item on the political agenda. The enhanced co-operation between several ministries, especially those related to water management and nature conservation and fisheries, resulted in mutual participation in field projects.

Reconsideration

In 1995, the Dutch government decided that the financial boundary conditions for reaching the environmental targets had to be rearranged, so that the deadlines for realisation of these targets were postponed to 2030–2040. Despite the recent financial profits at the level of the national government, the postponement of this deadline was not reconsidered in 2000 (Fig. 3) indicating a change in political interest at that time in favour of national well fare or well being, instead of nature. This illustrates that the willingness of society to take its responsibility for investments in the (natural) environment is an important part of the problem as well (Fig. 4). Although today the income 'per capita' is steadily growing in the Netherlands, the willingness to take care and responsibility for our environment seems to stabilise or even decline. Protection against flooding and improvement of infrastructure in terms of railroads and motor ways, resulting in more pollution and increased loss of important natural areas, are at present more important issues on the political agenda than during the previous decade. Obviously, the well being of men and women is more easily expressed in Euro's and employability than in ecological goods and services.

Another remarkable change in policy occurring since the publication of the Fourth Report on Water Management in 1999 is that (ecological) 'restoration' is no longer the main item, as this term has been replaced by (ecological) 'resilience'. This change in approach from restoration to resilience seems to indicate the present-day tendency to grind down nature instead of applying 'sustainability' and applying the 'precautionary principle'. A magnifying problem is that a suitable and practically applicable definition of 'resilience' is not available. Resilience is connected to sustainability and in politics both terms seem to be very useful, but do not necessarily contain a solid scientific and practical meaning, partly due to the fact that the development of new concepts from fundamental research is stagnating. In recent Dutch policy reports, it is stated that no additional restoration projects will be formulated and carried out, except those that have been listed already. This strategy implies that it will be very difficult in the future to evaluate and control the progress in the field of coastal restoration projects.

Useful principles in scientific concepts

Large scale processes of decisive importance in restoration projects

The success of ecological restoration will primarily be determined by the recognition of the operational environmental conditions and the (im)possibilities to influence or regulate these conditions. The main driving physical factors in the coastal zone are those governing freshwater discharge, tidal- and wind-induced mixing and transport of water and sediment. Other important factors are derived from these main driving forces such as the gradients in salinity, turbidity and nutrients, the distribution of contaminants, the sediment composition (e.g. mud content) and the mobility of the soil. From this listing, it is clear that restoration projects in open or half open coastal systems are very difficult to achieve without taking additional measures to create the desired new physical and chemical conditions. Intentionally positive measures may result in unforeseen side effects at the same locality, in the neighbourhood or even remote from the locality under consideration, resulting in a zero or even negative net result; e.g. the negative side effects for nature of the storm-surge barrier in the Oosterschelde, designed for safety and nature conservation, may be seen in this way. It means that once started with physical interference, the water managers often have to continue this interference to maintain the desired physical conditions.

Boundary conditions for development of coastal habitats

When focusing on the functioning of estuarine systems, it is possible to present a general and simple picture showing the overall and integrative effects of the factors nutrients, light and primary production under estuarine conditions. For instance, Figure 5 shows the relation between river discharge and turbidity (light extinction) and the resulting primary production in the water-phase of the Ems estuary. Some negative aspects of eutrophication are the replacement of e.g. seagrasses by green algae like *Enteromorpha* and *Ulva* spec. and the accumulation of large amounts of green algae, causing noxious smell and related dying off of benthic animals due to de-oxygenation of the soil (see e.g. Sfriso & Marcomini, 1997, 1999 and Sfriso et al., 1992).

The relationship between average salinity and the potential number of species in estuarine areas below mean high water is shown in the old and strongly generalised 'figure of Remane' (1934; Fig. 6). In the former Delta area (before the 'delta works'), this relation was confirmed by several scientists, as reviewed by Schmidt van Dorp (1979). Although the 'figure of Remane' suggests that a rather high or very low salinity might be the optimum situation from the point of view of species diversity, it has to be taken into account that in the zone with intermediate salinity ranges important species occur, typical for these brackish areas. These brackish areas are very important in the life cycle of many saltwater or freshwater species, be it for spawning, for juvenile phases or as transmigration zone between freshwater and the sea. A number of marine plant species, like seagrasses (*Zostera* spp) and pioneer salt marsh species (e.g. *Salicornia* spec.), benefit from a period with decreased salinities, compared to the average salinity for germination and settling of seedlings. The figure of Remane does not take into account the effect of the normal fluctuations and the extremes in salinity on the species composition of water bodies. De Jonge (1974) investigated the cor-

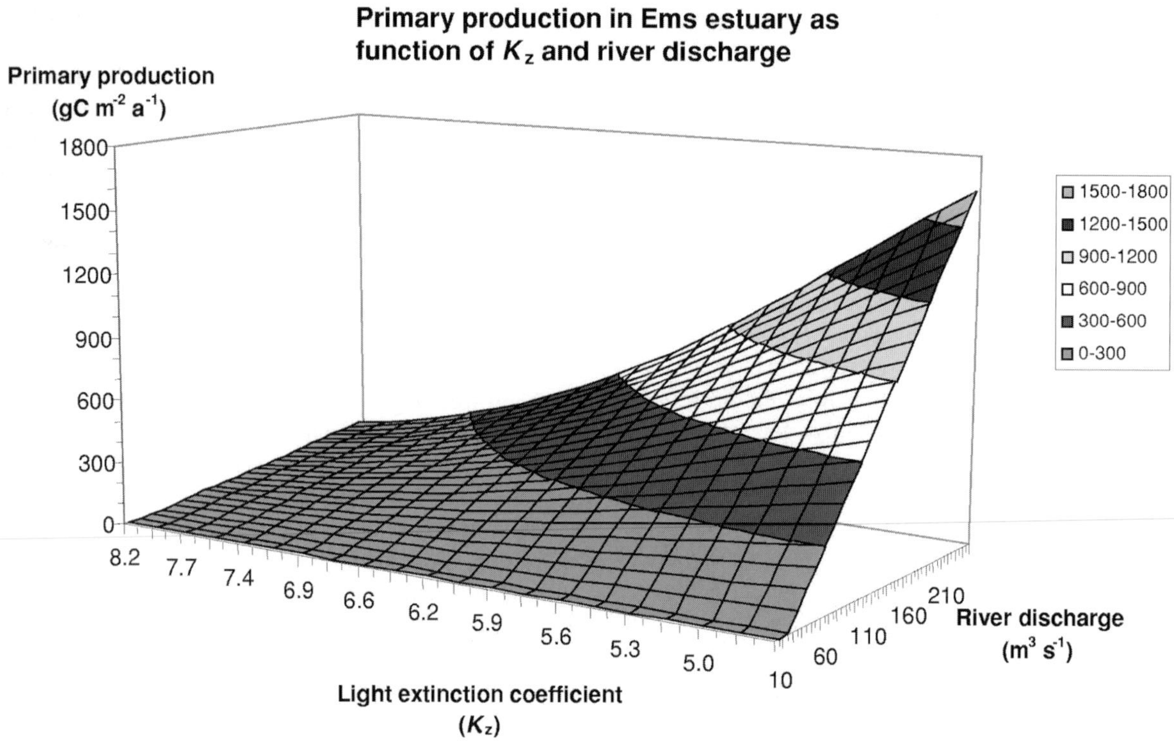

Figure 5. Integrative effect of turbidity (expressed in light extinction coefficient K_Z), the nutrient supply (given in load per unit of time) and the annual primary production (expressed in g C m^{-2} yr^{-2}).

relations between variations as well as extremes for a number of brackish coastal inland waters in the south-western part of the Netherlands. Two results of this analyses shown (Fig. 7) that different combinations of average salinity and fluctuations in salinity lead to different communities.

Another subject is the effect of the innumerable number of contaminants present in Dutch coastal wetlands. Recently, it is assumed that these pollutants may have led to (sub)lethal effects or unexpected phenomena like amplified algal blooming because of anorexia behaviour of affected zooplankton by contaminants (Scholten et al., 1996). Additional problems in restoration projects may be caused by the sluicing out of e.g. herbicides to coastal areas during very short time periods within the growing season, killing algae and shortening the growing season for e.g. eelgrass. Shortening of the growing season of eelgrass may reduce the seed production and consequently the reproduction capacity of the (annual) populations. The ample use of herbicides during the short period of the potato harvest may serve as a good example. The combination of the freshwater sluicing schemes with the high water dispersion capacity (dilution effect) in our coastal areas,

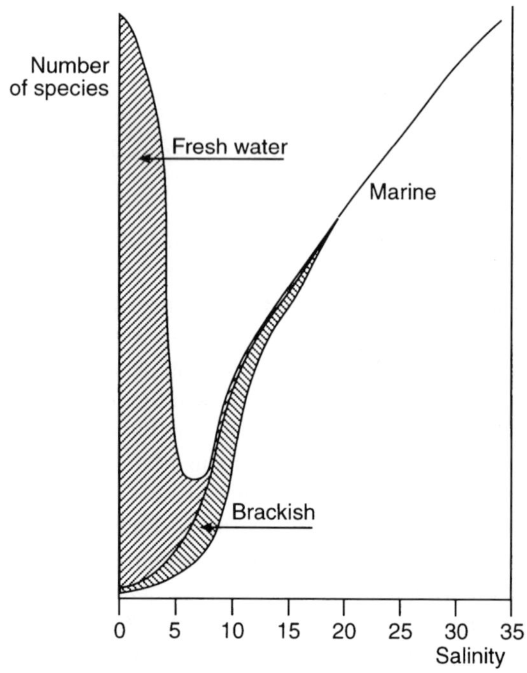

Figure 6. Relation between the number of species and the mean salinity (in ‰)as generalized for estuarine systems by Remane (1934).

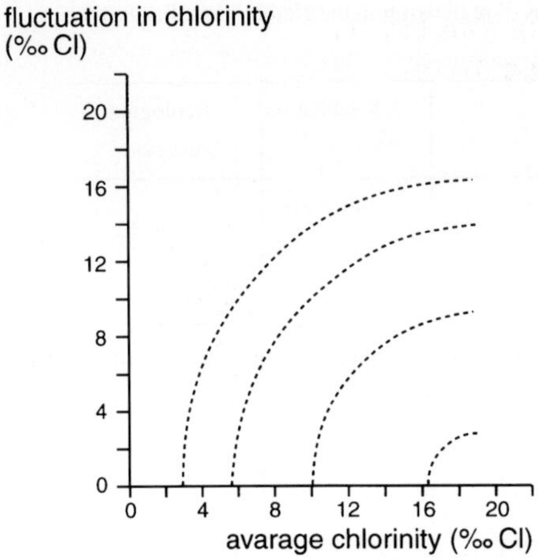

Figure 7. Boundaries between species assemblages of oligohalinicum, α-mesohalinicum and β-mesohalinicum, polyhalinicum and euhalinicum in isolated coastal brackish inland waterbodies based on mean salinity of the water and its fluctation (De Jonge, 1974).

strongly hampers the assessment of the possible negative effects of herbicides to the development of algae and seagrasses under natural conditions.

Historical references

Defining historical references in the context of ecological restoration measures causes quite often large problems and dilemma's. An implicitly as well as explicitly used reference is the situation prior to the great engineering works in the Netherlands, and the large scale introduction of artificial fertilisers and xenobiotic compounds, some eighty years ago. The period 1920–1930 represents the 'good old days', when the world population was only 2 billion people, and stands for a more or less ideal situation in terms of human pressure on ecosystems. Firstly, no actual data of a past situation older then several decades are present. Due to lack of data such a reconstruction is open to endless discussions. Secondly, the overall situation in the past is hardly comparable with the present situation. The number of human inhabitants in the Netherlands has doubled, mobility has increased tremendously as well as welfare. Per capita we use and need more space for living then 70 year ago. Former remote areas are now intensively used for recreational purposes. Large areas have been embanked since the 1920s. These developments are, in fact, irreversible and in our present considerations we have to take this into account.

Two examples from our experience may illustrate these statements. The first example deals with the question of the 'desired' area of salt marshes in Dutch coastal areas. At present in most tidal areas, hardly any space is available for natural salt marsh formation. The present salt marsh area in the SW Netherlands is mainly due to the accretion capacity following the introduction of the new species *Spartina anglica*, around 1920. In the Wadden Sea, the present salt marsh area is both the result of intensive accretion works carried out along the coastline of Groningen and Friesland, as well as of the construction of sand dams on the islands resulting in sheltered areas. Without human interference the total area of 'natural' salt marshes should even have been much less than in the actual situation. What reference should be chosen: the 1920 situation, before the construction of sand dams and the introduction of Spartina anglica, or the present situation?

The second example concerns, in short, the situation around the breeding habitats of coastal birds like Kentish plover and little tern, whose very dynamic breeding areas have vanished in the last decades due to our needs for safety for the human population and recreation facilities. What period should be chosen here as a reference for restoration measures?

Importance of serendipity

Scientists usually consider conceptual knowledge on system functioning and accompanied system structuring as a good basis for ecological management measures. However, there is also a dilemma which is based on recent developments in theories and computer experiments on chaos and self organisation (e.g. Holling, 1986; Kay & Schneider, 1994; Schneider & Kay, 1994). It has been indicated that, even when we are anyway able to re-create lost environmental conditions, it still must be questioned whether lost communities actually will re-settle according to the structure known from historical records, or that unknown new communities will come forward. These insights indicate that it might be better not to focus too much on complete restoration of water bodies. Focus should be on the restoration of boundary conditions, physical and chemical conditions and processes and human management, that may lead to the development of specific habitats and hence may contribute to the increase of the diversity in natural ecological communities. This approach comprises sufficient pre-

Table 1. Examples of estuarine ecological restoration plans and status of realization in the Netherlands. For further explanation see text

Brackish coastal areas	Examples	Item	Realization	Ecological success
Large river transition areas	Haringvliet	fish passage	under way	
Open systems	Voordelta	system development	yes	+
	(Maasvlakte)	seawater inundation	yes	+
	Green beach	seawater inundation	yes	just finished
	Terschelling			
	Kroonspolder			
	(Vlieland)			
Half open brackish/ marine systems	Wadden Sea	Eelgrass beds	under way	
		Mussel beds	-	
	Eems estuary	-	-	
	Dollard	fish passage	no	
	Oosterschelde	river input	no	
	Westerschelde	dredging	yes	
	Zwin			
Brackish inland systems	Zeeuwse Inlagen			
	De Bol (Texel)	rehabilitat. brackish	no	
	Roggesloot (Texel)	env.	yes	+/−
	Polder Breebaart	fish passage by	under way	
	(Dollard)	siphon		
		rehabilitat. brackish		
		env		
Lakes	IJsselmeer	salinity gradient	no	
		'Afsluitdijk'		
	Amstelmeer	fish passage		
	Lauwersmeer	salinity gradient		
	Veerse Meer	salinity gradient		
	Grevelingen	reduction nutrient		
		effects		
		river input		
Locations	Den Helder	fish passage	no	
	Oostoever	fish passage	no	
	Termunterzijl	fish passage	no	
	Roptazijl	fish passage	no	
	Zwarte Haan	fish passage	no	
	Harlingen	fish passage	no	

tension to optimise the conditions for development of natural communities.

Another reason for hesitation is due to practical experience based on several small scale attempts to restore parts of ecosystems, and the very few that really have proven to be successful (see Table 1). A serious problem, often not enough recognised, is that the realisation of actual restoration projects has to be followed by maintenance work of the newly developed or newly created situation. On the longer time scale, this usually needs much more investment in terms of energy and currency than the original action. This problem is enlarged owing to the relatively long period required for the 'recreation' of new habitats, which is generally much longer than the political life span and the period the public is prepared to wait for expected results.

Use of Geographical Information Systems

Although the settlement of specific species in tidal wetlands is hardly predictable, still some progress has recently been made. Powerful instruments to explore the possibilities for habitat development are Geographical Information Systems (GIS). The use of this instrument enables the researcher to test the available knowledge about boundary conditions of a species or habitat in a spatial way, and thus to test whether the available conceptual knowledge about boundary conditions may suffice.

To explore the possibilities for the occurrence of intertidal eelgrass beds and intertidal mussel beds in the Dutch Wadden Sea, habitat maps in GIS have been developed recently (de Jonge et al., 1996; de Jong, 1999). These maps are based on the main environmental driving factors for potential occurrence of the two seagrass species and the communities they represent. These factors are wave action, current velocity, sediment composition and emersion time, and for intertidal mussel beds also distance to the low water mark; actually salinity should be taken into account as well, but reliable maps were not available at that time. Figure 8 shows the example for intertidal eelgrass beds (cf. de Jonge et al., 2000).

Evaluation of executed and planned restoration projects

Protection by jurisdiction

From an historical perspective, the best management measures taken so far have been statutory regulations, meant to protect an entire area. The entire international Wadden Sea area is subject to the Ramsar convention and also to the EU Habitat directive and EU Wild Birds directive, which are, or will be implemented by national legislation. In order to do so the entire Dutch Wadden Sea is under the Dutch Nature Conservation Act (article 17). It has further been agreed (Key Planning Decision) that the starting point for the entire area is 'sustainable development', which means

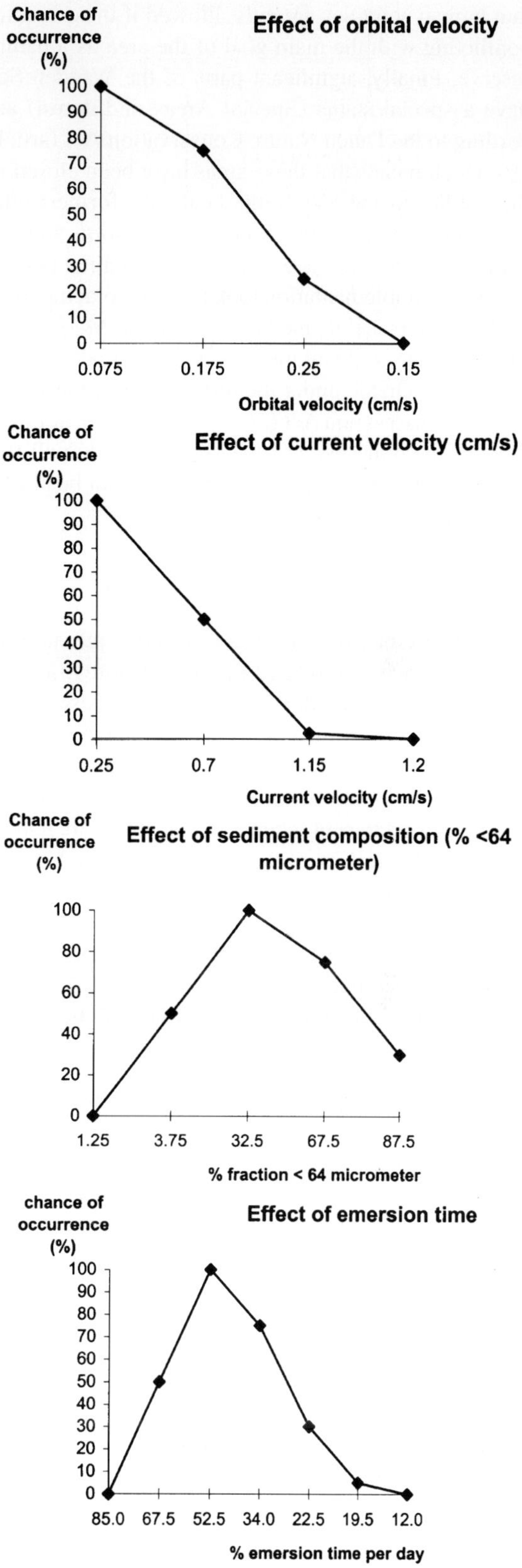

Figure 8. Potential chance of eelgrass occurrence for the factors current velocity, wave action, sediment composition and emersion time (cf. De Jonge et al., 2000).

that human activities are only allowed if these are not conflicting with the main goal of the area as a nature reserve. Finally, significant parts of the Wadden Sea have a special status (Special Areas of Interest) according to the Dutch Nature Conservation Act (article 19) which means that these areas have been closed to the public. In the SW Netherlands all (former) tidal areas are subject to the Ramsar convention and the Wild Birds directive as well as the EU-Habitat Directive. So far implementation took place by placing most of the Oosterschelde, the Haringvliet, the Markiezaatmeer and the Volkerakmeer as well as small parts of the Westerschelde under the jurisdiction of the Dutch Nature Conservation Act (article 17). In the near future most of the other parts of the Westerschelde will follow, and only the shipping channels will not be placed under the law.

Thematic evaluation of executed measures

Remarkably, nearly all coastal restoration measures in the Netherlands are related to previously realised engineering works or side effects related to these works. To discuss the results of measures, the restoration attempts are thematically grouped according to the basic problem. Some of these measures are related to changed management schemes while others are related to direct physical interference. From a recent policy report, it is clear that in addition to emission reducing measures in the period 1984–1989 now mainly physical measures will be taken, within the field of physical coastal engineering.

Five main themes of restoration will be distinguished, concerning:

(A) abrupt transitions between marine and freshwater due to flood prevention works and changed hydraulic management;
(B) counteraction of restrictions of physical processes; examples are the fixation of coastal dunes, the 'cliff fixation' of salt marshes, the construction of the Oosterschelde Barrier and the dredging of shipping channels in the Westerschelde;
(C) normative aspects in compensation measures;
(D) the reduction of noxious seasonally occurring developments, particularly the blooming or proliferation of macroalgae;
(E) compensation measures outside the tidal system.

Abrupt transitions between marine and freshwater

Abrupt transitions from freshwater to marine water have been created by the building of dikes and modern sluices that prevent salt intrusion. Prevention of salt intrusion has been an important item during the realisation of the 'delta works', because it was recognised that high quality freshwater should be one of the most important prerequisites for future development of the rapidly growing human population in the Netherlands. Both the large scale and small scale engineering works bore many (unforeseen) consequences.

In the first place, a reduction was achieved in the number of localities with small ebb sluices (sluices draining freely during low tide). In the Dutch Wadden Sea 17 transitions occur between freshwater and seawater (Fig. 9), of which only one represents the full gradient between freshwater and seawater (Ems estuary). In all other cases, either a pumping system or sluices are used for freshwater discharge. In the Delta area, the situation is similar, comprising one full gradient transition (Westerschelde), one transition with a very short gradient (Nieuwe Waterweg) and further pumps, and sluices equipped with systems to prevent exchange of fresh and saltwater (in the Oosterschelde). Due to the reduction in number of the classic ebb sluices in the Dutch as well as in the German Wadden Sea, remarkable elements belonging to the brackish water communities were lost from many places. The negative impact of overall environmental deterioration could be added to the loss of salinity gradients, as was shown by Michaelis et al. (1992): from 25 typical brackish water species only circa 50% were really present in the estuary with the best developed macrobenthic fauna. These data indicate that the smaller the system, the more variable its salinity gradient and consequently the more vulnerable the biotic communities are, which can be considered a very important notion in habitat restoration schemes (Fig. 10).

A second consequence was that a number of (diadromous) fish species (Table 2) were no longer able to migrate between the freshwater, and the marine environment. This has led to not well documented but significant declines in the local populations of many of these fish species (cf. de Groot, 2001). Out of the 11 diadromous fish species known from The Netherlands (Table 2) four are already extinct, three others are heavily under threat, two others are vulnerable and only three fish species are still doing well. Due to the creation of abrupt transitions between the sea and the

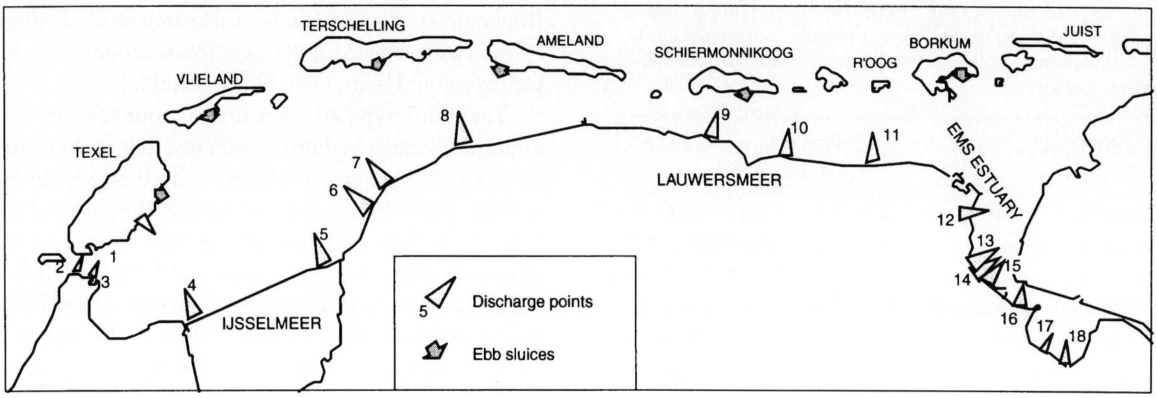

Figure 9. Map with the main transitions between freshwater and seawater in the Wadden Sea area.

freshwater there has also been a strong reduction in the flounder population (*Platichthus flesus*) in the Wadden Sea and the IJsselmeer since the closure of the former Zuiderzee (Rijnsdorp & Vethaak, 1989). Diseases occurring in the flounder population were also thought to be partly caused by the strict separation of freshwater and seawater along the coast, so that normal migration via salinity gradients was not possible any longer (Vethaak et al., 1995).

To mitigate the problems connected to the abrupt transitions between the marine environment and the freshwater environment, three types of actions were carried out. The first type of restoration measures is the construction of local facilities to enable species to reach either the marine or the freshwater environment. These facilities comprise fish channels and fish passages by siphons, the latter being used to lure organisms from one basin to the other. One example of these passages is found on the island of Texel near Eijerland, where it has been built to support migration of the Three spined stickleback (*Gasterosteus aculeatus*) which is an important prey species for the Spoonbill (*Platalea leucorodia*). Another facility on the island of Schouwen (Zeeland) is meant to improve the migration of young eel (*Anguilla anguilla*) from the Oosterschelde to the ditches on the island. Typically, these measures have been taken to support the population of a specific bird species, or to enhance the commercial and recreational fishery respectively, and not to improve the entire deteriorated aquatic community.

The second type of restoration measures is the implementation of a freshwater discharge scheme that enables fish to pass sluices, e.g. the sluices separating lake IJsselmeer from the Wadden Sea. Since 1938, the discharge schemes allow eel to reach the

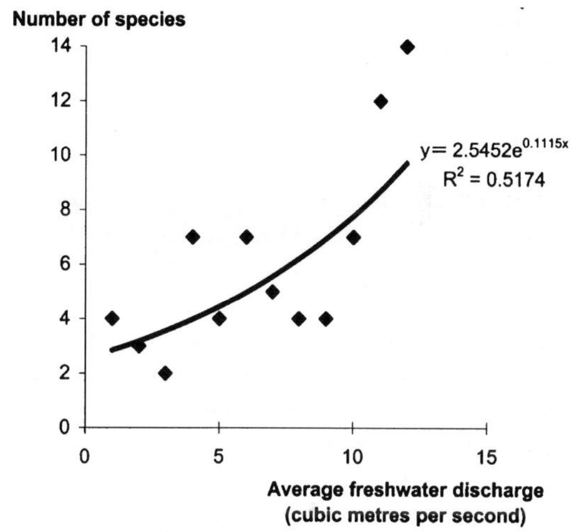

Figure 10. The number of benthic brackish water macrofauna species shown in relation to the mean annual freshwater discharge of a number of estuaries and tidal outlets (based on data of Michaelis et al., 1992).

freshwater lake, a measure exclusively meant to maintain a sizable eel population, to sustain the local eel fishery and associated fish industry. Since 1991, a more migration friendly freshwater discharge scheme was successfully introduced in the sluices of IJsselmeer near Den Oever(4). This scheme implies that during the flood period when the water masses seaward and landward of the sluice are at the same level, the sluice doors will be kept open longer than usual. In doing this brackish water can penetrate the freshwater reservoir. As long as the current velocities are low, organisms can actively pass the sluice. This result in a lengthening of the migration period per tidal period, mainly for fishes. The first positive effects of the new sluicing schemes were a strong increase in the flounder population (Dekker, 1994) and a de-

Table 2. Diadromous fish species. Given are common species and species under threat or extinct from the Netherlands due to environmental stress (e.g. removal of transition areas between river and sea)

Species	Position (cf. Jager, 1999)
Eel	–
Anguilla anguilla	(in NL vulnerable)
Three spined stickleback	–
Gasterosteus aculeatus	
Smelt	–
Osmerus eperlanus	–
Flounder	–
Platichthys flesus	
River lamprey	vulnerable
Lampetra fluviatilis	(in NL threatened)
Sea trout	susceptible
Salmo trutta	(in NL threatened)
Sea lamprey	threatened
Petromyzon marinus	(in NL threatened)
Twait	vulnerable
Alosa fallax	(in NL vulnerable)
Salmon	critical (under immediate threat of extinction)
Salmo salar	(in NL extinct)
Houting	critical (under immediate threat of extinction)
Coregonus lavaretus oxyrhinchus	(in NL extinct)
Allis shad	critical (under immediate threat of extinction)
Alosa alosa	(in NL extinct)
Sturgeon	extinct
Acipenser sturio	(in NL extinct)

crease in the diseases of the flounder (Vethaak et al., 1995). At Lauwersoog (Lauwersmeer), a lift-lock is used to enable fish migration. Further, a lot of plans are available for stimulating fish passage (Wadden Sea area: Den Helder(1), Oostoever(2), Termunterzijl(16), Roptazijl(7), Zwarte Haan(8), Harlingen(6), Dollard (Westerwoldsche Aa)(18) (see for locations Fig. 9); Delta region: Haringvliet, Oosterschelde).

The third type of measures comprises the restoration of entire salinity gradients, but as yet only some small scale projects have been executed. On the western part of the island of Terschelling a gradient between a dune area and the sea has been restored physically (Green beach project). On the island of Vlieland, the physical contact between an intertidal area reclaimed in the period 1905–1922 (Kroon's polders) has been restored in 1996, and surveyed in 1999. The levees around a number of reclaimed areas along the Wadden Sea have been deliberately breached locally, to enable seawater flooding during high tides, enabling a creek ecosystem with specific brackish water species to re-develop in the area. A similar process was initiated in some other places (Paessener lannen (Wadden Sea), Sieperdaschor (Westerschelde)) where naturally breached dikes were not repaired, so that the sea gained free entrance to the former, embanked polders (see also Bakker et al., 2001). The introduction of a reduced tidal movement in the polder Breebaart in the Ems estuary area (between discharge points 16 and 17 in Fig. 9) is part of an ongoing restoration project. The area is only 60 ha and the connection between the polder and the estuary will be realised by a tube through the dike. Further, a fish passage has been planned so that diadromous species can migrate again between the sea and the polder ditches connected to the main waterway (Doeglas, 1999).

So far, large scale restorations of salinity gradients have not been carried out, although the item has been discussed now for already ten years. In the Wadden Sea area plans have been presented for the area near Den Helder (Amstelmeer near Balgzand) and IJsselmeer, Lauwersmeer and Westerwoldsche Aa (preparation of management plan). Recently for the Haringvliet, a large former transition area between the river and the sea in the SW Netherlands (Fig. 1; Haringvliet south of Nieuwe Waterweg), a decision has been made to re-open the sluices in such a way that both the salinity gradient as well as the tidal movement will be partly re-established (Sonneveldt et al., 1997).

It is a well known fact that species diversity in enclosed and in open estuarine systems is increasing parallel with increasing salinities (Fig. 6 and de Jonge, 1974). Despite this fact, an estuarine environment experiencing an increase in salinity may nevertheless be valued negatively, due to the loss of specific brackish water elements. One of these typical brackish water

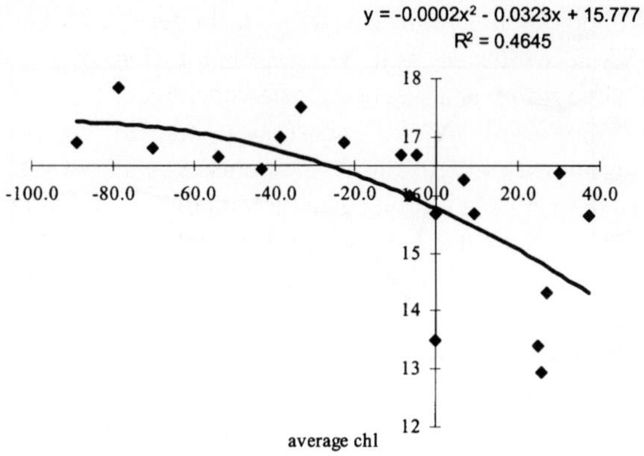

Figure 11. Relative (%) changes in surface area of eelgrass in the Grevelingen from year to year plotted as a function of salinity (Cl^-) in the period 1971–1998.

communities is the eelgrass community that has nearly vanished from the Grevelingenmeer. After the closure in 1971, the eelgrass stands significantly increased concomitantly with decreasing salinity (Fig. 11) until 1978 when 44 km^2 were covered by this species. This increase was explained (Anonymous, 1996) by the very low turbidity, absence of strong currents and low nitrogen concentrations. In 1979–1980, the salinity of the water body increased within a very short time period, resulting in stratification and die off of eelgrass rhizomes. Since 1980, the eelgrass beds increased again in size but from 1990 onwards the situation slowly deteriorated, and the surface area covered by eelgrass decreased steadily to near zero km^2 in 1998. The available data strongly suggest that this dramatic decline is related to the increased salinity over the last decade (Wijgergangs & de Jong, 1999; Kamermans et al., 1999). Together with the eelgrass beds, accompanying small fish species, fish eating birds and plant eating birds almost disappeared from the Grevelingenmeer. Despite the entire loss of this characteristic ecosystem, the final evaluation of the ecological situation in the Grevelingenmeer is judged positively (Anonymous, 1996). Similar problems with the eelgrass community can be observed in the Oosterschelde, apparently also caused by high salinity problems or related factors.

It should be noticed that the situation in the Grevelingenmeer and Oosterschelde is complex: the loss and importance of the extensive eelgrass beds and other estuarine aspects has to be weighed against the observed increase in species diversity related to higher (marine) and more stable salinities. In this discussion it is important to realise that a unique brackish/estuarine water community might become extinct while the increased species diversity does not represent a unique development, because most of the new species also can be found 'next door' in the North Sea.

The realisation of technical facilities to transport organisms from one to the other environment is a very first and artificial step in the improvement of specific estuarine populations. It is a typical engineering solution generally focused on the improvement of the population of a very specific part of the ecosystem, mostly fish species, sometimes a bird species. From the perspective of full development of the characteristic communities that we may expect in the brackish water environment or even in terms of sustainable development such an action is not satisfying. Changing water discharge schemes, including re-designing of e.g. sluices seems a better way to further improve and rehabilitate the many lost natural transition zones between the river and the sea. It may enable both upstream and downstream migration of organisms like fishes and the recovery of typical brackish water elements, like mussel beds, oyster beds and the lost eelgrass community which existence is favoured by a seasonal variation in salinity (van Katwijk, 2000).

From a scientific point of view, it is at least remarkable that the arguments for the existing restoration plans of estuarine gradients are derived from only a very small selection of species, such as migratory fish and birds. There has not been given any attention to the well documented brackish water communities in the former Zuiderzee (Redeke, 1922, 1936), the Wadden Sea (van Goor, 1919) and the south-western part of the Netherlands and to the existing knowledge about the crucial role of salinity levels and fluctuations. Based on all this unused knowledge, criteria could have been developed to serve as a sound scientific basis for adapted freshwater discharge schemes or the maintenance or the attainment of a certain salinity level. There is a big difference between a gradient that is relatively stable in time and in space, in contrast to a gradient that is always present in time but largely shifts in area, or one that is only present in certain periods of the year. The first, stable gradient-type offers opportunities for both migrating species and mobile and sessile brackish species, the second, only temporarily stable gradient-type only offers an opportunity for mobile and migrating species, while the sediment is almost a 'desert', and the third, unstable and unreliable gradient-type may only be suitable for migrating species.

Counteracting human induced restricted physical processes

Many human activities in coastal areas lead to constraints of natural processes. Some examples are the fixation of coastal dunes and the 'cliff fixation' of salt marshes. The consequences of the construction of the Oosterschelde Barrier also may be judged in this way. The dredging programs in the Westerschelde and Ems estuary led to an amplification of some of the physical processes (e.g. de Jonge, 1983).

Slufters

One of the consequences of the fixation of the coastal dunes (Fig. 1 between Westerschelde and Wadden Sea) is that the development of tidal dune inlets (the so-called 'slufters') is restricted to near zero. Nowadays 'slufters' are considered as very valuable and the remaining ones (Texel, Zwin, Zwarte Polder) are strongly protected. During the last years new 'slufters have been created: along the storm-surge barrier (ca 15 yr old), near Oostvoorne (ca 10 yr old) and near Bergen (ca 3 yr old), or planned (Schouwen). However, as a 'landscape element' the 'slufter' is the result of a moving coastline, occurring both in retreating and developing dune systems; both the development and the maintenance is a difficult enterprise in a completely fixed coastline as in the Netherlands. The existing 'slufters' require serious management measures, including regular dredging of the entrance channel to prevent silting up. So far, the results of the recently created tidal dune inlets are not very satisfactorily. The future will teach us whether it is possible to re-create 'slufters' also in a stable dune system.

Salt marshes

Erosion of salt marsh edges is a serious problem in the Wadden Sea and in the estuaries in the SW Netherlands. This erosion is, in fact, a natural phenomenon as in a natural estuary salt marshes come and go continuously due to natural processes. At present, the Dutch estuaries, including the Dutch Wadden Sea, are so 'designed' due to the land reclamations in the past, that hardly no salt marsh formation may occur any longer. In the SW Netherlands, most salt marshes originate from the mass introduction of *Spartina anglica* in the 1920s. As a consequence, all salt marshes there are of a similar age and in a similar stage of development. In the Dutch Wadden Sea, the salt marshes along the coast of Groningen and Friesland in fact only exist as the result of the massive effort to maintain them

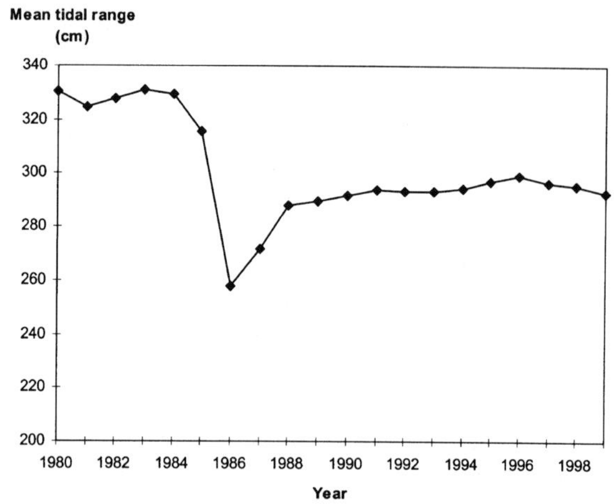

Figure 12. Reduction of the tidal range in the Oosterschelde near Yerseke due to the construction of the engineering works in the period 1977–1987 (data from Rijkswaterstaat).

there. This leads to the dilemma what salt marsh area is desired in an area. A sound historical or scientific base for a solid standard cannot be set. It is more or less a matter of perception what society, or better what nature conservation organisations want to have, and what this may cost.

Another aspect, connected to this problem, is the cyclic process in salt marsh formation. In a sound estuary, salt marshes come and go, they silt up and erode. In a natural situation, all stages of salt marshes, primary salt marsh, low and mature salt marsh, are present somewhere in the estuary. When a salt marsh becomes too highly elevated, it becomes vulnerable to erosion and after erosion it gives way for new salt marsh formation. Nowadays society hardly accepts this natural process of 'wax and wane', because so many salt marshes have been lost due to land reclamation. As a consequence, salt marshes tend to be protected and will grow higher and higher; Cliff fixation of salt marshes is considered a desirable option in many places to counteract erosion processes. Finally the marshes reach a height due to which they hardly can be considered as a salt marsh anymore. They should erode, or being rejuvenated by excavation measures. However, this is in conflict with the nature conservation aim of protecting the salt marshes as much as possible.

Oosterschelde Barrier

The Oosterschelde engineering works (1977–1987) concerned the construction of a storm-surge barrier and two auxiliary compartmentalisation dams. The ecological consequences of the construction of the dams are illustrative for what might happen when the boundary conditions for a system are changed. From the ecological research carried out (cf. Nienhuis et al., 1994 and references therein) it is clear that the changes in abiotic factors predominantly have dictated the new conditions for the biotic communities. These abiotic changes were changes in hydrodynamic conditions and related geomorphologic developments as well as changes in water quality. Primary consequences included a 13% reduction of the tidal range (Fig. 12), a 30% reduction of the tidal volume and reductions in current velocities up to 90%. These primary effects were the beginnings of numerous derived consequences.

An important physical result was the shift in erosion-sedimentation processes in the littoral zone from a more or less internal stable system to local erosion dominance, resulting in the morphological degradation of the sand and mud flats. They are gradually decreasing in height. The expectation is that they may even disappear largely below low water in the future. The change is mainly caused by the unbalance between the present hydrology and the given morphology: the tidal channels are now too wide and too deep for the present, reduced tidal prism, compared to the period previous to the storm-surge barrier. The only way to stop this process is by filling the channels up to a level that they are in balance again with the tidal volume. This requires several hundreds of millions of cubic metres of sand, which should be a very costly solution. Due to the storm-surge barrier the Oosterschelde changed from a sediment exporting system into a system that needs to import sediment. The disturbed balance between local erosion and sedimentation results in a decrease of the tidal flat area of about 30–35 ha yr^{-1}, a process which is still going on. Due to the decreasing elevation of the tidal flats the erosion of the salt marsh area increased generally to about 5–6 ha yr^{-1}, while the sedimentation rate decreased with about 50% (van Berchum & Wattel, 1997).

Due to the building of the auxiliary dams the salinity gradient disappeared completely which has caused a 'lagoonisation' of the former estuary into a bay now. Consequently, the salinity increased from 16.2 to 16.9‰ Cl^- and in summer salinity levels rise to 18‰ Cl^-. The lack of river input resulted in a decrease of nutrient concentrations by 20–60%. There is even a tendency into a 'reversed' nutrient gradient in that the concentrations near the seaside inlet are now higher than at the rear end of the basin, so that the Oosterschelde now mirrors North Sea conditions. The turbidity of the water decreased significantly, and primary production is now generally nutrient limited instead of light limited. In terms of primary productivity, the Oosterschelde has not shown large overall changes (Wetsteyn & Kromkamp, 1994; Klepper et al., 1994; Scholten & van der Tol, 1994). Considering benthic primary production, the importance of microphytobenthos increased from 18.10^6 kg C yr^{-1} to 26.10^6 kg C yr^{-1} (pre- and post-barrier period, respectively). In the present situation, about 25% of this production is generated in the sublittoral areas (Stapel & de Jong, 1997). This is possibly related to improved growth conditions due to decreased light extinction and less resuspension pressure due to reduced current velocities.

Ecological changes in terms of diversity have also been observed in Oosterschelde estuary. The change from an estuarine system into a tidal bay without significant salinity variations had large consequences. There is a general dominance of flagellates over diatoms now. Large heavy diatom species disappeared and were replaced by smaller diatom species and flagellates. The former summer species now occur in spring while the importance of the spring species diminished. Among the planctonic grazers the importance of the species *Acartia tonsa* was replaced by *Acartia clausi*. Further, the rotifer *Synchaeta* spp became an important species (Bakker, 1994).

Diadromous fishes decreased in numbers because of the lack of river water input, and the almost impossibility for migrating fish to pass the sluices in the Oosterschelde dams. The wader birds decreased in number from 181 000 to 125 000 while the ducks increased from 57 500 to 98 600. Although the less turbid water and the decreased hydrodynamics are in favour of seagrasses, both species steadily decline, due to the high salinity of the water (Wijgergangs & de Jong, 1999).The shellfish-fishery has negatively influenced predator prey relationships of oystercatchers and blue mussels, because scarcity in mussels on the intertidal flats changed the diet of this wader in favour of the edible cockle. In the pre-barrier system the diet consisted for 50% of mussels and 50% of cockles, while in the post-barrier system the diet was 20% blue mussels and 80% edible cockles (Meire et al., 1994).

The ecological developments sketched above are investigated in order to get insight in the actual rate and direction of the changes. There are no actual plans to counteract ongoing developments, and an important reason for this reserved attitude is that most negatively valued changes can only be tackled in a basin wide approach, making it quite difficult to fully understand and predict the consequences of possible interventions. Further, the economic costs of basin wide interventions will be very high.

Dredging in the Westerschelde and Ems estuary

In order to allow continuous freight traffic to and from the big harbour cities, the channels in the Westerschelde and the Ems estuary have to be dredged at many places. The (in)direct effects of dredging have been studied extensively. For the Ems estuary it has been demonstrated that dredging leads to increased turbidity and to shifts in primary production (de Jonge, 1983, 2000). For the Westerschelde, recently a new dredging programme has started, leading to further deepening of the channels. As a consequence, the tidal prism increased, leading to effects opposite those observed in the Oosterschelde. On the one hand, sand flats tend to increase in surface and to rise to a level where development of pioneer salt marsh may occur. It is doubtful whether these salt marshes may grow further to mature marshes, as they are not well protected against severe storms. Further, at salt marsh elevation the importance for benthic animals is rather small, and consequently this counts also for feeding wader birds. On the other hand, mud flats and salt marshes connected to the mainland are liable to severe erosion, often down to the compact peat layer underneath. Under these circumstances, the living conditions for benthic animals decrease strongly. Finally, there is a tendency that, owing to increased exposure, the areas with high energy sand ripples increase. Altogether a negative, downward spiral is to be observed, when considering the geomorphologic and ecological changes in Westerschelde ecosystems.

Apart from stopping the dredging program, ways to counteract these negative processes were thought to be either creation of new sand and mud flats and salt marshes or enlargement of the total estuarine area. The first option was considered as less realistic, because a gain in one type of ecosystem would mean a loss in another type; e.g. increase of salt marsh area would mean decrease in mud flat area. Thus, a plan has been developed to enlarge the total estuarine area by breaching the levees of a number of polders, enclosed in former centuries. This plan, however, encountered severe prohibited objections from the local people, and it had to be withdrawn. New ways to counteract the negative effects of dredging are again sought in the estuary itself, although there are only very limited possibilities in this highly anthropogenically influenced basin. In fact, only solutions to preserve the present salt marsh area are investigated at the moment. The lesson to be drawn from this 'failure' is that a proposal for a serious sacrifice of agricultural land has to be prepared properly, including sound communication and promotion, otherwise it will 'die in glory'.

Normative aspects in compensation measures

In order to meet desired natural developments, quite often standards and criteria are developed based on species that represent only a part of an ecosystem. Examples are the standards for surface areas of salt marshes in the Wadden Sea, and numbers of some breeding birds (e.g. plovers and terns) and seals. To meet these standards, sometimes very large efforts are to be made. Contradictory to these large efforts, a sound scientific basis for the standards is rare. The standards are generally based on feelings about 'how it should be' or 'how it might have been'. The consequence is that quite often standards for old situations are used in the present world, resulting in huge conflicts between the aims of the policy maker and the environmental manager on the one hand and the society, the ordinary citizen, on the other hand (see also the preceding paragraph). As long as these standards are fulfilled in a more or less 'remote' area like the Wadden Sea, the problems are relatively small and in fact mostly of financial nature. However, the plans for the restoration of almost barren breeding habitats for Kentish plovers and Little terns in the original beach areas meets large opposition, evoked by the amenity function of these areas for humans. Consequently, these plans are doomed to fail. However, new possibilities are created such as artificial islands in the newly formed lakes in the SW Netherlands, and in so-called 'inlagen', brackish coastal inland waters surrounded by wetland. These wetlands, however, require strong annual management measures, otherwise the breeding habitats will change into unsuitable vegetated areas within a few years. In fact, in a heavily populated country like the Netherlands, there is hardly any breeding place left for plovers and terns. The dilemma is whether to proceed with this type of expensive 'poultry-farming' or to focus on other potentially more successful restoration possibilities.

Reduction of noxious blooming of macroalgae

The problem of annual noxious blooms of macroalgae is rather specific for the Veerse Meer, a brackish and nutrient rich lake in the SW Netherlands. Due to the high nutrient levels in the water *Ulva* species are very abundant in this lake, and during summer sea lettuce is piling up in certain recreational areas, causing odour nuisance to the public. To counteract this problem, a plan has been made to build a new sluice in the Zandkreekdam, at the east entrance of the lake, and to flush the lake with saline water from the Oosterschelde. It is expected that this will lead to a decrease in nutrient concentrations and an increase in salinity, and consequently to less inconvenience caused by *Ulva*.

Compensation outside the tidal system

As mentioned before, it is often quite difficult to take restoration measures in a tidal system. A measure at one place might cause unforeseen and unwanted processes at another place, while it further is almost impossible to take appropriate measures due to the open tidal nature of the physically dominated estuaries. In such cases compensation for lost habitat is sought outside the specific tidal system. 'Managed retreat' might be considered as a new management option, although the creation of brackish inland areas close to, or even connected to the seawalls are already common practice. An example is the nature restoration scheme 'Tureluur' ('Redshank'), a large reconstruction scheme under execution in the south-western part of the Netherlands (Zeeland). The aim of this scheme is to create brackish wetland areas, comprising a salt-tolerant vegetation which will offer ample feeding and breeding opportunities to many bird species. The projects are inspired by similar activities in the past, which, although not aimed at nature development but at safety against flooding, led to the development of interesting wetlands for birds and vegetation (the so-called 'inlagen' and 'karrevelden' e.g. on the island of Tholen and in Zeeuws Vlaanderen in the SW Netherlands).

Concluding remarks

Processes

Coastal ecosystems are the result of complex interactions of large-scale and small-scale processes, implying a holistic approach in scientific investigations. Consequently, restoration of these systems primarily has to be realised by influencing the basic processes. This is in our opinion the only way to preserve or regain in a sustainable way ecological elements like species composition. Focusing only at one particular species (e.g. breeding terns) or a specific habitat (e.g. salt marsh) easily ends up in a sort of 'bio-agriculture' where the natural processes have been mainly expelled. This statement, however, does not exclude the species or mono-habitat oriented approach, but this has to be used with care.

Based on the available knowledge and experiences, it may be concluded that large interventions in estuarine and coastal processes are precarious, as they might have severe and often unexpected side-effects. Examples were discussed in preceding paragraphs: e.g. the reduction in tidal range of only 10% in the Oosterschelde resulted in a swing from stationary into eroding tidal areas; an increase of only 2‰ in salinity in the Grevelingenmeer resulted in the total disappearance of the eelgrass ecosystem including the accompanying fishes and birds; and a deepening of the shallow parts in the main shipping channel in the Westerschelde with about 2 m resulted in significant geomorphological changes in the sand and mud flats.

In general, coastal restoration should focus on the redirection of processes towards a desired status by stimulating certain processes or process parameters. Monitoring of the results and, if necessary, gradual readjusting of the governing factors, is an essential part of this approach. Physical, chemical and biological processes form the basis of restoration measures of coastal habitats, and this means that a thorough knowledge of these processes is essential. This includes basic concepts like serendipity ('chaos theory') which tells us that aspects like 'chance' and 'succession' also play an important part in restoration endeavours.

The spatial scale of restoration projects is also very important. It was shown, for instance, that large salinity fluctuations in small ecosystems have adverse effects on the long-term survival of many species, and that the spatial scale of a system is of utmost importance in maintaining specific ecological communities or community elements. This illustrates that rehabilitation success very much depends on the spatial scale of a system, because larger systems are over time more stable than small systems.

induced by civil-engineering works. Hydrobiologia 282/283: 575–592.

Oenema, O., 1988. Early diagenesis in recent fine-grained sediments in the Eastern Scheldt. PhD Thesis, Utrecht University.

Redeke, H. C., 1922. Flora en fauna der Zuiderzee. Monographie van een brakwatergebied (flora and fauna of the Zuiderzee; monography of a brackish water area). Ned. Dierk. Ver., Den Helder: 1–406 (in Dutch).

Redeke, H. C., 1936. Flora en fauna der Zuiderzee. Supplement (flora and fauna of the Zuiderzee; annex). Ned. Dierk. Ver., Den Helder: 1–176 (in Dutch).

Reise, K., K. Kolbe & V. N. de Jonge, 1994. Makroalgen und Seegrasbestände im Wattenmeer. In Lozán, J. L., E. Rachor, K. Reise, P. von Westernhagen & W. Lenz (eds), Warnsignale aus dem Wattenmeer – Wissenschaftliche Fakten. Blackwell, Oxford: 90–100 (in German).

Rijnsdorp, A. D. & A. D. Vethaak, 1989. Beschrijving van de populaties van bot (*Platichthys flesus*) in de Noordzee en het Nederlandse kust- en binnenwater (description of the population of *Platichthys flesus* in the North Sea, the Dutch coast and in inland waters). In: Ecologisch Profiel Vissen, Den Haag, RWS, DGW: 26 pp. (in Dutch).

Remane, A., 1934. Die Brackwasserfauna. Verh. Ddeutsch. Zool. Ges. 36: 34–74 (in German).

Rilov, G., A. Gasith, S. M. Evans & Y. Benayahu, 2000. Unregulated use of TBT-based antifouling paints in Israel (eastern Mediterranean): high contamination and imposex levels in two species of marine gastropods. Mar. Ecol. Prog. Ser. 192: 229–238.

Saeijs, H. L. F, 1982. Changing estuaries; a review and new strategy for management and design in coastal engineering. Rijkswaterstaat Communications no.32/1982. Government Publishing Office – The Hague. ISBN 90-12-03921-5.

Schmidt van Dorp, A. D., 1979. Literatuuronderzoek naar de soortenrijkdom van het makrozoobenthos in relatie tot het zoutgehalte. [literature review of species richness of macrozoobenthos in relation to salinity] DIHO, rapporten en verslagen nr. 1979–5 (in Dutch).

Schneider, E. D. & J. J. Kay, 1994. Complexity and thermodynamics: towards a new ecology. Futures 24: 626–647.

Scholten, H. & M. W. M. van der Tol, 1994. SMOES: a simulation model for the Oosterschelde ecosystem. Part II: Calibration and validation. Hydrobiologia 282/ 283: 453–474.

Scholten, M. C.Th., E. M. Foekema & R. G. Jak, 1996. Toxic anorexia of daphnids: a critical factor in the development of eutrophication problems. In Kramer, P. R. G., D. A. Jonkers & L. van Liere (eds), Proceedings of the Workshop 'Interactions of Nutrients and Toxicants in the Foodchain of Aquatic Ecosystems'. RIVM report no. 703715001: 49–53.

Selten, F. M., R. J. Haarsma & J. D. Opsteegh, 1999. On the mechanism of North Atlantic decadal variability. J. Climate 12: 1956–1973.

Sfriso, A. & A. Marcomini, 1997. Macrophyte production in a shallow coastal lagoon. Part I: coupling with chemico-physical parameters and nutrient concentrations in waters. Mar. Envir. Res. 44: 351–375.

Sfriso, A. & A. Marcomini, 1997. Macrophyte production in a shallow coastal lagoon. Part II: coupling with sediments, SPM and tissue carbon, nitrogen and phosphorus concentrations. Mar. Envir. Res. 47: 285–309.

Sfriso, A., B. Pavoni, A. Marcomini & A. A. Orio, 1992. Macroalgae, nutrients cycles and pollutants in the lagoon of Venice. Estuaries 15: 517–528.

Sommer, U., 1996. Nutrient competition experiments with periphyton from the Baltic Sea. Mar. Ecol. Prog. Ser. 140: 161–167.

Sonneveldt, H. L. A., N. M. de Rooij, A. J. Wijdeveld & R. C. J. Bakkum, 1997. MER beheer Haringvlietsluizen (EIA sluices Haringvliet). Report Delft Hydraulics WL, T1695 (in Dutch).

Stapel, J. & D. J. de Jong, 1998. Ontwikkelingen in de biomassa van het microfytobenthos in de Oosterschelde en Westerschelde in de periode 1981–1996 (development of biomass of microphytobenthos in the Oosterschelde and Westerschelde, 1981–1996). Rapport RIKZ-98.023: 57 pp. (in Dutch).

Van Berchum, A. M. & G. Wattel, 1997. De Oosterschelde, van estuarium naar zeearm (the Oosterschelde, from estuary to lagoon). Bekkenrapportage 1991–1996. Rapport RIKZ-97.034: 96 pp. (in Dutch).

Van Ginkel, R., 1995, Groen-zwart, Texels in het hart. Beschouwingen over een eilandcultuur (a description of the island culture of Texel). Amsterdam, Het Spinhuis: 107–130, footnotes: 133–139 (in Dutch).

Van Goor, A. C. J., 1919. Het zeegras (*Zostera marina* L.) en zijn beteekenis voor het leven der visschen (eelgrass and its importance in fish lifehistory). Rapp. Verh. Rijksinst. Viss. Onderz. I: 415–498 (in Dutch).

Van Katwijk, M. M., 2000. Possibilities for restoration of *Zostera marina* beds in the Dutch Wadden Sea. PhD Thesis Nijmegen: 151 pp.

Van Katwijk, M. M., D. C. R. Hermus, D. J. de Jong, R. M. Asmus & V. N. de Jonge, 2000. Habitat Suitability of the Wadden Sea for restoration of *Zostera marina* beds. Helgol. mar. Res. 54: 117–128.

Vethaak, A. D., J. G. Jol, J. Jungman & M. B. Meijer, 1995. Visziekten in de Waddenzee in 1994 vergeleken met de situatie in 1988 (fish diseases in the Wadden Sea in 1994 compared to 1988). Report RIKZ-95.039: 31 pp. (in Dutch).

WCED, 1987. Our Common Future. The Report of the World Commission on Environment and Development Oxford University Press: 400 pp.

Wijgergangs, L. J. M. & M. M. van Katwijk, 1993. Zeegrassterfte in het Grevelingenmeer; een studie naar demogelijke oorzaken van de afname van het Groot zeegras sinds eind jaren tachtig (die off of eelgrass in the Grevelingenmeer; possible causes of the decline since the end of the 1980s). Vakgroep Oecologie, werkgroep Aquatische Oecologie, KU Nijmegen: 60 pp. (in Dutch).

Wijgergangs, L. J. M. & D. J de Jong, 1999. Een ecologisch profiel van zeegras en de verspreiding in Nederland (an ecological profile of seagrasses and the distribution in the Netherlands). KUN/RIKZ-rapport: 75 pp. (in Dutch).

Hydrobiologia **478:** 29–51, 2002.
P.H. Nienhuis & R.D. Gulati (eds), Ecological Restoration of Aquatic and Semi-Aquatic Ecosystems in the Netherlands (NW Europe).

Restoration of salt marshes in the Netherlands

J.P. Bakker[1], P. Esselink[2], K.S. Dijkema[3], W.E. van Duin[3] & D.J. de Jong[4]
[1]*Laboratory of Plant Ecology, University of Groningen, P.O. Box 14, 9750 AA Haren, The Netherlands*
[2]*Koeman & Bijkerk bv, Ecological Research and Consultancy, P.O. Box 14, 9750 AA Haren, The Netherlands*
[3]*ALTERRA, Green World Research, P.O. Box 167, 1790 AD Den Burg, Texel, The Netherlands*
[4]*National Institute for Coastal and Marine Management/RIKZ, P.O. Box 8039, 4330 EA Middelburg, The Netherlands*

Key words: salt marsh, summerpolder, restoration, vegetation, seed dispersal, The Netherlands

Abstract

The conquest of land from the sea has been a long tradition in the Netherlands. When salt marshes were high enough, they were embanked when it was economically feasible, and transformed into intensively exploited agricultural land. This resulted in the transformation of halophytic communities to glycophytic communities. Often as an alternative, a low levee, a summerdike was built, which greatly reduced the flooding frequency of the landward summerpolder, hence creating a sedimentation deficit therein. Such summerpolders now cover 1200 ha in the Netherlands, 2100 ha in NW-Germany and small areas in England. Due to continuous embankments, the present salt-marsh area is relatively small with respect to the tidal basins. Discussions have been started how to increase the salt-marsh area. Two options will be discussed, firstly de-embankment of summerpolders and maintenance of the protective seawall, secondly increase of the effects of saline seepage behind the seawall by top soil removal. Both options include the restoration of salt-marsh communities (target communities) in intensively agriculturally exploited sites that have been salt marshes before. From the few examples abroad and experiments it is discussed (1) to which extent the sedimentation deficit in summerpolders could be compensated for, (2) if the soil seed bank is likely to contribute to re-establishment of salt-marsh communities, (3) if the dispersal of propagules of halophytic plants will be possible by hydrochory when the summerdike is breached, (4) to what extent is dispersal by endozoochory through waterfowl important in case re-establishment in a saline seepage area behind the seawall without open connection to the sea is envisaged. Two case studies of de-embanked summerpolders in the Netherlands revealed that the sedimentation deficit can be counteracted by rapid sedimentation, provided enough transport is possible from the foreshore. Dispersal by incoming tidal water from the nearby salt-marsh source area into the target area is possible for many salt-marsh plant species. The rate of success seems to depend on the relative position of source area and target area. A case study in a saline seepage area after top soil removal in the Netherlands, showed that the number of viable seeds dispersed by droppings from waterfowl is limited. Hence the possibilities for restoration of inland halophytic plant communities seem much lower than after de-embankment of summerpolders.

Introduction

Coastal salt marshes are valued as important habitats for their characteristic plant and animal life by many ecologists (Adam, 1990; Allen & Pye, 1992). The present area of coastal salt marshes in Europe along the Atlantic coast and the Baltic amounts to 175 800 ha (Dijkema, 1984; Dijkema, 1990; Allen & Pye, 1992). Salt marshes are either of natural origin, or have been artificially created by Man as a result of sedimentation works. Both natural and artificial salt marshes have often been embanked. The new seawall enabled better coastal protection and intensification of agricultural exploitation of the embanked polder. Most salt marshes have always been exploited agriculturally. Hence, untouched salt marshes are hardly found. Haymaking has gradually decreased and nowadays the most widespread use is livestock grazing. Seventy per-

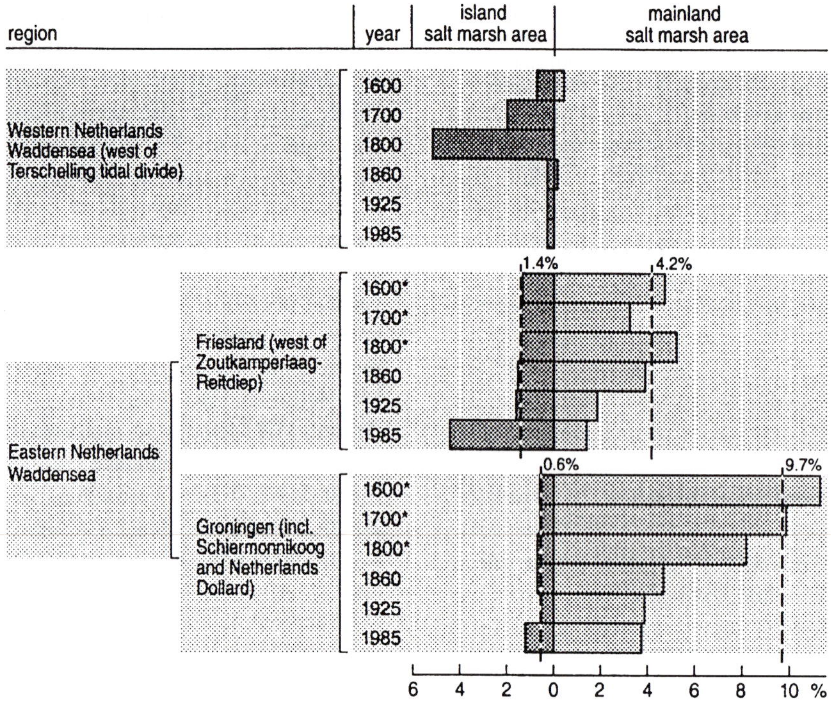

Figure 2. Salt-marsh area in the Netherlands Wadden Sea after 1600, in percentages of the tidal area. The former island of Huisduinen has been included. Data for 1985 including the Slufter on Texel, but excluding summer polders and mainland pioneer vegetation. ——- = mean salt-marsh percentage for the years 1600, 1700 and 1800. (After Dijkema, 1987.)

of embankment were roughly balanced in that period. The mainland salt marshes of the Dutch Wadden Sea area cover about 50 km^2 below this standard nowadays (Fig. 2).

Apart from the area, salt marshes suffer from another problem. The presence of the seawalls as landward boundary of salt marshes and the dependence of the existing marshes on human intervention to maintain the boundary with the intertidal flats, has stabilized the mainland marshes so that they have lost much of their dynamic character. Because vertical sediment accretion rates in these salt marshes are about 10 mm yr^{-1}, equivalent to at least four times the local contemporary increase of MHT, the spatial fixation means that the succession will inevitably lead to marsh maturation: extension of middle- and high-marsh communities at the expense of pioneer and low-marsh communities. The overall process is shown in Figure 4 for two salt-marsh complexes in the Dutch Wadden Sea coast that presently are protected by accretion works on their seaward boundary. In both areas construction of sedimentation fields reached its greatest extent in area in the mid-1960s, and the high relative contribution to the total marsh area of over 90% of pioneer salt marsh by 1960 indicates the young age of the marshes (see Esselink, 2000).

Historical development: Man and the coastal environment

The pristine coastal environment, i.e. before human impact has been described by Behre (1979), see also Bakker et al., (1997). A continuous sea-level rise has occurred since the last glacial period. This transgressive process continued until the mid-Holocene, after which the coastline stabilised more or less at its present position. The first time that the salt marshes became inhabitable, as a result of a decreased risk of flooding during a regression interlude, colonisation of the salt marshes started, by about 600 B.C. The first colonists came from the northwestern part of Lower Saxony between the Ems and Weser river. People from the Pleistocene Drenthe Plateau contributed to the salt-marsh occupation probably at a later stage, after 400 B.C. (Taayke, 1996). In response to increased risk of flooding during the next ingression, people started to build their dwellings on artificial mounds. It

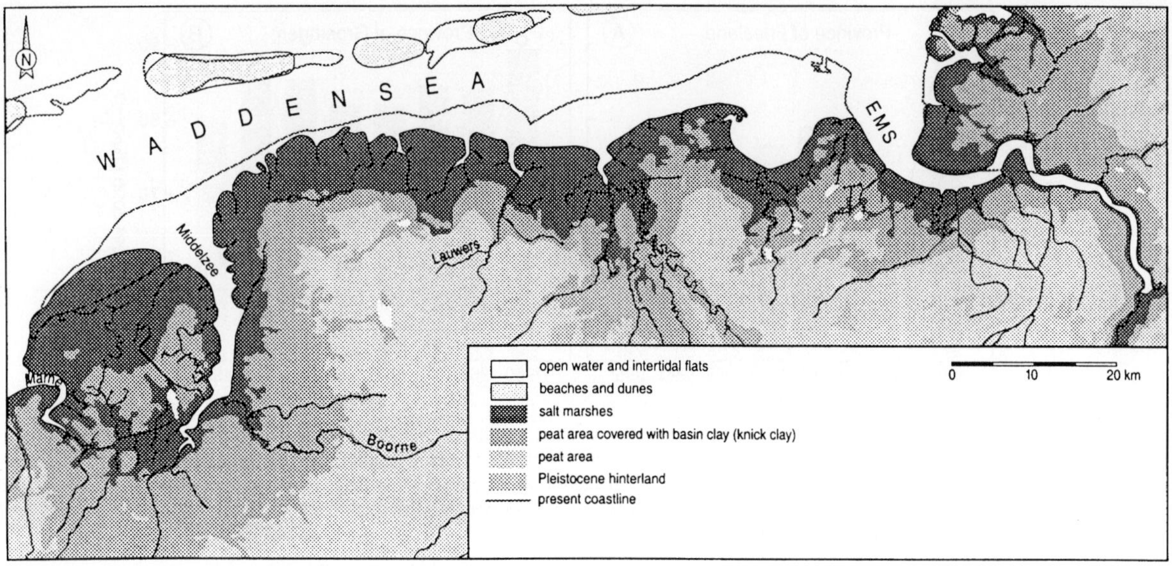

Figure 3. Paleographical reconstruction of the eastern Dutch Wadden Sea and the adjacent German sector by approximately A.D. 600, before human influence became an important factor in the geological development of the coastal area. The position of the barrier islands is inexact. The present mainland coastline is depicted for reference. (After Knol, 1993.)

is assumed that first a salt-marsh area was grazed by livestock. After further sedimentation the area became suited for arable field, and later for permanent settlement. During the arable field phase, low dikes might have been constructed to protect the crop from inundating salt water. Such dikes, enclosing some hectares of arable field, in what may be called the first summer-polders, have been found in Friesland already from the 1st century B.C. (Bazelmans et al., 1998). The mounds gradually grew higher through the accumulation of waste and dung. The macrofossils contain excellent records of former settlement and land use.

It has been possible to reconstruct the salt-marsh plant communities from the time of the mound settlements in the northern part of the Netherlands and adjacent Germany from palaeo-ecological samples (van Zeist, 1974; Behre, 1979, 1985). According to the macrofossils found, it must be assumed that the salt marshes were heavily exploited (Waterbolk, 1976; Bakker et al., 1993). Moreover, *Atriplex portulacoides*, *Artemisia maritima* and *Elymus athericus*, dominant species on ungrazed salt marshes, were rarely found as macrofossils. But Cappers (1994) points out that remains of *Atriplex portulacoides* and *Artemisia maritima* are not easily preserved and hence incorporated in the subfossil assemblage of the mounds. Nevertheless, it seems likely that by the year 0 A.D. no pristine mainland salt marshes were left.

At first, the only defence against inundations were for centuries the mounds and low dikes. The construction of dikes was probably a response to increased risk of flooding and relative sea-level rise (i.e. land subsidence) which was at least partly due to human impact. Much of the flooding of the land was enhanced by the exploitation of the peatland behind the coast. Levelling, drainage and subsequent oxidation after aeration led to subsidence, and the land became vulnerable to flooding (Oost & de Boer, 1994). Along the coast peat has been dug for fuel for at least 2000 years, and later on also for salt production. For this purpose and also for agriculture, the area was drained. Large parts of the mainland were flooded as a result of mining of peat. In the period 800–1300, the Zuiderzee (now IJsselmeer), Middelzee, Lauwerszee and Dollard were formed (Fig.1, sites 5, 4 and 2, respectively) in the northern part of the Netherlands (Oost & de Boer, 1994). Large floods also occurred in SW-Netherlands in the 14th, 15th and 16th century and created for instance the Braakman and Saeftinge as well as the large tidal area in the eastern part of the Oosterschelde (Fig. 1, sites 1 and 8, respectively) (Beeftink et al., 1982).

Large-scale counter measures against flooding started around 1000 A.D. in the northern Netherlands, when seawalls were constructed to enclose sheltered areas and to reclaim lost land (Waterbolk, 1976).

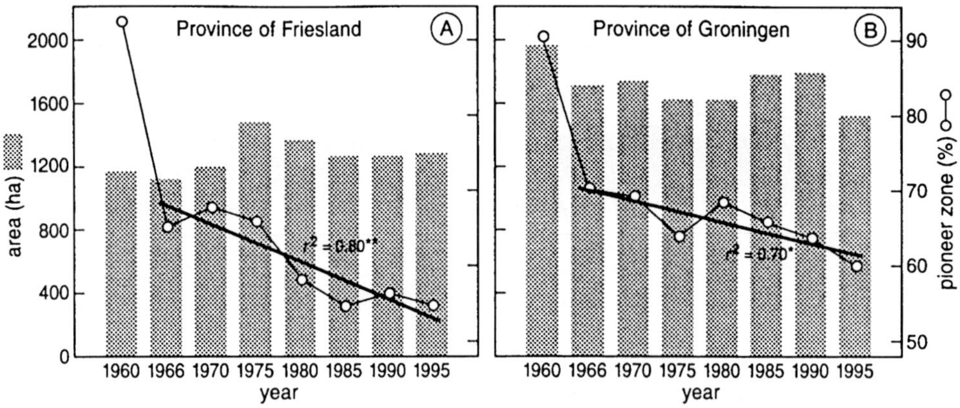

Figure 4. Development of the area of artificial salt marshes (left vertical axis), and the decline of the relative contribution of pioneer salt marsh (right vertical axis) on two stretches of the Dutch mainland coast in the Wadden Sea. Sedimentation fields surrounded by brushwood groynes reached their greatest extent during the mid-1960s. At present, sedimentation works are primarily maintained to preserve existing marshes. The total area did not show a statistical trend on either of the stretches between 1966 and 1995, but the relative share of the pioneer zone declined significantly. $^*P<0.05$, $^*P<0.01$. (After Esselink 2000.)

Seawall building started in adjacent Germany in the 11th century (Behre, 1985). Also for SW-Netherlands, the first seawalls were mentioned in the 11th century (Beeftink et al., 1982).

Restoration of salt marshes and salt-marsh communities

We define restoration as an attempt to bring back destroyed habitats to an original state. Here it deals with changing the present state of agriculturally intensified sites into a (more valuable) former state. It may be carried out at either the original site, or as a replacement at a different site. However, in the latter case mitigation might be a better term.

The effects of restoration management can only be measured after targets have been set. The definition of targets and monitoring the effects of ecological restoration should be strongly interrelated. Monitoring is a *sine qua non* for the evaluation of the effects of ecological restoration. If the target is met within a certain time span, the ecological relationships in the ecosystem under study are apparently clear enough to enable reliable predictions. When the target is not met, the ecological relationships, including the effects of restoration, are not yet clear. The important conclusion should be that further research is necessary to unravel the mechanisms in the ecosystem under study in order to increase the predictive power for future ecological restoration in similar ecosystems.

The targets can be both abiotic and biotic. The abiotic target is to restore saline conditions. Moreover, an interaction between tidal amplitude and soil level should be restored to such an extent that zonations can develop characteristic of different levels with respect to MHT. These zonations should harbour typical plant and animal communities, the biotic target. The plant target community concept can easily be adopted in countries with a tradition in plant community classification. Otherwise the concept of target species can be used and derived from reference sites (Bakker et al., 2000). Targets can imply prerequisites for the establishment of plant (and animal) species, but their very establishment is the ultimate target, as it integrates the suitability of abiotic conditions and plant traits.

The area of mainland marshes in the Wadden Sea has strongly decreased during the last century (Dijkema, 1987). Moreover, the proportion of pioneer salt marsh is decreasing as a result of maturation within the presently fixed salt-marsh area (Fig.4; Esselink, 2000). Both the total salt-marsh area and that of the pioneer zone could be enlarged by further artificial extension of sedimentation fields into the intertidal area. Another possibility has been practised in the newly embanked areas of Hauke-Haienkoog and Beltringharderkoog, Germany, where sluices were built to maintain some tidal influence. This alternative cannot be considered as very successful, since extensive salt-marsh areas can not be maintained, due to the decrease in tidal amplitudes by nearly 100%. This results in a shift of the salt-marsh area: former salt marshes are transformed into fresh marshes, and former intertidal flats become salt marshes as shown in the Beltringharderkoog (Wolfram et al., 2000). A shift of vegetation zones towards lower parts of the salt marsh for the Oosterschelde inlet in the SW Netherlands after reduction of the tidal amplitude by 10% was reported (de Jong & van der Pluym, 1994; de Leeuw et al., 1994). Both methods imply a decrease of the area of intertidal flats and may, therefore, conflict with other nature conservation interests like foraging area for wading birds. Another option for restoration is to aim at a long-term accretion-erosion cycle in the most seaward fringe of a marsh (see Esselink, 2000). Finally, options may be realised at the landward side of the salt marshes, and enlarge salt marshes at the expense of earlier embanked marshes. Minor options for restoration of halophytic communities seem the possibilities to increase salinity at the landward side of the seawall by (1) using saline seepage water high in the soil profile by top soil removal or (2) the inlet of seawater through sluices. The best option is landward retreat of the marshes by de-embankment of summerpolders. In this option, the entire tidal amplitude will be maintained enabling salt-marsh development by the interaction of abiotic and biotic components.

The theory: possible abiotic and biotic constraints

Before revealing case studies, possible mechanisms on restoration of salt marshes and salt-marsh plant communities will be discussed. Embanked summerpolders have experienced a low number of annual inundations only during storm periods. This implies that salinity has strongly decreased and that hardly any sedimentation took place since the embankment of the summerpolder. Moreover, drainage of the polder may have caused some shrinkage of the clay soil. Therefore, the summerpolder lies at a lower level with respect to present MHT than the foreland salt marsh. De-embankment will quickly increase the salinity, but to what extent can the sedimentation deficit be counteracted? Restoration of the abiotic conditions is an important prerequisite for restoration of characteristic communities.

Summerpolders have been applied with artificial fertilisers and sown to productive pastures or arable fields. The summerpolders only suffer from sea water inundations with extreme weather conditions: ranging from a few times per year to no inundations during several subsequent years. Therefore, halophytic plant communities have disappeared. The polders are quickly drained by sluices after inundations. Halophytic communities affected by saline seepage behind the seawall have been reduced or disappeared as a result of drainage, fertiliser application, sowing for productive pastures or arable fields. The first restoration projects are currently in progress by removal of the top soil. This measure reduces both the high nutrient level and the effects of drainage.

There might be constraints for sessile plants in the restoration of salt marshes. Are salt-marsh plants still available in the community pool as persistent seeds in the soil seed bank? If not, can diaspores then, after removal of the summerdike, easily be dispersed by the incoming tidal water from the foreland salt-marsh source area? In the case of saline-seepage areas, the question is how diaspores can reach the target area, since the area is not open to tidal currents. Species characteristic of the halophytic communities are further referred to as target species. We should try to get an impression of the possibilities of establishment of target species from the local species pool (Zobel et al., 1998).

Sedimentation and biogeochemistry

In coastal systems with dominant sedimentation processes, minerogenic coasts, embanked former salt marshes are situated at a lower level with respect to MHT than foreland salt marshes with continuous sedimentation: since banks block the input of inorganic sediment (Warren & Niering, 1990). Differences in soil level will be larger when the polder is severely drained for agricultural purposes. This will also hold for coastal systems with accumulation of peat, organogenic coasts, where strong subsidence can take place (Roman et al., 1995). In all embanked polders, drainage accompanied by ditching lowers the groundwater table, allows sediment aeration, subsequent decomposition of organic matter, and hence shrinkage (Roman et al., 1984). Salt-marsh plant communities will be transformed into freshwater communities. Measurements in north-eastern American coastal systems suggested that the restoration of seawater flow will likely extend flooding depth and duration, lower the redox potential, increase cation exchange, and subsequently increase NH_4, Fe(II) and PO_4 mobilisation (Portnoy & Giblin, 1997A). Microcosm experiments in the greenhouse confirmed these findings. Seawater reintroductions may cause porewater sulphides to increase, likely reducing the success of re-vegetation (Portnoy & Giblin, 1997B).

In organogenic coastal areas such as the Baltic with tidal ranges of about 20 cm in Denmark (Vestergaard, 1997) to zero tidal range towards the east, removal of the summerdike might result in permanent stagnant pools containing slightly brackish water without any vegetation after die-off of the freshwater community. These events were indeed found at Karrendorfer Wiesen near Greifswald where a summerdike surrounding an area of about 350 ha was completely removed in 1993 (Bernhardt et al., 1996; Müller-Motzfeld, 1997).

Another experiment was carried out in south-east England, where a seawall was breached in 1995 and seawater was allowed to flow onto a site of 21 ha that has been a polder for at least 150 years. The mean annual rate of sedimentation at the salt marshes near Tollesbury marsh is estimated 4–5 mm yr^{-1} during the last 5 years. The first 2 years after breaching the seawall a sedimentation of approximately 47 mm yr^{-1} was estimated (Boorman & Hazelden, 1996). Further sedimentation and particularly the establisment of salt-marsh plant communities have to be awaited (Boorman, 1999).

As only little knowledge exists on sedimentation in former summerpolders, we might learn from experiences in salt-marsh systems where management has changed. In the Dollard Bay, part of the Ems Estuary at the border between Germany and the Netherlands (Fig.1), abandonment of the drainage system in the artificial marshes led to an increase of variations in elevation, and abiotic diversity. After designation to a nature reserve in 1981, grazing with cattle continued, but at reduced stocking rates (0.5–1 animal ha^{-1}), whereas the maintenance of the drainage system was abandoned by 1984. Many ditches were filled in, which resulted in decreased drainage of the marshes and the development of temporarily waterlogged depressions. Vertical accretion rates, derived from levelling programmes carried out in 1984 and 1991/1992, were generally correlated negatively with (a) the initial marsh elevation in 1984, (b) the distance from the tidal flats, (c) distance from main creeks, and (d) in many cases, the distance from minor creeks (Esselink et al., 1998). As a consequence of a gradient in grazing intensity, vegetation density decreased from the outer marsh towards the seawall, and was probably one of the causes in the aforementioned accretion patterns. After the abandonment of drainage, the number of levees along minor creeks (former ditches) increased, as did the elevation differences at many already existing levees (Esselink et al., 1998). The levee development was more pronounced towards the seawall. This could be explained by the greater differences in vegetation structure between the levees and the marsh interiors towards the seawall, and indirectly by the gradient in grazing intensity in the marshes (Esselink et al., 2002). The possible effect of grazing on the sedimentation rate was also demonstrated in the mainland marshes of the Leybucht, Germany. After 4 years of measurements a rate of sedimentation of 21 mm yr^{-1} was found in ungrazed sites at MHT +40 cm, and 16–17 mm yr^{-1} in grazed sites (Erchinger et al., 1996). This has resulted to differences of a few cm near the seawall and up to 20 cm near the intertidal flats in the grazed and ungrazed marsh after 17 years (van Wijnen, 1999).

A possible consequence of the abandonment of the drainage system in the Dollard was a negative effect on the overall vertical accretion rates. These ranged from 6.6 mm yr^{-1} to 11.4 mm yr^{-1} among different study sections. Taking into account the sheltered position of the marshes in the Dollard, these values are rather low in comparison with other artificial marshes in the Wadden Sea, which show vertical accretion rates of well above 10 mm yr^{-1} (Dieckmann, 1988; Dijkema et al., 1990; Erchinger et al., 1996). All these marshes still have a well-maintained drainage system, which will prevent the development of badly drained and sparsely vegetated depressions (Jakobsen, 1954; Dijkema, 1983). Though negative accretions could be measured locally in the Dollard (Esselink et al., 1998), on average, vertical accretion rates in these depressions were positive, and ranged from 5.6 mm yr^{-1} to 9.2 mm yr^{-1}. These values exceed the average increase of MHT, which indicates that the development of the waterlogged depressions may not be a threat to the marshes. Maintenance of the artificial drainage system was also suspended in a cattle-grazing trial in the Leybucht marshes, Germany, but it was resumed a few years later, since the impact of trampling by cattle on the sward was considered too great (Erchinger et al., 1996).

The rise of MHT level of 3.7 mm yr^{-1} from 1960 to 1990 was independent of the relative sea-level rise. Moreover, there are no predictions of future MHT levels. The MHT rise has to be considered in relation to the level of both the salt-marsh zone itself and to the tidal flat just in front of it. It seems that the potential loss of salt-marsh area through erosion from the seaward edge is not related with the sedimentation processes in the salt marsh itself, but is determined by sedimentation processes in the pioneer zone in front

Table 2. Longevity index of salt-marsh species ranging from 0 to 1 (according to Bekker et al., 1998), in parentheses the number of records (Thompson et al., 1997)

Agrostis stolonifera	0.42 (84)
Armeria maritima	0 (10)
Artemisia maritima	– (0)
Aster tripolium	0.75 (20)
Atriplex littoralis	– (0)
Atriplex portulacoides	– (0)
Atriplex prostrata	1 (15)
Carex extensa	– (0)
Carex distans	– (0)
Elymus athericus	0 (5)
Festuca rubra	0.12 (119)
Glaux maritima	0 (10)
Juncus gerardi	0.63 (8)
Juncus maritimus	1 (1)
Limonium vulgare	0 (5)
Plantago coronopus	0.33 (3)
Plantago maritima	0.11 (9)
Potentilla anserina	0.17 (18)
Puccinellia maritima	0.39 (23)
Salicornia europaea	0.71 (27)
Spartina anglica	1 (1)
Spergularia maritima	0.33 (3)
Spergularia salina	0.58 (12)
Suaeda maritima	0.09 (11)
Triglochin maritima	0 (5)

of the marsh (Dijkema et al., 1988; Boorman et al., 1989; Houwing et al., 1999). Sea-edge erosion may result in large-scale cliff formation. If cliff erosion is not prevented by groyness artificial salt marshes may disappear in the longer term (Dijkema, 1994).

Seed bank

Only a few authors published on the seed banks of salt marshes (Milton, 1939; Ungar & Riehl; Jefferies et al., 1981; Ungar, 1988; Bakker, 1989; Hutchings & Russell, 1989; Bernhardt, 1992; Bakker et al. 2001). Data from a number of authors have been summarised by Adam (1990). Many seed bank data were compiled in the seed bank database of the NW European flora (Thompson et al., 1997). The database enables to calculate a longevity index (Bekker et al., 1998; Thompson et al., 1998). Although salt marshes are underrepresented in the seed bank database, the data available suggest that only very few salt-marsh species may have a long-term persistent seed bank (Table 2). In general, seed densities are low in salt marshes (Adam, 1990). *Spergularia salina* is the exception; it is the second species with respect to seed densities in the seed bank database with over 375 000 seeds m^{-2} in 0–5 cm soil (Ungar, 1988).

The viable soil seed banks in a summerpolder in Noord-Friesland Buitendijks, The Netherlands (Fig. 1) and in the adjacent high salt marsh were estimated in April 1997. Inundations with seawater did not occur during the two previous winter periods. The vegetation of the salt marsh was dominated by *Plantago maritima*, *Spergularia maritima*, *Puccinellia maritima*, *Glaux maritima*, and *Suaeda maritima*. The vegetation in the summerpolder was characterised by *Elymus repens*, *Alopecurus geniculatus*, *Potentilla anserina*, *Leontodon autumnalis*, *Poa trivialis* and *Agrostis stolonifera*. The soil seed bank was quantified by taking twelve soil cores, diameter 7 cm, of the top soil 0–5 cm. This procedure was repeated in ten replicates. The samples were concentrated and grown in the glasshouse, according to the method described by ter Heerdt et al. (1996).

The composition of the soil seed banks on the salt marsh and the summerpolder differed strongly. The comparison of the number of viable seeds at the salt marsh and the summerpolder revealed that seeds of the halophytic species were found in much lower numbers in the summerpolder (Table 3; Bakker et al., 2001). Small amounts of seeds of *Spergularia* spp., *Atriplex prostrata* and *Puccinellia maritima* were found in the summerpolder. These seeds might have been transported by previous inundation events. The data suggest that the return of many target species can not rely on the soil seed bank of the summerpolder.

Dispersal of propagules

Dispersal of propagules by tidal currents was estimated by Huiskes et al. (1995) in SW-Netherlands. Most propagules were trapped in floating nets, particularly seeds of *Spartina anglica*, *Triglochin maritima*, *Limonium vulgare* and *Elymus athericus*. Seeds of *Salicornia* spp., *Aster tripolium* and *Spergularia* spp. were more caught in standing nets. These propagules were apparently mostly dispersed in the watercolumn. This also holds for seedlings of *Salicornia* spp., *Triglochin maritima*, *Aster tripolium* and *Limonium vulgare*. Other salt-marsh species were trapped in low numbers. To which extent dispersal of propagules

Table 3. Number of viable seeds m^{-2} in the 0–5 cm soil layer in the salt marsh and summerpolder of Noord-Friesland Buitendijks (site 6 in Fig. 1). *Spergularia maritima* and *S. salina* might have been incorrectly identified. Species are arranged according to salinity class: 0 =fresh; 1 = slightly brackish; 2 = halophyte (adapted from Scherfose, 1990). The log ratio was calculated as log (number of seeds in the salt marsh + 1)/log (number of seeds in the summerpolder + 1) (after Bekker et al. 2001)

	Salinity class	Salt marsh	Summer polder	Log ratio
Glaux maritima	2	16942	0	4.2
Plantago maritima	2	1062	0	3.0
Suaeda maritima	2	2405	65	1.8
Spergularia maritima	2	116773	4420	1.4
Spergularia salina	2	5521	282	1.3
Salicornia spec.	2	1257	65	1.3
Atriplex prostrata	2	14082	2513	0.7
Puccinellia maritima	2	520	412	0.1
Agrostis stolonifera	1	1018	650	−0.2
Polygonum aviculare	1	390	845	−0.3
Potentilla anserina	1	5438	54898	−1.0
Elymus repens	1	0	87	−1.9
Alopecurus geniculatus	1	43	15859	−2.6
Leontodon autumnalis	1	0	1473	−3.2
Poa annua	0	0	152	−2.2
Rumex crispus	0	0	325	−2.5
Matricaria maritima	0	0	347	−2.5
Taraxacum spec.	0	0	758	−2.9
Bellis perennis	0	0	1083	−3.0
Poa trivialis	0	0	2123	−3.3

by tidal currents implies an exchange between different salt marshes is not clear. Comparatively few propagules were transported into the marsh from elsewhere by incoming tide, whereas a significantly higher number of propagules was transported out of the marsh with ebb currents (Huiskes et al., 1995). The viability of the seeds was not tested in this experiment. Koutstaal et al. (1987) found that the floating capacity in basins of several halophytic species amounted to weeks and that most species did not loose their germination capacity. This suggests a potentially good dispersal capacity.

Large-scale sampling of driftline material in over 60 stations along the entire coast (200 km) of Niedersachsen, Germany, revealed a close relationship between the species composition of the driftline and its origin. Along the exposed coasts mainly halophytic species were found, whereas species characteristic of brackish reeds were deposited in bays and estuaries (Persicke et al., 1999). Apparently driftline material did not travel long distances. More detailed studies demonstrated a close relationship between the composition of the vegetation, quantified by relevés, of the foreland salt marsh and the driftline material. Discrepancies were only found for species with a proportion less than 1% in the driftline (Gerlach et al., 1999). Seeds formed only a small part of the driftline: mainly seeds of *Atriplex* spp. *Elymus* spp., *Aster tripolium*, *Phragmites australis*, *Triglochin maritima* were found (Persicke et al., 1999). Their viability was not tested. Waisel (1972) reported seeds of *Atriplex portulacoides*, *Limonium vulgare*, *Puccinellia maritima*, *Salicornia* spp., *Suaeda maritima*, *Triglochin maritima*, *Elymus athericus* and *Atriplex prostrata* in driftline material in the U.K., but again their viability was not tested. Driftline litter does include viable seeds, as shown for 13 salt-marsh species at an island salt marsh in the Netherlands (Bakker et al., 1985).

Driftline material deposited at the summerdike of the polder in Noord-Friesland Buitendijks, The Netherlands (Fig. 1) was sampled in order to estimate potential dispersal towards the summerpolder from plant species thriving at the salt-marsh foreland. Samples deposited during the previous autumn and from a freshly deposited driftline were collected in April 1997. From both driftlines ten replicate samples of 20 × 20 cm were grown in the glasshouse. The larger amounts of viable propagules in autumn as compared to the spring are striking. Probably seeds from most species dropped onto the soil during the winter period. We conclude that propagules of eight salt-marsh species potentially can be dispersed by tidal currents (Table 4; Bakker et al. 2001). After a storm that inundated the summerpolder in February 1999, viable seeds of these species were indeed found in the 0–1 cm soil layer of the polder (Table 4). *Puccinellia maritima* emerged from vegetative parts that can float. This phenomenon was also reported by Adam (1990). Most species are also found in the salt-marsh foreland, with the exception of *Elymus athericus* and *Artemisia maritima*. Like *Matricaria maritima* and *Polygonum aviculare*, they grow at the foot of the summerdike.

At the aforementioned Karrendorfer Wiesen in the Baltic, Germany, where the summerdike was removed, diaspore dispersal into the former summerpolder was estimated by using traps. Seeds of *Spergularia salina* and *Juncus ambiguus* were found during the entire season, therefore probably dispersed by wind according to Bernhardt (1998). However, windtunnel experiments suggest that the vector wind dispersal

Table 4. Plant species composition at the salt-marsh foreland of Noord-Friesland Buitendijks (site 6 in Fig.1), expressed as the frequency of occurrence in 100 subplots of 0.4 m^2. *Spergularia maritima* and *S. salina* might have been incorrectly identified. Mean number of viable seeds in 10 replicates of 20 × 20 cm driftline material, deposited in autumn 1996 and spring 1997. Mean number of viable seeds in 10 replicates of 100 cm^2 in the 0–1 cm topsoil after inundation in February 1999 (after Bakker et al., 2001)

	Low	High	Mean number of propagules m^{-2}		Mean number of viable seeds in 0–1 cm m^{-2} February 1999 summerpolder	
	salt marsh		Autumn 1996	Spring 1997	high	low
Atriplex prostrata	2	44	6658	555	2	5
Suaeda maritima	85	61	3353	288	4	4
Plantago maritima	1	90	1738	53	14	7
Potentilla anserina	0	29	403	13	62	187
Polygonum aviculare	0	0	380	35	1	7
Spergularia maritima	34	83	153	0	0	0
Glaux maritima	0	68	118	15	1	0
Matricaria maritima	0	0	78	5	91	12
Puccinellia maritima	43	79	55	63	0	0
Elymus athericus	0	0	55	3	1	0
Agrostis stolonifera	0	22	50	13	159	858
Aster tripolium	2	8	23	3	2	1
Salicornia spec.	100	26	13	10	22	17
Artemisia maritima	0	0	5	8	0	1
Spergularia salina	0	0	0	0	88	60
Juncus gerardi	0	0	0	0	7	0
Scirpus maritimus	0	0	0	0	2	0

may be overestimated (Strykstra et al., 1998). Seeds of *Agrostis stolonifera* were only found after ripening of the seeds, and were supposed to be transported by water. Propagules of some target species of brackish marshes were found in driftline material such as *Atriplex prostrata*, *Triglochin maritima*, *Spergularia salina*, *Puccinellia distans*, *Scirpus maritimus*. Establishment in the open soil, after die-off of the dominant *Elymus repens* stands was documented so far for *Spergularia salina* and *Puccinellia distans* (Bernhardt, 1998).

Dispersal of propagules by animals is well-known. Seeds and fruits of halophytic plant species may be transported internally or externally by birds (de Vries, 1961; Olney, 1963; de Vlaming & Proctor, 1968; Proctor, 1968; Siira, 1970). Birds may thus provide a potential vector for long-distance dispersal (Adam, 1990). Bakker et al. (1985) found viable seeds of salt-marsh species in both geese pellets and cattle dung patches at a salt marsh in the Netherlands. Livestock grazing is practised in many European salt marshes and may hence play a role in short-distance dispersal of halophytic species.

Expectations for plants

From the aforementioned we might derive possibilities and constraints for the restoration of halophytic plant communities in former summerpolders with open connection to the sea after breaching or complete removal of the summerdike, and in polders behind the seawall after top soil removal and increase of the effect of saline seepage water.

The rate of sedimentation is uncertain, it may be higher than expected because of the low position with respect to MHT (see Tollesbury), or lower because of poor drainage (see the Dollard). Differences in soil level may increase. After inundation by salt water, the glycophytic vegetation will die off. It should be studied what the repercussions will be for evapotranspiration from the then bare soil with possibly high salinity in the topsoil and its effect on seedling establishment. Will long periods of inundation result in algal blooms that on their turn might affect seedling establishment? To which extent will seedlings survive herbivory? It is expected that the seed bank in the summerpolder will hardly contribute to the establishment of the target community. Propagules have to be dispersed from elsewhere. Salt-marsh communities at a large distance are not likely to contribute much to dispersal by tidal currents. Dispersal by water from adjacent salt marshes seems possible for many species. Hence sowing or planting of target species is no option. Dispersal by wind is not expected for most species. Long-distance or short-distance dispersal by birds might occur for some species.

In summary, restoration of halophytic plant communities might be successful in former summerpolders. They might be restored in polders behind the seawall, when the target species are nearby in the polder area, otherwise restoration seems unlikely. These expectations will be tested in a few case studies available in the Netherlands.

Expectations for animals

In the previous section it was suggested that birds may play a role in the dipersal of propagules. This implies that restoration of salt marshes is expected to be immediately successful for birds, as they have little dispersal problems. The success should be measured at a somewhat longer term, including possibilities for nesting, and food availability for chicks.

The occurrence of birds depends on the grazing management (Bakker et al., 1997). This is true for established marshes, but will also be the case for marshes to be restored. The effect of grazing on the function of salt marshes as an area for breeding birds was studied in the German Wadden Sea area. An increase in the number of breeding oystercatchers (*Haematopus ostralegus*) was observed, which was positively related to the increased grazing of artificial salt marshes, in contrast the number of redshanks (*Tringa totanus*) decreased. The avifauna was in this case negatively influenced by the frequency of flooding, the trampling of nests and young, and the increased stocking rate; the vegetation composition and structure was considered less important (Schultz, 1987). However, in salt marshes in Great Britain the occurrence of redshank densities were positively related with the extent of *Elymus athericus* community. This relation could be explained by variation in vegetation structure. Heavily grazed plots, dominated by *Elymus athericus* community supported the most structurally diverse vegetation and the highest breeding densities. In contrast, ungrazed plots of similar habitat contained tall, uniform vegetation and supported significantly lower breeding densities (Norris et al., 1997). In the brackish Dollard marshes in the Netherlands, redshank breeding densities were also higher in *Elymus repens* stands than in other plant comunities. Due to a decreasing stocking rate the tall *Phragmites australis* community increases adjacent to the intertidal flats resulting in a decrease of breeding area. At other parts of the marsh redshank numbers declined, probably due to a decrease in the area of more marginal habitat, and a delay of sufficient vegetation cover to nest in spring, induced by the increased exploitation of the marsh by barnacle geese (Esselink, 2000).

Winter- and spring-staging geese prefer short canopy where they can forage on their preferred plant species (van der Wal et al., 2000). During succession from bare soil to older salt marshes the canopy gets taller, hence geese can no longer forage at the older stages of a productivity gradient (van de Koppel et al., 1996). Grazing by cattle facilitates for geese at older successional stages (Olff et al., 1997).

Salt marshes feature characteristic halobiontic invertebrates when plants are allowed to set flowers. This is especially the case during the first years after the cessation of livestock grazing. However, both heavy grazing and long-term cessation of grazing result in stands dominated by a single plant species and, therefore, are negative for the invertebrate fauna (Andresen et al., 1990).

Based on this evidence we expect that grazing regimes will play an important role in the development of animal life in the restoration of salt marshes.

Successful restoration projects

Peazemerlannen

The two summerpolders near Peazemerlannen (53° 21′ N, 5° 45′ E) (Fig. 1, site 7) resulted from artificially enhanced sedimentation by brushwood groynes and improvement of the drainage by ditching. The youngest one was embanked after 1930 and exploited as pasture. The summerdike of this polder breached at one point in April 1973. The dike of the older summerpolder remained intact. The following winter season had many storms and attempts to repair the summerdike were suspended. In 1979, another part of the dike eroded, and from that period onwards the dike was gone over a distance of about 500 m, giving free access to the sea in an area of about 100 ha. The de-embanked summerpolder is no longer grazed by livestock. Levelling with respect to MHT in 1964 revealed a relatively flat topography ranging from MHT −36 cm to MHT +34 cm. The first studies in this unintentionally restored salt marsh were carried out in 1995. It was clear that a new salt marsh had developed. At that time, the former summerpolder featured levees with *Elymus athericus* stands, a low salt marsh with *Puccinellia maritima* stands, and waterlogged depressions with *Salicornia* spp. stands, ranging from MHT −21 cm to MHT +109 cm (van Duin et al., 1997). Apparently sedimentation took place and differences in soil level increased.

Long-term net accretion (result of sedimentation and shrinkage of the soil in dry summer periods) was estimated by comparing the 1964 levelling with the repeated levelling in 1996. Hence the mean rate of accretion over 23 years (1973–1996) could be estimated. Differential Global Positioning System enabled levelling at the same coordinates with a 1–2 cm error. Another long-term estimation of net accretion was carried out by ^{137}Cs-measurements in 1996, indicating the Chernobyl deposition in 1986 (Ehlers et al., 1993). Short-term estimations of net accretion were carried out by measuring the soil level at intervals of two months by sedimentation-erosion-tables or -bars (SEB) (Boumans & Day, 1993). Sedimentation resulting from single tides was measured by putting filters on the soil.

The mean net accretion was similar as measured by the different methods (Table 5). The relatively high rate of sedimentation measured by the long-term repeated levelling can most likely be attributed to the high rate of sedimentation during the first years, but unfortunately no data are available from the different periods after breaching of the summerdike. Both at 10 cm and 22 cm depth clear peaks of ^{137}Cs were found. The mean accretion for the entire salt marsh amounted to 16 mm yr^{-1} (van Duin et al., 1997). The accretion along the entire northern mainland coast is about 10 mm yr^{-1} (Dijkema et al., 1990). Hence, no problems with the sediment budget could be expected in the restored salt marsh. After storm tides up to 10 mm fresh sediment was deposited at some filters.

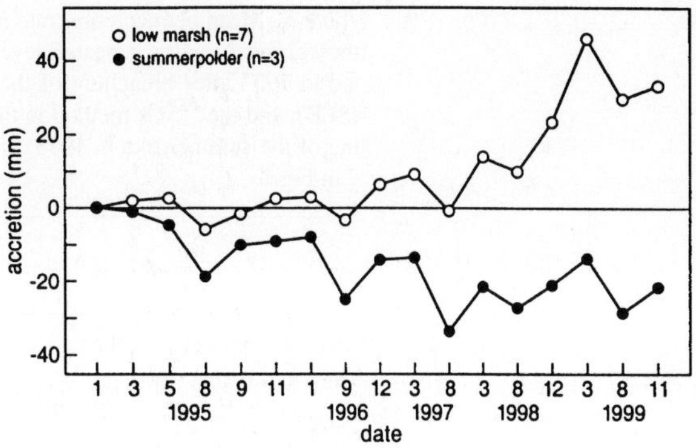

Figure 5. Changes in soil surface level in the low marsh (formerly summerpolder) and the (intact) summerpolder of the Peazemerlannen (site 7 in Fig. 1), measured with the Sedimentation-Erosion-Bar from January 1995 until November 1999.

This suggests that much sediment is deposited during storm tides, and not at a regular basis. The importance of storm tides was also stressed by others (Stumpf, 1983; Dankers et al., 1984; Reed, 1989). The situation close to the opening in the summerdike deviates. Here *Salicornia* spp. build thin stands covering only 50% at the level of MHT +17 cm, and *Puccinellia maritima* gets co-dominant (but only covering 40%) at MHT +28 cm. Because of the low plant cover, sediments are trapped at a low rate. Therefore, this zone is referred to as the pioneer zone, resembling the pioneer zone of mainland salt marshes that is regarded the problem zone in case of sea-level rise (Dijkema et al., 1990; Dijkema, 1997; Houwing et al., 1999). Because this pioneer zone is very exposed to waves and protection by brushwood groynes no longer exists, the area is very vulnerable. Erosion and cliff formation can be observed. When the difference in soil level between intertidal flats and pioneer zone is levelled off, this problem will be solved.

Because it seldom gets flooded, the summerpolder of course shows no net accretion measured by repeated levelling (Table 5). The negative figures from the SEB measurements indicate shrinking of the soil. The Peazemerlannen area shows the importance of de-embankment or coastal retreat for coastal protection. Instead of a relatively low-lying summerpolder behind the summerdike, a higher restored salt marsh can develop in front of the seawall (Fig. 5). However, it is important that enough sediment is imported from outside the local system, otherwise erosion and cliff formation will occur.

The vegetation at the restored salt marsh shows a pattern of levees, harbouring the *Elymus athericus*

Table 5. Mean net accretion rate (mm yr^{-1}) (number of replicates in parentheses) measured by repeated levelling (1964 and 1996; sedimentation started in 1973 after breaching of the summerdike), sedimentation-erosion-bar (SEB), and the ^{137}Cs method in the new salt marsh (developed after breaching of the summerdike in 1973) and summerpolder in Peazemerlannen (site 7 in Fig.1)

	Repeated levelling	Sedimentation-Erosion-Bar	^{137}Cs
Period of measurements	1973–1996	1995–1997	1986–1996
New salt marsh (average)	16.1 (115)	8.1(25)	
Levee			
Elymus athericus stands			
(MHT +36–109 cm))	22.5 (47)	10.8 (6)	
Low salt marsh			
(*Puccinellia maritima* stands			
(MHT +14-49 cm)	15.6 (29)	8.1(11)	10 (2)
Depression			
Salicornia spp. stands			
(MHT −18 to +24 cm)	10.9 (24)	4.1 (6)	
Pioneerzone			
behind opening in summerdike			
(MHT +7–33 cm)	5.4 (15)	12.3 (2)	
Summerpolder	−0.3 (89)	−7 (3)	

Table 6. Mean cover percentages in depression, low marsh, and creek bank levees in Peazemerlannen (site 6 in Fig.1) since 1995

	Depression (*n*=3)			Low marsh (*n*=3)			Creek bank levee (*n*=2)		
Year	1995	1996	1998	1995	1996	1998	1995	1996	1998
Total cover (%)	53	70	55	90	100	90	100	95	90
Aster tripolium	<1	1	3	27	17	15	15	1	<1
Atriplex portulacoides	–	<1	–	–	1	<1	15	15	3
Atriplex prostrata	–	<1	–	<1	7	5	10	1	<1
Elymus athericus	–	–	–	–	–	–	55	75	88
Glaux maritima	–	–	–	–	–	<1	–	–	–
Plantago maritima	–	–	–	–	–	<1	–	–	–
Puccinellia maritima	<1	2	20	67	87	85	20	<1	–
Salicornia spp.	53	21	15	13	<1	–	–	–	–
Spartina anglica	<1	–	–	<1	<1	<1	–	–	–
Spergularia maritima	–	–	–	–	<1	<1	–	–	–
Suaeda maritima	2	50	20	23	5	2	<1	<1	–

community, low salt marsh and depressions with the *Salicornia* spp. and *Puccinellia maritima* communities in 1992 (Fig. 6). We assume that the dispersal by propagules took place by tidal currents from the adjacent salt marsh. After the breach in the seawall, the artificial drainage system was abandoned. This resulted in stagnant water and locally die-off of the vegetation. Since that time, the soil level rose and the drainage improved in most depressions. The effect of this can also be observed in the species composition that was monitored since 1995 (Table 6). The species composition in the permanent plots in the low marsh

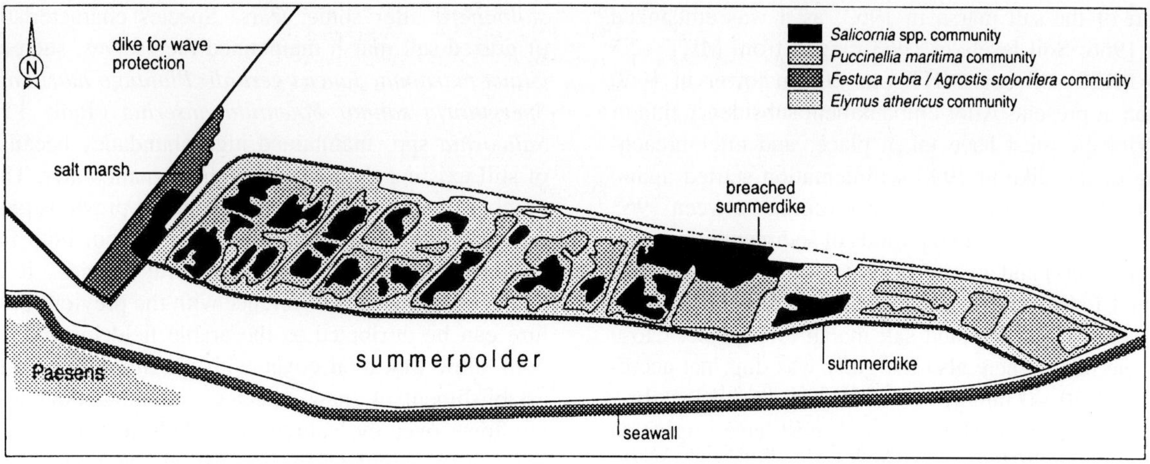

Figure 6. Vegetation map of the Peazemerlannen (site 7 in Fig. 1) (produced in 1995, but based on aerial photographs from 1992).

seems to be stable. However, in dry years *Atriplex prostrata* can profit very quickly from the poor condition of *Puccinellia maritima* and invades the area. The species composition in depressions and creek bank levees is still changing. In the depressions *Salicornia* spp. and *Suaeda maritima* are being replaced by *Puccinellia maritima*, and at the levees *Elymus athericus* is taking over from *Puccinellia maritima* and *Aster tripolium*. These data suggest that after about 25 years the salt-marsh restoration with respect to plant species replacement still has not stabilised completely, and succession is continuing. We expect that *Elymus atehricus* stands will dominate eventually when the area remains ungrazed (Andresen et al., 1990). Species such as *Glaux maritima*, *Plantago maritima* and *Spergularia maritima* have very low abundancies, whereas *Festuca rubra*, *Artemisia maritima*, *Triglochin maritima* and *Limonium vulgare* are absent from the permanent plots. Although a source area, the salt marsh (Fig. 6) is nearby, incoming tidal water does not flow over this source area. Unfortunately no data are available before 1995 when rapid changes must have taken place.

From Selenapolder to Sieperdaschor

The Selenapolder of about 100 ha (51° 22′ N, 4° 13′ E) (Fig. 1, site 10) was embanked from the large salt marsh Saeftinge in 1966. It is a narrow summerpolder 200–500 m width and 3500 m long. Before 1966, ditching took already place in order to enhance the drainage. Levelling in 1963 showed only slight differences in soil level (Stikvoort & de Winder, 1998). It is known that about 25% of the new summerpolder was never ploughed and drained by ditches. The agricultural history until 1975 is not documented. Until 1979, about 60% was exploited as arable field, and 40% was pasture. From 1979 onwards, 65% was pasture. Although no data are available prior to 1990, it is assumed that in the pastures *Agrostis stolonifera*, *Elymus repens*, and *E. athericus* were important (Stikvoort & de Winder, 1998).

The summerpolder suffered from two breaches of the summerdike that were repaired immediately. After the third breach in February 1990 the connection with the brackish Westerschelde remained open and a 500 m long breach developed. In 1993, the enormous amounts of water transport by tidal currents eroded a small road and forced the managers to dig a gully of about 500 m from the former summerdike into the new marsh. The area became a nature reserve later in 1993 and its name was changed into Sieperdaschor. Since 1990, about 35% of the former summerpolder was abandoned, and 65% was grazed by various and changing stocking rates. Vegetation changes immediately after the start of the unintentional restoration of the salt marsh were not recorded, but in the autumn of 1993 permanent plots were established. The position of the permanent plots enables comparison of (1) now ungrazed, previous arable field, (2) now ungrazed, previous pasture, (3) initially grazed, ungrazed since 1997, previous pasture, (4) now grazed, previous pasture, and (5) now grazed, previous arable field until 1979, afterwards pasture.

Between 1963 and 1994, a net accretion between 10 and 15 mm yr^{-1} was documented (Krijger, 1993). However, it should be realised that the polder was still part of the salt marsh in 1963, as it was embanked in 1966. Soil levels in 1963 ranged from MHT −55 to MHT +35 cm. MHT was 20 cm lower in 1960 than at present. After embankment subsidence due to shrinkage must have taken place, and after breaching of the dike in 1990 sedimentation started again. Therefore, the differences in levelling between 1963 and 1994 included two periods of sedimentation (total eight years) and an intermediate period of subsidence (total 13 years). Nevertheless, it seems that overall no erosion of the foreland salt marsh took place. Close to the inlet, where also the gully was dug, net accretion of 30–40 mm yr^{-1} was recorded, and locally a levee developed. Most material must have originated from the gully. In the remainder of the new salt marsh, an accretion 5–10 mm yr^{-1} was measured. Soil level ranged from MHT −55 cm to MHT +5 cm in 1994. As pointed out above, a similar range was found in 1963, a few years before embankment. The ditches in the western part of the marsh initially could not drain the large amounts of incoming water. This resulted in large areas with stagnant water. Gradually new creeks were formed and the area of stagnant water has decreased. In 1990, 70% of the new marsh was bare soil, especially the former arable fields. In 1995, 30% was bare soil, particularly the places with stagnant water.

After initially low cover, probably after die-off of the original vegetation, *Aster tripolium* and later *Puccinellia maritima* reached high cover abundance, but without grazing *Phragmites australis* took over within 10 years in the previous pasture and might be the climax (Table 7). In the previous arable field, the cover is already higher in the fourth year, as the area started with bare soil. Here *Scirpus maritimus* and *Phragmites australis* are taking over without grazing after the high cover of *Aster tripolium* and *Puccinellia maritima* (Table 8). No climax seemed to be reached here yet. Starting from a previous pasture, grazing decreased *Elymus athericus*, and *Aster tripolium* did not get high abundancies. Grazing benefitted *Juncus gerardi* and *Spergularia salina*. The cessation of grazing since 1997 resulted in an increase of *Elymus athericus* and *Scirpus maritimus* (Table 9). This area will further change during the following years. Intensive grazing in a former pasture increased the dominance of *Puccinellia maritima*, and decreased *Elymus athericus* and *Agrostis stolonifera*. In the short canopy, *Salicornia* spp., *Glaux maritima* and *Spergularia salina* were able to maintain themselves (Table 10). Grazing with low stocking rates in the previous arable field revealed co-dominance of *Puccinellia maritima* and *Agrostis stolonifera* after some years. Species characteristic of grazed salt marsh maintained themselves, such as *Glaux maritima*, *Juncus gerardi*, *Plantago maritima*, *Spergularia salina*, *Potentilla anserina* (Table 11). *Salicornia* spp. maintained high abundancy because of still existing open places with stagnant water. The number of species is higher than in the previous pasture and increased during the study period, both for salt-marsh species and more ruderal species. It is unclear whether this difference with the previous pasture can be attributed to the arable field history, as both show low total cover with possibilities for the establishment of 'new' species. Incoming tidal water flows over the source area. This might explain the nearly complete species list in the target area within a few years, in contrast to the aforementioned Peazemerlannen.

Restoration projects that are less successful

The Emmapolder study site (53° 28′ N, 6° 44′ E) (Fig. 1, site 3) is about 20 ha. It was embanked by a seawall in 1943. The polder is characterised by seepage. The chloride-content of shallow groundwater at 200 cm below soil surface locally amounts to 900 mg l^{-1} (Houthuesen, 1994). In 1997, the top 30 cm was removed from the entire area, and a pool with stagnant brackish water was created, apart from two creek-like new gullies, where also a thicker soil layer was removed. The pool immediately attracted waterfowl. The topsoil was not completely removed from the area, but partly stored in rows all over the area in 1997. Hence, arable weeds established from the stored soil. The remainder of the stored soil was removed in 1998. To prevent complaints from the adjacent farmer for contaminating his land with arable weeds, the area was sown with *Lolium perenne* and *Trifolium repens*, except for a small strip bordering the pool. The area is grazed with low stocking density by sheep, and the aftermath is cut.

Vegetation recordings in 10 plots 2 × 2 m^2, subdivided in 10 subplots of 0.4 m^2 each, enabled to quantify the frequency of plant species occurring in 1999. The majority of the sown area was dominated by *Trifolium repens* and *Lolium perenne*, whereas arable weeds complete the vegetation (Table 12). In the small unsown part arable weed species are most frequent, but in the open canopy also many other species were found and among them some species characteristic of brackish conditions such as frequent *Juncus*

Table 7. Mean (*n*=2) cover percentages Selenapolder (site 10 in Fig. 1) at now ungrazed area, previously pasture

Year	1993	1994	1995	1996	1997	1998	1999
Total cover	13	60	90	83	95	100	100
Salicornia spp.	1	15	1	1	–	–	–
Aster tripolium	13	48	65	50	53	5	–
Puccinellia maritima	1	6	26	30	33	50	–
Scirpus maritimus	1	1	5	4	4	11	–
Phragmites australis	–	1	3	10	48	50	95
Agrostis stolonifera	–	3	2	1	–	–	1
Elymus repens	–	–	–	–	–	1	1
Elymus athericus	–	–	–	–	–	–	18

Table 8. Mean (*n*=3) cover percentages Selenapolder (site 10 in Fig. 1) at now ungrazed area, previously arable field

Year	1993	1994	1995	1996	1997	1998	1999
Total cover	37	54	64	87	99	100	100
Salicornia spp.	4	1	6	7	–	–	–
Aster tripolium	14	15	12	30	39	21	1
Puccinellia maritima	29	17	20	47	57	60	11
Spergularia maritima	1	–	–	–	–	–	–
Agrostis stolonifera	7	22	27	10	5	13	40
Atriplex prostrata	–	1	2	9	32	20	9
Scirpus maritimus	–	1	6	8	14	19	19
Glaux maritima	–	–	–	1	–	–	–
Phragmites australis	–	–	–	–	1	2	33
Elymus repens	–	–	–	–	–	13	–

ambiguus, and rare *Aster tripolium*, *Scirpus maritimus* and *Puccinellia distans* (Table 12). Geese foraging at the salt marsh and later forage or rest in the polder, may disperse seeds attached to their feet or via dung. The first 90 min after landing in the polder geese droppings will contain seeds from the salt-marsh vegetation. Pilot studies revealed that very little viable seeds of target species are found in geese droppings in the Emmapolder.

Evaluation and perspectives

In general, the sedimentation rate after breaching the summerdike, is fast enough to cope with the time-lag of several decades caused by the embankment of summerpolders. Of course, this holds for the small summerpolders of the case studies described. Careful measurements are needed to study the sediment budget between the salt-marsh foreland and larger summer-polders to be de-embanked as in Noord-Friesland Buitendijks (Fig. 1), where eventually about 1200 ha of salt marsh is planned to be restored. At the small scale, an increase of abiotic differentiation has been found, such as the development of levees and the origin of depressions. Especially the latter is characteristic of more naturally drained mainland salt-marshes (Esselink et al., 1998).

Restoration of salt marshes or halophytic plant communities cannot rely on a viable soil seed bank, that might reflect the vegetation prior to embankment. This is caused by the fact that the majority of salt-marsh plant species have a transient or short-term persistent seed bank. It should be noted that our knowledge of the longevity of seeds of salt-marsh plant species is underrepresented as compared to many other plant communities: further studies are needed.

Table 9. Mean (n=4) cover percentages Selenapolder (site 10 in Fig. 1) at initially heavily grazed, ungrazed since 1997, previously pasture

Year	1993	1994	1995	1996	1997	1998	1999
Total cover	69	39	47	72	90	90	85
Salicornia spp.	1	1	1	10	2	1	–
Aster tripolium	11	7	11	11	29	8	9
Puccinellia maritima	18	11	14	31	36	45	23
Spergularia maritima	1	–	–	–	–	–	–
Festuca rubra	1	–	–	–	–	3	8
Agrostis stolonifera	19	11	14	11	3	16	15
Juncus gerardi	1	1	1	5	3	2	3
Spergularia salina	–	1	1	1	–	1	1
Elymus athericus	26	13	9	16	27	37	27
Atriplex prostrata	–	–	1	3	3	3	3
Scirpus maritimus	–	1	1	1	2	3	27
Phragmites australis	–	1	–	–	–	1	–
Ranunculus sceleratus	–	1	–	–	–	–	–
Cirsium arvense	–	2	2	–	–	1	3
Stellaria media	–	1	–	–	–	–	–
Puccinellia distans	–	–	2	–	–	–	–
Rumex crispus	–	–	1	1	–	–	–
Taraxacum spp.	–	–	–	–	1	–	–
Elymus repens	–	–	–	–	–	1	1
Glaux maritima	–	–	–	–	–	–	1

Table 10. Mean (n=5) cover percentages Selenapolder (site 10 in Fig. 1) at now heavily grazed, previously pasture area

Year	1993	1994	1995	1996	1997	1998	1999
Total cover	77	69	56	54	62	66	71
Salicornia spp.	2	1	1	1	1	1	2
Aster tripolium	4	9	6	1	8	12	9
Puccinellia maritima	43	42	44	54	56	58	63
Glaux maritima	1	1	1	1	1	1	2
Juncus gerardi	1	–	–	–	–	–	1
Spergularia maritima	1	–	–	–	–	–	–
Elymus athericus	26	1	–	–	–	–	–
Atriplex prostrata	–	1	1	–	1	1	1
Agrostis stolonifera	34	25	11	–	–	–	1
Festuca rubra	–	2	1	1	–	–	–
Suaeda maritima	–	–	1	–	–	–	–
Scirpus maritimus	–	–	–	–	–	–	1

Many salt-marsh plant species have established at the restored salt marshes of the aforementioned case studies. However, it is striking that even after about 25 years, some species are lacking locally. Because of lack of monitoring, no information is available of the period of establishment. We do not know whether species that did not establish were not dispersed at all, or did not establish although they were dispersed. We should know the potential of dispersal of salt-marsh species: can they float for a long time, are

Table 11. Mean (n=12) cover percentages in Selenapolder (site 10 in Fig. 1) in now lightly grazed, previously arable field until 1979, afterwards pasture

Year	1993	1994	1995	1996	1997	1998	1999
Total cover	75	73	65	69	70	64	73
Salicornia spp.	9	5	2	26	17	7	15
Aster tripolium	6	7	7	5	12	13	7
Puccinellia maritima	15	12	17	12	30	24	24
Festuca rubra	1	1	–	1	–	–	–
Glaux maritima	1	1	1	1	1	2	2
Juncus gerardi	1	1	1	1	2	1	1
Spergularia maritima	1	–	–	1	1	–	–
Elymus athericus	1	1	2	1	2	1	3
Atriplex prostrata	1	1	1	6	2	1	1
Scirpus maritimus	1	1	6	1	2	3	3
Agrostis stolonifera	52	48	37	16	14	17	25
Plantago maritima	1	2	2	1	1	1	3
Spergularia salina	–	1	1	1	5	3	4
Potentilla anserina	–	1	1	1	1	1	1
Cirsium arvense	–	2	1	1	1	1	1
Trifolium repens	–	1	1	–	–	–	–
Phragmites australis	–	–	1	1	1	1	1
Trifolium fragiferum	–	–	–	1	1	1	1
Elymus repens	–	–	–	1	–	5	3
Triglochin maritima	–	–	–	–	1	1	1
Juncus bufonius	–	–	–	–	–	2	1
Atriplex littoralis	–	–	–	–	1	–	–
Juncus maritimus	–	–	–	–	1	1	–
Chenopodium album	–	–	–	–	1	–	–
Taraxacum spp.	–	–	–	–	–	–	1
Rumex crispus	–	–	–	–	–	–	1
Plantago coronopus	–	–	–	–	–	–	1

they able to pass the alimentary tract of waterfowl and maintain their viability? This knowledge could explain why some species do not establish. In general, dispersal is often potential, propagules may arrive either in a non-salt-marsh environment or be carried across the marsh to the driftline (Adam, 1990). The discrepancy between realised dispersal and subsequent establishment remains to be quantified.

Although comparisons are not easy because of differences in the history prior to restoration, and the management during restoration of restored salt marshes during the first 10 years after breaching the summerdike, some generalisations seem possible. The development at former arable fields takes more time than at previous pastures. Livestock grazing is necessary to maintain a species-rich salt-marsh plant community and the fauna depending on it. No grazing will eventually result in a tall canopy, the dominant species depending on the salinity, and its characteristic fauna (see also Bakker et al., 1997; Esselink et al., 2002).

Restoration of brackish plant communities behind the seawall can only be replenished by dispersal of propagules by waterfowl. The first results do not seem very promising. Its success may depend on the presence of the target communities in adjacent areas behind the seawall. In the case of open connections, dispersal by water may play an important role.

How long the aforementioned processes will take is not known. The oldest case study started in 1973. However, it should be noted that the case studies such as Peazemerlannen and Selenapolder were not planned. This is the reason that they were not monitored from the beginning and certainly not before the

Table 12. Frequency of occurrence in 100 plots of 0.4 m^2 in 1999 in sown and unsown parts of the Emmapolder (site 3 in Fig.1)

	Frequency of occurrence	
	Sown	Unsown
Trifolium repens	100	20
Lolium perenne	97	6
Festuca rubra	29	–
Agrostis stolonifera	27	50
Cirsium vulgare	13	17
Elymus repens	13	–
Matricaria recutita	11	97
Holcus lanatus	9	–
Epilobium hirsutum	5	26
Rumex crispus	3	–
Sonchus oleraceus	3	76
Papaver dubium	2	1
Alopecurus geniculatus	2	19
Bromus hordeaceus	2	–
Chamerion angustifolium	1	52
Poa pratensis	1	–
Stellaria media	1	15
Juncus bufonius/ambiguus	–	97
Poa annua	–	61
Ranunculus sceleratus	–	49
Persicaria maculosa	–	45
Chenopodium spec.	–	29
Polygonum aviculare	–	24
Plantago major	–	10
Aster tripolium	–	7
Scirpus maritimus	–	5
Phragmites australis	–	2
Puccinellia distans	–	1
Potentilla anserina	–	1

start of the 'experiment'. Monitoring started in the mid 1990s. Firstly, it is important to continue current monitoring. Secondly, monitoring should be started before the start of new projects. Part of the large amounts of money available for acquirement of the land, and subsequent lay-out, should be allocated to monitoring and research behind the mechanism of restoration of salt marshes and brackish plant communities.

Possibilities for restoration with respect to their area, only consider the mainland salt marshes in the Wadden Sea, as shown in Figure 2 (Dijkema, 1987). However, the configuration of the tidal basins in the 17th century differs strongly from the present one, as a result from shortening the coastline. Moreover, a discrepancy exists between such a standard of naturalness along the mainland coast featuring artificial salt marshes.

The present seawalls are expected to be strong enough for protection of the hinterland. This implies that no further embankments are necessary for coastal protection. The need for extension of agricultural areas has diminished, especially in the framework of European Union agricultural policies. This means that the restoration of salt marshes and halophytic plant communities has become an option in the Netherlands within a few years. Creation of new marshes by dumping of dredging material from elsewhere is no option, because a salt marsh is the outcome of the interactions of physical and biological processes and not a dumping site. It should be taken into account that the present mainland salt marshes in the Wadden Sea are artificial, and do not harbour meandering creeks, creek bank levees and depressions. It could be argued that since the estuarine salt marshes in SW-Netherlands do feature these natural geomorphological phenomena, it should be studied how their area could be enlarged naturally, but the possibilities seem restricted. The area of natural back-barrier marshes at the Wadden Sea islands, is of no concern with respect to restoration as they cover more than ever before as indicated in Figure 2 (Dijkema, 1987). However, not only the total area matters, but from an ecological point of view also the occurrence of the completeness of the entire series from pioneer community to old salt marsh is important. As most salt marshes in the Netherlands senescence, this issue will be relevant in the near future in the discussion about restoration of salt marshes. Most salt marshes in SW-Netherlands are about the same age due to the massive planting of *Spartina anglica* in the 1920s. Also mainland salt marshes in the Wadden Sea area suffer from a similar phenomenon, as sedimentation works have been abandoned recently. Only island salt marshes will include young marshes as long as they can extend eastward. Apart from maintaining or increasing the total area of salt marshes, we should think about cycling from pioneer to eroding old salt marshes within the present area.

It is important to know whether it is politically feasible to restore salt marshes including their characteristic communities. Long discussions have resulted in the choice for de-embankment of summerpolders in front of the seawall along the mainland coast of Friesland, Noord-Friesland Buitendijks (Fig. 1) (about

1200 ha). Where no summerpolders are present along the coastline, as in the other parts of the Netherlands, there will be no local inland removal of the seawall, but only restoration of saline seepage behind the seawall along the mainland coast of Groningen (small-scale projects) and along the southcoast of the island of Schouwen (about 700 ha) in SW-Netherlands (Fig. 1). The traditional conquest of land from the sea and the relatively recent disaster of 1953 in SW-Netherlands, makes removal of the present seawall no issue in the near future.

References

Adam, P., 1990. Saltmarsh Ecology. Cambridge University Press, Cambridge.

Allen, J. R. L. & K. Pye, 1992. Coastal salt marshes: their nature and importance. In Allen, J. R. L. & K. Pye (eds), Salt Marshes, Morphodynamics, Conservation and Engineering Significance. Cambridge University Press, Cambridge: 1–18.

Andresen, H., J. P. Bakker, M. Brongers, B. Heydemann & U. Irmler, 1990. Long-term chnages of salt marsh communities by cattle grazing. Vegetatio 89: 137–148.

Armstrong, W., E. J. Wright, S. Lythe & T. J. Gaynard, 1985. Plant zonation and the effects of the spring-neap tidal cycle on soil aeration in a Humber salt marsh. J. Ecol. 73: 323–339.

Bakker, J. P., 1989. Nature Management by Grazing and Cutting. Kluwer Academic Publishers, Dordrecht.

Bakker, J. P., M. Dijkstra & P. T. Russchen, 1985. Dispersal, germination and early establishment of halophytes and glycophytes on a grazed and abandoned salt marsh gradient. New Phytol. 101: 291–308.

Bakker, J .P., J. De Leeuw, K. S. Dijkema, P. C. Leendertse, H. H. T. Prins & J. Rozema, 1993. Salt marshes along the coast of the Netherlands. Hydrobiologia 265: 73–95.

Bakker, J. P., P. Esselink, R. van der Wal & K. S. Dijkema, 1997. Options for restoration and management of coastal salt marshes in Europe. In Urbanska, L. M., N. R. Webb & P. J. Edwards (eds), Restoration Ecology and Sustainable Development. Cambridge University Press, Cambridge: 286–322.

Bakker, J. P., A. P. Grootjans, M. Hermy & P. Poschlod, 2000. How to define targets for ecological restoration? Appl. Veg. Sci. 3: 3–6.

Bakker, J. P., G. van den Brink, G. L. Verweij & P. Esselink, 2001. Zaadvoorrad en dispersie bij een proefverkweldering in Noord-Fryslân Bûtendijks. De Levende Natuur. 102: 19–23 (in Dutch).

Bazelmans, J., D. Gerrets & P. Vos, 1998. Zoden aan de dijk; kleinschalige dijkbouw in het Friesland van de Romeinse tijd. Noorderbreedte 22: 18–21 (in Dutch).

Beeftink, W. G., 1977. The coastal salt marshes of Western and Northern Europe: an ecological and phytosociological approach. In Chapman, V. J. (ed.), Ecosystems of the World I. Wet Coastal Systems. Elsevier, Amsterdam: 109–155.

Beeftink, W. G., P. De Bruin, S. Jelgersma, J. A. Trimpe-Burger, P. J. van der Feen, J. Visser & W. J. Wolff, 1982. Geschiedenis. In Duursma, E. K., H. Engel & T. J. M. Martens (eds), De Nederlandse Delta. Koninklijke Nederlandse Akademie van Wetenschappen, Amsterdam: 15–77 (in Dutch).

Behre, K. E., 1979. Zur Rekonstruktion ehemaliger Pflanzengesellschaften an der Deutschen Nordseeküste. In Wilmanns, O. & R. Tüxen (eds), Werden und Vergehen von Pflanzengesellschaften. Cramer, Vaduz: 181–214.

Behre, K. E., 1985. Die ursprungliche Vegetation in den deutschen Marschgebieten und deren Veränderung durch prähistorische Besiedlung und Meeresspiegelbewegungen. Verh. Gesellsch. für Ökologie 13: 85–96.

Bekker, R. M., J. P. Bakker, U. Grandin, R. Kalamees, P. Milberg, P. Poschlod, K. Thompson & J. H. Willems, 1998. Seed shape and vertical distribution in the soil: indicators for seed longevity. Funct. Ecol. 12: 834–842.

Bernhardt, K. G., 1992. Besiedlungsstrategieen an sandigen Extremstandorten im Tidebereich. Flora 187: 272–281.

Bernhardt, K. G., 1998. Auswirkungen der Widerzulassung von Überflutungsdynamik auf die Vegetation und Pflanzenpopulation im Grünlandkomplex 'Karrendorfer Wiesen' bei Greifswald. Schriftenreihe für Landschaftspflege und Naturschutz 56: 251–262.

Bernhardt, K. F., J. Tesmer, C. Ruth & H. Schurbohm, 1996. Die Vegetation der Karrendorfer Wiesen – Inventarisierung des Zustandes 1994–1995. Natur un Naturschutz in Mecklenburg-Vorpommern 32: 85–100.

Boorman, L. A., 1999. Salt marshes – present functioning and future change. Mangroves and Salt Marshes 3: 227–241.

Boorman, L. A., J. D. Goss-Custard & S. McGrorty, 1989. Climatic change, rising sea level and the British coast. HMSO, London. Natural Environm. Res. Council. ITE Research Publication 1: 1–24.

Boorman, L. & J. Hazelden, 1996. New marshes for old: saltmarsh creation in Sussex, England. Ocean Challenge 6: 34–37.

Boumans, R. M. J. & J. W. Day Jr, 1993. High precision measurements of sediment elevation in shallow coastal areas using a sedimentation-erosion-table. Estuaries 16: 375–380.

Cappers, R. T. J., 1994. Botanical macro-remains of vascular plants of the Heveskesklooster terp (The Netherlands) as tools to characterize the past environment. Palaeohistoria 36: 107–167.

De Jong, D. J. & A. M. van der Pluijm, 1994. Consequences of a tidal reduction for the salt-marsh vegetation in the Oosterschelde estuary (The Netherlands). Hydrobiologia 282/283: 317–333.

De Jong, F., J. F. Bakker, C. J. M. van Berkel, K. Dahl, N. M. J. A. Dankers, C. Gättje, H. Marencic & P. Potel, 2000. 1999 Wadden Sea Quality Status Report. Common Wadden Sea Secretariat, Wilhelmshaven.

De Leeuw, J., L. A. Apon, P. J. Herman, W. De Munck & W. G. Beeftink, 1994. The response of salt marsh vegetation to tidal reduction caused by the Oosterschelde storm-surge barrier. Hydrobiologia 282/283: 335–353.

De Vlaming, V. & V. Proctor, 1968. Dispersal of aquatic organisms: viability of seeds recovered from droppings of captive kildeer and mallard ducks. Am. J. Bot. 55: 20–26.

De Vries, V., 1961. Vegetatiestudie op de Westpunt van Vlieland. PhD.Thesis, University of Amsterdam (in Dutch).

Dankers, N., M. Binsbergen, K. Zegers, R. Laane & M. Van Der Loeff, 1984. Transportation of water, particulate and dissolved organic matter between a salt marsh and the Ems-Dollard Estuary, The Netherlands. Estuar. coast. shelf Sci. 19: 143–165.

Dieckmann, R., 1988. Entwicklung der Vörländer an der nordfriesischen Festlandküste. Wasser & Boden 40: 146–150.

Dijkema, K. S., 1983. Use and management of mainland marshes and Halligen. In Dijkema, K. S. & W. J. Wolff (eds), Flora and Vegetation of the Wadden Sea Islands and Coastal Areas. Report 9 of the Wadden Sea Working Group. Rotterdam, Balkema: 302–312.

Dijkema, K. S., 1984. Salt Marshes in Europe. Council of Europe, Strasbourg.

Dijkema, K. S., 1987. Changes in salt-marsh area in the Netherlands Wadden Sea after 1600. In Huiskes, A. H. L., C. W. P. M. Blom & J. Rozema (eds), Vegetation Between Land and Sea. Dr W. Junk Publishers, Dordrecht: 42–49.

Dijkema, K. S., 1990. Salt and brackish marshes around the Baltic Sea and adjacent parts of the North Sea; their development and management. Biol. Conserv. 51: 191–209.

Dijkema, K. S., 1994. Auswirkung des Meeresspiegelanstieges auf die Salzwiesen. In Lozán, J. J., E. Rachor, K. Reise, H. van Westernhagen & W. Lenz (eds), Warnsignale aus dem Wattenmeer. Blackwell, Berlin: 196–200.

Dijkema, K. S., 1997. Impact prognosis for salt marshes from subsidence by gas extraction in the Wadden Sea. J. Coast. Res. 13: 1294–1304.

Dijkema, K. S., J. Van Den Bergs, J. H. Bossinade, P. Bouwsema, R. J. De Glopper & J. W. T. M. Van Meegen, 1988. Effecten van rijshouten dammen op de opslibbing en de omvang van de vegetatiezones in de Friese en Groninger landaanwinningswerken. Rijkswaterstaat Directie Groningen. Nota GRAN 1988–2010, RIN-report 88/66, RIJP-report 1988-33 Cbw, 1–119 (in Dutch).

Dijkema, K. S., J. H. Bossinade, P. Bouwsema & R. J. De Glopper, 1990. Salt marshes in the Netherlands Wadden Sea: rising high tide levels and accretion enhancement. In Beukema, J. J., W. J. Wolff & J. J. W. H. Brouns (eds), Expected Effects of Climatic Change on Marine Coastal Ecosystems. Kluwer Academic Publishers, Dordrecht: 173–188.

Ehlers, J., K. Nagorny, P. Schmidt, B. Stieve & K. Zietlow, 1993. Storm surge deposits in North Sea salt marshes dated by ^{134}Cs and ^{137}Cs determination. J. Coast. Res. 6: 698–701.

Erchinger, H. F., H.-G. Coldewey & C. Meyer, 1996. Interdisciplinäre Erforschung dis deichvorlandes in Forschunpsvorhaben 'Erosionsfestigkeit von Hellern'. Die Küste 58: 1–45.

Esselink, P., 2000. Nature management of coastal salt marshes; interactions between anthropogenic influences and natural dynamics. PhD.Thesis, University of Groningen.

Esselink, P., K. S. Dijkema, S. Reents & G. Hageman, 1998. Vertical accretion and marsh-profile development in man-made tidal marshes after abandonment. J. Coast. Res. 14: 570–582.

Esselink, P., K. S. Dijkema & L. F. M. Fresco, 2002. Vegetation change in a man-made salt marsh affected by a reduction in both grazing and drainage. Appl. Veg. Sci. (in press).

Gerlach, A., T. Brüning & K. Brüning, 1999. Untersuchung zur Zusammensetzung und Herkunft von Getreibsel ('Teek') an der niedersächsischen Nordseeküste. Internal Report Carl von Ossietzky-Universität Oldenburg.

Houthuesen, Y., 1994. Mogelijkheid voor het creëren van een brak-zout milieu in de Emmapolder. Internal Report, Vereniging Natuurmonumenten, 's-Graveland (in Dutch).

Houwing, E. J., W. E. Van Duin, Y. Smit-Van Der Waaij, K. S. Dijkema & J. H. J. Terwindt, 1999. Biological and abiotic factors influencing the settlement and survival of Salicornia dolichostachya in the intertidal pioneer zone. Mangroves and Salt Marshes 3: 197–206.

Huiskes, A. H. L., B. P. Koutstaal, P. M. J. Herman, W. G. Beeftink, M. M. Markusse & W. De Munck, 1995. Seed dispersal of halophytes in tidal salt marshes. J. Ecol. 83: 559–567.

Hutchings, M. J. & P. J. Russell, 1989. The seed regeneration dynamics of an emergent salt marsh. J. Ecol. 77: 615–637.

Jakobson, B., 1954. The tidal area in South-Western Jutland and the process of the salt marsh formation. Geografisk Tidsskrift 53: 49–61.

Jefferies, R. L., A. J. Davy & T. Rudmik, 1981. Population biology of the salt marsh annual Salicornia europaea agg. J. Ecol. 69: 17–31.

Knol, E., 1993. De Nederlandse kustlanden in de Vroege Middeleeuwen. PhD. Thesis, Free University, Amsterdam (in Dutch).

Koutstaal, B. P., M. M. Markusse & W. De Munck, 1987. Aspects of seed dispersal by tidal movements. In Huiskes, A. H. L., C. W. P. M. Blom & J. Rozema (eds), Vegetation Between Land and Sea. Dr W. Junk Publishers, Dordrecht: 226–233.

Krijger, G. M., 1993. Het Verdronken Land van Saeftinghe komt weer boven water. DGW, werkdocument GWWS-93.838x, Middelburg (in Dutch).

Milton, W. E. J., 1939. The occurrence of buried viable seeds in soils at different elevations and on a salt marsh. J. Ecol. 27: 149–159.

Müller-Motzfeld, G., 1997. Renaturierung eines Überflutungsgrünlandes an der Ostseeküste. Schriftenreihe für Landschaftpflege und Naturschutz 54: 239–263.

Norris, K., T. Cook, B. O'Dowd & C. Durdin, 1997. The density of redshank Tringa totanus breeding on the salt-marshes of the Wash in relation to habitat and its grazing management. J. appl. Ecol. 34: 999–1013.

Olff, H., J. De Leeuw, J. P. Bakker, R. J. Platerink, H. J. van Wijnen & W. de Munck, 1997. Vegetation succession and herbivory in a salt marsh: changes induced by sea-level rise and silt deposition along an elevational gradient. J. Ecol. 85: 799–814.

Olney, P. J. S., 1963. The food and feeding habits of teal Anas crecca L. Proc. Zool. Soc. London 140: 169–210.

Oost, A. P. & P. L. de Boer, 1994. Sedimentology and development of barrier islands, ebb-tidal deltas, inlets and backbarrier areas of the Dutch Wadden Sea. Senckenbergiana maritima 24: 65–115.

Persicke, U., A. Gerlach & W. Heiber, 1999. Zur botanische Zusammensetzung von Treibsel der niedersächsischen Deichvorländer und Deichabschnitte. Drosera '99: 23–34.

Portnoy, J. W. & A. E. Giblin, 1997A. Effects of historic tidal restrictions on salt marsh sediment chemistry. Biogeochemistry 36: 275–303.

Portnoy, J. W. & A. E. Giblin, 1997B. Biogeochemical effects of seawater restoration to diked salt marshes. Ecol. Appl. 7: 1054–1063.

Proctor, V. W., 1968. Long-distance dispersal of seeds by retention in the digestive tract of birds. Science 160: 321–322.

Reed, D. J., 1989. Patterns of sediment deposition in subsiding coastal salt marshes, Terrebonne Bay Louisiana: the role of winter storms. Estuaries 12: 222–227.

Roman, C. T., W. A. Niering & R. S. Warren, 1984. Salt marsh vegetation change in response to tidal restriction. Envir. Manage. 8: 141–150.

Roman, C. T., R. W. Garvine & J. W. Portnoy, 1995. Hydrologic modelling as a predictive basis for ecological restoration of salt marshes. Envir. Manage. 19: 559–566.

Scherfose, V., 1990. Salz-Zeigerwerte von Gefässpflanzen der Salzmarschen, Tideröhrichte und Salzwassertümpel and der Deutschen Nord- und Ostseeküste. Jahresbericht 1987. Forschungsstelle Küste Norderney 39: 31–83.

Schultz, W., 1987. Einfluss der Beweidung von Salzwiesen auf die Vogelfauna. In Kempf, N., J. Lamp & P. Prokosch (eds), Salzwiesen: geformt von Küstenschutz, Landwirtschaft oder Natur? WWF-Deutschland, Husum: 255–270.

Siira, J., 1970. Studies on the ecology of the seashore meadows of the Bothnian Bay with special reference to the Liminka area. Aquilo Ser. Bot. 9: 1–109.

Stapel, J. & D. J. de Jong, 1998. Sedimentatieveranderingen op het schor bij Waarde en het Verdronken Land van Saeftinge, Westerschelde (ZW Nederland). Report RIKZ-98.022, Middelburg (in Dutch).

Stikvoort, E. & B. De Winder, 1998. Sieperdaschor, van polder naar schor. Interim-evaluatie 1990–1996. RIKZ-report 98.002, Middelburg (in Dutch).

Stratingh, G. A. & C. A. Venema, 1855. De Dollard. Oomkens, Zoon & Schierbeek, Groningen (in Dutch).

Strykstra, R. J., D. M. Pegtel & A. Bergsma, 1998. Dispersal distance and achene quality of the rare anemochorous species *Arnica montana* L.: implications for conservation. Acta Bot. Neerl. 47: 45–56.

Stumpf, R. P., 1983. The process of sedimentation on the surface of a salt marsh. Est. coast. shelf Sci. 17: 495–508.

Taayke, E., 1996. Die einheimische Keramik der nördlichen Niederlande, 600 v. Chr. Bis 300 n. Chr., Teil V: Übersicht und Schlussfolgerungen. Berichten van de Rijksdienst voor het Oudheidkundig Bodemonderzoek (BROB) 42: 163–208.

Ter Heerdt, G. N. J., G. L. Verweij, R. M. Bekker & J. P. Bakker, 1996. An improved method for seed bank analysis: seedling emergence after removing the soil by sieving. Funct. Ecol. 10: 144–151.

Thompson, K., J. P. Bakker & R. M. Bekker, 1997. Soil Seed Banks of NW Europe: Methodology, Density and Longevity. Cambridge University Press, Cambridge.

Thompson, K., J. P. Bakker, R. M. Bekker & J. G. Hodgson, 1998. Ecological correlates of seed persistence in soil in the NW European flora. J. Ecol. 86: 163–169.

Ungar, I. A. & T. E. Riehl, 1980. The effect of feed reserves on species composition in zonal halophyte communities. Bot. Gaz. 141: 447–452.

Ungar, I. A., 1988. A significant seed bank for Spergularia marina (Caryophyllaceae). Ohio J. Sci. 88: 200–202.

Van de Koppel, J., J. Huisman, C. F. R. van der Wal & H. Olff, 1996. Patterns of herbivory along a productivity gradient: an empirical and theoretical investigation. Ecology 77: 736–745.

Van der Wal, R., S. van Lieshout, D. Bos & R. H. Drent, 2000. Are spring staging brent geese evicted by vegetation succession? Ecography 23: 60–69.

Van Duin, W. E., K. S. Dijkema & J. Zegers, 1997. Veranderingen in bodemhoogte (opslibbing, erosie en inklink) in de Peazemerlannen. IBN-report 326, Wageningen (in Dutch).

Van Zeist, W., 1974. Palaeobotanical studies of settlement sites in the coastal area of the Netherlands. Palaeohistoria 16: 223–371.

Van Wijnen, H. J., 1999. Nitrogen dynamics and vegetation succession in salt marshes. PhD.Thesis, University of Groningen.

Verhoeven, A. G., 1938. Fixation des terres alluviales. Congrès International Géographie. Amsterdam. Excursion A: Zeeland: 101–137.

Vestergaard, P., 1997. Possible impact of sea-level rise on some habitat types at the Baltic coast of Denmark. J. Coast. Conserv. 3: 103–112.

Waisel, Y., 1972. Biology of Halophytes. Academic Press, New York.

Waterbolk, H. T., 1976. Oude bewoning in het Waddengebied. In Abrahamse, J., W. Joenje & N. Van Leeuwen-Seelt (eds), Waddenzee. Landelijke Vereniging tot Behoud van de Waddenzee, Harlingen / 's-Graveland: Natuurmonumenten, 's-Graveland: 210–222 (in Dutch).

Warren, R. S. & W. A. Niering, 1989. Vegetation change on a Northeast tidal marsh: interaction of sea-level rise and marsh accretion. Ecology 74: 96–103.

Westhoff, V., 1985. Nature management in coastal areas of Western Europe. Vegetatio 62: 523–532.

Wolfram, C., U. Hörcher, D. Lorenzen, R. Neuhaus, E. Aegerter & K. Dierssen, (....). Vegetation succession in a salt-water lagoon in the polder Beltringharder Koog/German Wadden Sea. In White, P. S., L. Mincina & J. Lepš (eds), Proceedings of the 41st IAVS Symposium, Opulus Press, Uppsala.

Zobel, M., E. Van Der Maarel & C. Dupré, 1998. Species pool: the concept, its determination and significance for community restoration. Appl. Veg. Sci. 1: 55–66.

Hydrobiologia **478:** 53–72, 2002.
P.H. Nienhuis & R.D. Gulati (eds), Ecological Restoration of Aquatic and Semi-Aquatic Ecosystems in the Netherlands (NW Europe).

Ecological rehabilitation of the lowland basin of the river Rhine (NW Europe)

P.H. Nienhuis[1,3], A.D. Buijse[2,3], R.S.E.W. Leuven[1,3], A.J.M. Smits[1,3], R.J.W. de Nooij[1,3] & E.M. Samborska[1]

[1]*Department of Environmental Studies, Faculty of Science, University of Nijmegen, P.O. Box 9010, 6500 GL Nijmegen, The Netherlands*

[2]*RIZA Institute for Inland Water Management and Waste Water Treatment, P.O. Box 17, 8200 AA Lelystad, The Netherlands*

[3]*Member of the Netherlands Centre for River Studies (NCR), P.O. Box 177, 2600 MH Delft, The Netherlands*

Key words: ecological rehabilitation, lowland river basin, Rhine

Abstract

In this paper, the status of ecological rehabilitation of the Dutch lowland basin of the river Rhine has been reviewed. The historical perspective, mainly with regard to river regulation measures in the past, is given. The lower river Rhine comprises a man-dominated, strongly regulated catchment, polluted water and sediments, and annihilated and deteriorated ecosystems. During the past 25 years, the water quality and, to a lesser extent, the sediment quality, has improved considerably, hence leading to improvement of biotic diversity. The rehabilitation of lost and disturbed ecosystems started some 15 years ago. The non-refutable boundary conditions for restoration projects are protection against flooding and transport by cargo ships. Holistic ecological theories to promote river catchment management are available, but the rehabilitation projects dominantly rely on local and regional planning policies. The use and application of ecological criteria are increasingly playing a role in rehabilitation schemes. Ecological rehabilitation mainly focuses on nature conservation and restoration strategies, i.e. exploiting the hydrodynamic and morphodynamic potentials of the flowing river, and introducing a semi-natural grazing regime by large herbivores. Particularly the creation of new secondary channels contributes to the restoration of riverine habitat diversity and heterogeneity, and hence to the species diversity. The assessment of a number of rehabilitation projects has led to the conclusion that no general statement can be given concerning 'success' or 'failure' of specific restoration measures. The overall balance of 14 assessed projects out of approximately 30 running projects is positive. The highly dynamic river Waal offers the best possibilities for the restoration of natural sandy levees and back swamps. It will be very difficult to rehabilitate the hardwood floodplain forests under the present management conditions. Further recovery of the river biota depends on continued decrease of pollution, an increase of morphological dynamics, and the development of the original natural habitats.

Introduction

A palaeographic reconstruction of the lowland basins of the rivers Rhine and Meuse, roughly 2000 years ago (Fig. 1; Zagwijn, 1986) shows wide and unbridled river valleys with freely meandering river branches and floodplains, comprising freshwater marshes and swamp forests, and debouching into an area of fen mires and raised bogs, and closer to the North Sea coastal marshes and tidal flats beyond or under tidal regime. Until about 1000 BP, man lived on the higher grounds beyond the impact of river floods, and hardly touched these vast and inaccessible wetlands. During the 11th century, man began to invade the fertile floodplains, and to reclaim wetlands for agriculture, followed by systematic embankment of vast areas. Intensive and exponential human use of European rivers, including the Rhine and Meuse, started more than

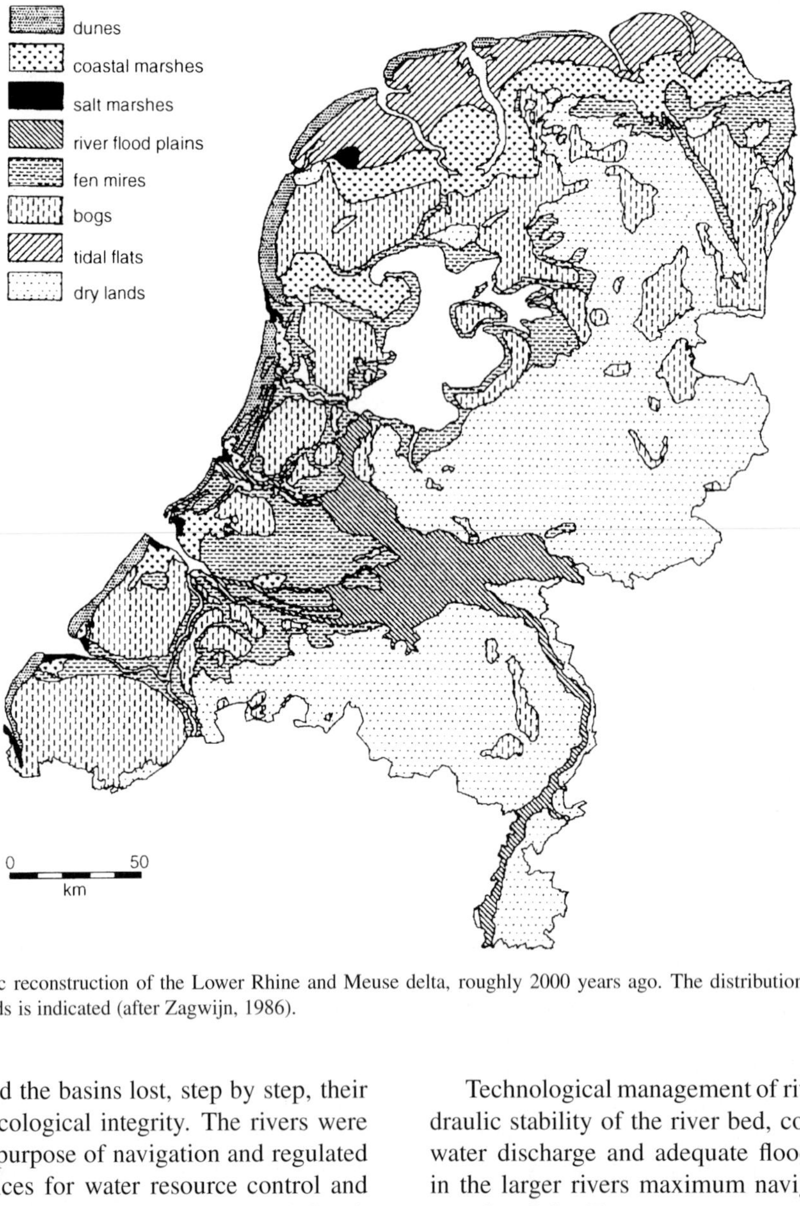

Figure 1. Paleographic reconstruction of the Lower Rhine and Meuse delta, roughly 2000 years ago. The distribution of the major wetland types in the Netherlands is indicated (after Zagwijn, 1986).

500 years ago, and the basins lost, step by step, their naturalness and ecological integrity. The rivers were canalised for the purpose of navigation and regulated by weirs and sluices for water resource control and flood defence, habitats were fragmented and flood-plain land was reclaimed for urban and industrial purposes. The basins were used as sewers to carry waste and drainage water away from the urban environment (Smits et al., 2000). From the early 1900s, major dam building activities started for both hydroelectric power and drinking water supply. At present, almost 80% of the total water discharge of the main European rivers is more or less strongly affected by flow regulation measures (Dynesius & Nilsson, 1994), while in a very densely populated country like the Netherlands this percentage is close to 100.

Technological management of rivers, aiming at hydraulic stability of the river bed, comprising efficient water discharge and adequate flood prevention, and in the larger rivers maximum navigability, ruled the agenda of the European water managers for centuries. Confronted with severe water pollution during the build-up phase after the Second World War, the cleaning-up of the rivers Rhine and Meuse started in the 1960s and 1970s. Considering the complexity of the remaining environmental problems in river catchments, it is becoming more and more difficult to gain positive and predictable results (Nienhuis et al., 1998).

The change from technological river management to ecological river management, in which values of landscape and nature and the sustainable use of water resources are the essential parameters, is a general EU-wide policy, in anticipation of the implementation

Table 1. Hydrological characteristics of the river Rhine

Length (km)	1320
Q (m^3 s^{-1}) at Lobith	2300
Qmin (m^3 s^{-1}) at Lobith	790
Qmax (m^3 s^{-1}) at Lobith	12 000
Average speed (m s^{-1}) at Lobith	0.5–1.5
Catchment (km^2)	185 000
Rain/melt	R / M

of the forthcoming EU Water Framework Directive. Among the other objectives, this legislation will focus on water management based on river basins. The best model for a single system of water management is the natural geographical and hydrological unit, the river basin. For each river basin district, a river basin management plan will need to be established and updated every six years. The plan is a detailed account of how the objectives set for the river basin, such as the chemical status, the ecological status, the conservation of protected areas, the socio-economic and legal status, are to be achieved within the time scale required (V&W, 2000).

The Rhine comprises a large trans-boundary river basin, having its origin in the mountains of Switzerland, fed by numerous smaller and larger tributaries while flowing through France, and Germany and debouching into the North Sea in the Netherlands (Table 1; Fig. 2). Rivers traverse different national borders, and the present day practice is that in most cases river management is restricted to within the national borders, and consequently only parts of the data of projects for environmental management and river rehabilitation are accessible in international journals (De Waal et al., 1995). The river Rhine is, however, an exception in this respect, owing to the enduring activities of the International Commission for the Protection of the Rhine, which celebrated its 50th anniversary in 2000 (ICPR, 2000).

In this paper, the status of ecological rehabilitation of the lowland basin of the river Rhine will be reviewed (Fig. 3). The question arises with regard to the use and application of ecological theory and empirical knowledge in the framework of river restoration projects. Other questions pertain to the successes and failures of ecological rehabilitation projects in the Netherlands. The paper starts with the historical perspective of river restoration in a Rhine-catchment context, in which drastic regulation measures and deterioration of water- and habitat quality are predominant. General goals for river rehabilitation are sketched, underpinned by ecological theories, and the contrasting every day practice of the river manager is also shown. The next section elaborates on the quality assessment of river rehabilitation projects, starting with the status of a number of Dutch projects, followed by some results of trans-boundary rehabilitation projects along the rivers Rhine and Meuse. The paper ends with some conclusions on the implementation of Dutch rehabilitation plans.

River rehabilitation: the historical perspective

River regulation

It is now well known that almost all large European rivers are heavily influenced by man. The deterioration of the river Rhine already started in the early Middle Ages. Already in the period 800–1200, considerable areas in the Rhine flood plains were deforested, owing to an increasing demand of wood for building and fuel purposes. Building of flood defence dikes were started from 1000 onwards (van Urk & Smit, 1989), and the canalisation started already in the late 14th century. In the early 18th century, almost all beech and oak forests were cut down, and replaced by grassland and cattle, the picture which we know so well nowadays. In the 19th century the Rhine changed at an accelerated speed: flood defences, flood control dams, barrages and hydroelectric power stations were built in the upper Rhine. Industrialization took place in the Middle and Lower Rhine areas: the construction of roads, human and industrial settlements, large factories for steel and chemicals. Canalisation is the most striking human disturbance of the riverbed. Nowadays the river Rhine is a completely man-manipulated river (Tittizer & Krebs, 1996). Downstream the completion of the Haringvlietdam in 1970 meant a physical barrier between the rivers Rhine and Meuse and the North Sea. The presence of this dam caused extremely detrimental effects on the ecology of the former estuary, and in a wider context also on the ecology of the entire rivers Rhine and Meuse, and the coastal strip along The Netherlands (Smit et al., 1994; Nienhuis, 1996).

The regulation of the river Rhine started centuries ago, was intensified in the 19th and 20th century, and was only very recently completed, with the construction of a storm surge barrier in the Nieuwe Waterweg (1996), in the mouth of the river Rhine, close to

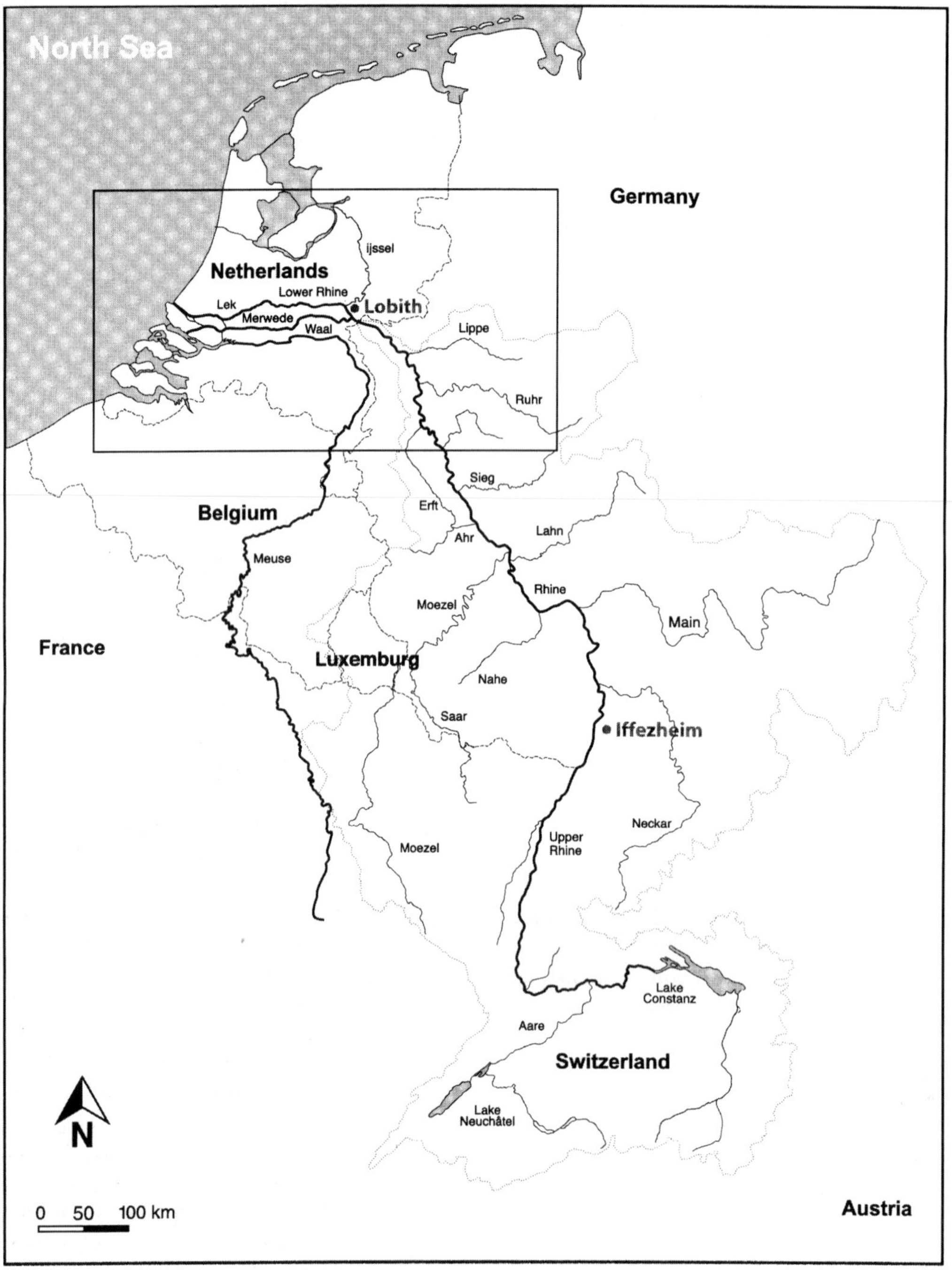

Figure 2. The catchment of the river Rhine and its tributaries. The Rhine enters the Netherlands at Lobith. The Lower Rhine in the Netherlands splits in several branches, IJssel, Nederrijn-Lek, and Waal-Merwede; downstream of Lek and Merwede the complicated Rhine–Meuse Delta is situated, comprising a.o. Nieuwe Waterweg, Hollands Diep and Haringvliet, as mentioned in the text. Insert: the Lower Rhine in Germany and in the Netherlands (see Fig. 3 for details). The river Meuse is indicated without tributaries.

the North Sea. The long-term effects of the building of these dams and weirs could not have been foreseen in earlier centuries, but the entire Rhine regulation has had great irreversible consequences for the present-day rehabilitation schemes. It is known that the reservoirs behind a dam trap roughly 95% of the

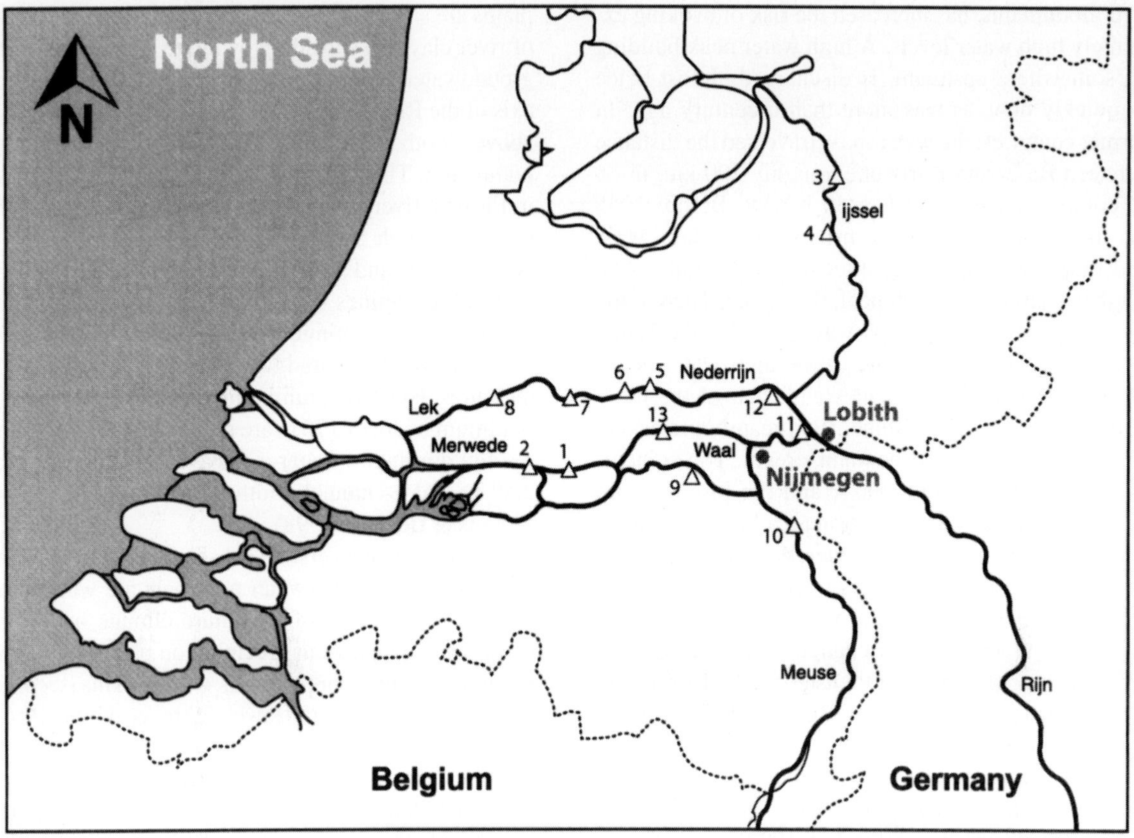

Figure 3. Ecological rehabilitation projects along the Dutch section of the Lower Rhine and Meuse discussed in this paper. 1. Opijnen floodplain and secondary channel and Gamerensche Waard secondary channel; 2. Loevestijn floodplain; 3. Duursche Waarden floodplain; 4. Ossewaard floodplain; 5.Blauwe Kamer floodplain; 6. Amerongse Bovenpolder floodplain; 7. Bosscherwaarden floodplain; 8. Grote Bol Willige-Langerakse Waard floodplain; 9. Liedensche Waard Batenburg floodplain; 10. Ossenkamp Gennep floodplain; 11. Millingerwaard floodplain; 12. Meinerswijk floodplain; 13. Beneden-Leeuwen secondary channel and floodplain.

sediment load coming from upstream. Further, owing to the shortening of the river course, the water gradually started to flow faster, and the faster currents eroded the river basin, leading to unrestrained scouring and incision of the river bed. In the period from 1828 to 1920, a lowering of the Rhine river bed in Germany of 3 m was measured, increasing to 7 m in recent years (Bernhart, 1990). Scouring of the summer-bed continues over the Dutch–German border, and only farther downstream in the Merwede and the Hollands Diep-Haringvliet river mouth, where the water mass slows down, eroded sand and silt settles down.

The problem has been 'solved' by the continuous dumping of hundreds of thousands of tons of gravel and sand downstream of the larger weirs in Germany. This process is expensive and certainly not sustainable. The dumping of sand and gravel has to be continued year after year, but still the bottom of the river is slowly lowering, e.g. at the Dutch–German border by 2 cm per year. Further erosion of the river bed will increase the instability of the river banks and groyne constructions. Recently, in 1996–1998, the Dutch water managers have armoured stretches of the river Rhine (Waal) by paving the bottom of the river with large blocks of stone, hopefully this will prevent further degradation of the river bed (RWS, 1996). Whether this experiment would be successful is difficult to say at present, but bridling of a dynamic, sediment transporting lowland river with a stone-strengthened bottom does nor appear to be a sustainable solution.

One of the main reasons for building the regulation works in the Rhine and its tributaries was a more effective control of the dangers from flooding (Dister et al., 1990). The irony of the fact is that the canalisation works, together with the enclosure of large areas of the river floodplains, has increased the risk of evoking extremely high water levels. A high water peak building up somewhere upstream, is discharged almost twice as quickly now, as was more than a century ago. In former centuries, the water mass traversed the distance between Basel and Karlsruhe, roughly 200 km, in 65 h, compared with less than 30 h now. Before 1955 the high water peak of the tributaries Neckar, Main and other rivers reached, after heavy rainfall in the southern part of the catchment, the main branch of the river Rhine before the high water peak of the Rhine itself. Nowadays the timing of the high water peaks of all the tributaries coincides with that of the maximum discharge of the Rhine, significantly increasing the high water peak of the main stream. The coincidence of the high water peaks, and the loss of more than 130 km^2 of water retention area in the German–French floodplains, has considerably diminished the protection against high water of the lower parts of the cities along the upper Rhine: the former flooding frequency of 1/200 per year changed into a chance of 1/50 per year (Dister, 1990; Dister et al., 1990). The major cities along the lower Rhine, Koblenz, Köln and Nijmegen now regularly experience downtown flooding. Owing to human interference, the amplitude between high and low water mark of the river has risen enormously during the last 1000 years, e.g. at the Dutch–German border from 1.5–2.5 m to 7–10 m at present (Knaapen & Rademakers, 1990).

Complexity of recent water management

Since the completion of the modern regulation works of the upper Rhine in 1977 (weir at Iffezheim; Fig. 2), the catchment experienced an unprecedented frequency of extremely high water peaks. The floods of 1993, 1995 and 1998 are the very recent examples of this. This recently recognised high water problem, induced by hydraulic regulation works, demands stringent solutions for restoring, i.e the water-carrying capacity of the Rhine catchment. Ever since 1978 the German water authorities are engaged in plans for creating retention basins along the Upper Rhine, large areas to be used as overflows during extreme high water. Along the Lower Rhine, in Nordrhein-Westfalen, the design and execution of retention reservoirs and the re-activation of old overflow areas are in full swing (ICPR, 1998a).

Dutch and German river managers are confronted with several problems simultaneously: the navigation channel proper is eroding and deepening, the floodplains are silting up, and the inland polders, consisting of river clay, are subsiding owing to a lowering of the groundwater table. Continuous navigation on the main axis of the Rhine, the Waal-Merwede, is a prerequisite above all other functions, except safety for the human population. The river Waal is one of the busiest rivers in Europe. Every year about 160 000 vessels cross the German–Dutch border, comprising a cargo capacity between 250 and 18 000 tonnes per vessel, carrying 140 million tonnes of freight every year. The Dutch government is continuously improving the navigability of the Waal, by dredging and improved regulation measures. By 2010 a minimum depth of 2.80 m, and a minimum width of 170 m are required. At the moment, the aim during low water is to have a channel depth of 2.50 m and a channel width of 150 m (RWS, 1996; Simons & Boeters, 1998).

Besides the irreversible consequences of the Rhine regulation works, the water managers are worried by the model scenarios of the future climate in Europe that predict a change in precipitation pattern, less rain in summer and more rain in winter. This scenario implies a threat to the lower sections of the Rhine (Middelkoop, 1997a,b). Because more than 80% of the Rhine catchment is situated outside of the Netherlands, the Dutch water managers can only to a limited extent undo the adverse effects on the river of changes in climate pattern. The planned solutions to the high water problem considered in the Netherlands, such as the excavation of the floodplains or the widening of the river bed, have only an additional function. Supposing that during heavy rainfall a high water peak is building up in Germany, the flood wave would be difficult to stop on its way to the sea, and despite the existence of wide floodplains along the river in the Netherlands, the height of the peak will not diminish drastically. The Dutch policy in river management was, therefore, until very recently designed at the quickest possible dispatch of the high water peak down to the sea, implementing the River Act of 1908. That is also the reason why no discharge-hindering obstacles are allowed in the river bed (Nienhuis & Leuven, 2001).

Triggered by the extreme floods of 1993 and 1995, and considering the predicted changes in river discharge in the near future (e.g. for the Rhine an increase in winter discharges and a decrease in summer discharges), the State Commission on Water Management (CWB21, 2000) advised a rigorous change in water management strategies. The main features of the planned strategy are to: (1)*retain* surplus water in the upstream stretches of the catchment as much as possible; (2) *store* surplus water temporarily in retention basins, if the strategy (1) does not work; (3) *discharge* surplus water if strategies (1) and (2) both fail. This *'retain, store, discharge'* sequence, of course, only works when water managers in the middle and upper stretches of the Rhine also concurrently apply the planned measures. Slowing down and reducing of high water peaks over time in Germany is of great importance for the situation in the Netherlands. Thus, for 'keeping their feet dry' the Dutch are dependent on the measures taken by the Germans. International cooperation to reach sustainable protection against flooding is of prime importance, and this is a high priority subject in the international negotiations (ICPR, 2000).

It should be stressed that unwanted flooding is a problem for the entire catchment of Rhine and Meuse. Measures to lower the water level during peak discharges in the Lower Rhine basin will forfeit their purpose if measures in the Upper Rhine are neglected. In a worst case scenario, at a discharge considerably exceeding the one in 1995, only 10% of the surplus water volume flowing down to the Netherlands will be retained in the planned retention reservoirs in Nordrhein-Westfalen. On the other hand, if the discharge peaks of the Rhine and its tributary, the Mosel, coincide, the retention capacity for river water in Nordrhein-Westfalen would be of greater importance for the protection of the Lower Rhine, than the retention reservoirs along the Upper Rhine (Dister et al., 1990; Havinga & Smits, 2000).

Water quality and biodiversity

From the early human settlements until the Middle Ages, the composition of the water in the river Rhine was hardly influenced by humans. The massive cutting down of trees in the Middle Ages, led to erosion, and changed the sediment load of the river. From 1920 to 1990, the human population in the Rhine catchment increased from 5 million to 45 million people, which caused an exponentially increased impact on water quality and habitat occupation. The maximum pollution of the river occurred in the 1960s and early 1970s. Since 1975, however, a positive turn in water quality and in the composition of species has been observed (Tittizer & Krebs, 1996). For example, the average oxygen concentration in the water at the German–Dutch border (Lobith), which had steadily dropped to a critical value of 4 mg l^{-1} (minimum value of 2 mg l^{-1}) in 1971–1972, and rose thereafter to an average of 9–11 mg l^{-1} from 1995 onwards. The chloride load carried by the Rhine at Lobith, originating from mining and industrial activities, rose steadily. In 1874, the yearly mean-salinity at the German–Dutch border was 13 mg Cl^- l^{-1}; in 1988 it was 150 mg l^{-1} with maximum levels up to 234 mg l^{-1} Cl^- (Den Hartog et al., 1989). Concentrations dropped slightly after 1990, but are still too high; at present the Rhine contains 10 times its natural salt content (RIWA, 2000).

Inorganic nutrients are considered as eutrophying substances. At Lobith the concentrations of nitrate drop steadily, but very slowly, from 4 mg l^{-1} in the 1970s to 3.5 mg l^{-1} in 1998. Ammonia concentrations drastically dropped from 1 mg l^{-1} in 1975 to 0.1 mg l^{-1} in 1998. Orthophosphate concentrations showed a steady decline from 0.4 mg l^{-1} in the early 1970s to less than 0.1 mg l^{-1} in 1998. According to the standards laid down by the quality objectives of the Netherlands, concentrations of specific herbicides, pesticides and poly-cyclic aromatic organic compounds are still too high. Almost all heavy metal concentrations in river water have dropped dramatically over the last 25 years; the remaining problem comprises the legacy from the past, contained in the floodplain sediments (RIWA, 2000).

Concomitant with the deteriorating water quality the number of invertebrate species in the river Rhine have also decreased. In the early 1970s, only 20% of the original biota persisted (ICPR, 1987). The diversity of river biota is currently increasing again, largely as a result of the return of original inhabitants. Also, the immigrant species have contributed to this increase. Many of these species come from brackish environments and southern areas (Admiraal et al., 1993). The occurrence of several brackish-water species in the freshwater stretch of the river Rhine is a result of the increasing salinity of the river, and concurrent rise of the temperature of the river-water by approximately 2–3 °C. In addition, the continuously increasing integration of river systems, which used to be separated, contributes to these changes, e.g. the Rhine–Main–Danube Canal linking the river Danube and the river Rhine, connects the Black Sea with the North Sea.

The salt and temperature tolerant species of invertebrates have a strong competitive advantage, compared with autochthonous Rhine biota. Between 1970 and 1980, the freshwater wood louse (*Asellus aquaticus*) was the most important invertebrate on the stones. From 1980 to 1985, this position was taken over by the dance fly larvae with the genera *Cricotopus* and *Dicrotendipes* as the most important represent-

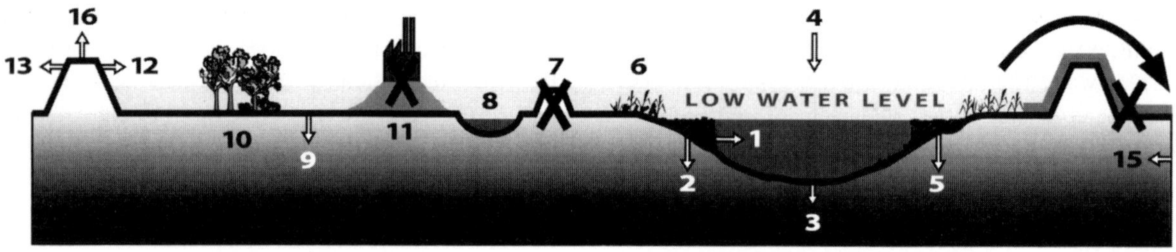

1 Narrowing of the main channel
2 Lowering of the groynes
3 Dredging
4 Redumping of sediment
5 Permanent layer
6 Natural bank
7 Removing summer embankment
8 Digging a secondary channel
9 Lowering of the embanked flood plain
10 Nature development
11 Removing of raised areas
12 Dike reinforcement
13 Dike repositioning
14 Retention (outside the high-water bed)
15 Obstructing lateral inflow
16 Dike raising

Figure 4. Proposed compensation measures in favour of flood protection and ecological rehabilitation along the Lower Rhine (V&W, 1998).

atives. In 1987 and 1988, an immigrant dominated for the first time: the tiger gammarid (*Gammarus tigrinus*). Since 1988, two species of basket mussels have been found in the river bed of the Dutch branches of the Rhine. These are the twisted form *Corbicula* cf. *fluminalis* (upstream) and the coarse ribbed *Corbicula* cf. *fluminea* (downstream; tolerates brackish water). Both species are immigrants. In 1990, the Caspian mudshrimp *Corophium curvispinum*, also an immigrant, was still a (sub)dominant species. This mudshrimp is a strong competitor to other organisms which live on stones, and consequently this species became dominant, while the populations of other organisms which live on stones, like the zebra mussel, *Dreissena polymorpha,* have greatly decreased. The mayfly *Ephoron virgo*, which occurred before in Dutch rivers, returned in 1991. In 1992, the Gammaridae *Echinogammarus ischnus,* a southern immigrant, was spotted in the Netherland for the first time (Bij de Vaate et al., 1996) .The floods of 1993 and 1995 brought an influx of diaspores and larvae from upstream areas, which increased biodiversity any further (G. van der Velde, Pers. Comm.).

Despite the existing international Rhine treaties (V&W, 1998) only recently is there a growing concern for the restoration and maintenance of river ecosystems, a policy which is strongly incident-driven. A striking example is the Sandoz fire in 1986 in Schweizerhalle, near Basel on the river Rhine. The fire evoked substantial chemical pollution and consequent massive mortality of fish and invertebrates in extensive stretches of the Rhine. This accident appeared to be the turning point in river management (Rhine Action Programme; ICPR, 1987): from that time onwards it was not only water quality that counted, but the integrity of Rhine ecosystems was at stake. Already in December 1986 the ministers of the Rhine states, joined in the International Commission for the Protection of the Rhine, decided that the ecosystem Rhine should be restored in the year 2000 to such a level that extinct species (e.g. the salmon) should have returned, that the quality of the water as drinking water should be guaranteed, and that the discharge of pollutants should have been drastically reduced (Schulte-Wulwer-Leidig, 1995). Further recovery of the river biota depends on continued repression of pollution, but also on the development of more natural habitats and an increase in morphological dynamics of the river catchment (Cals et al., 1998).

Goals for river rehabilitation

Ecological concepts for river rehabilitation

Significant advances have been made in stream ecology since the early 1980s, principally through the implementation of ecological theories like the 'river continuum' (Vannote et al., 1980) , the 'serial discontinuity'(Ward & Stanford, 1995) and the 'flood pulse'(Junk et al., 1989) concepts. These have stressed the need for holistic approaches in our effort to promote catchment management. The aim must be long-term sustainable management, and our vision for rehabilitation also needs to be in balance with the standards of today (de Waal et al., 1998). A considerable number of integrated river-concepts have been published over the years; Townsend & Hildrew (1994) and Townsend (1996) reviewed and evaluated a diversity of proposed ecological concepts; for a more thorough analysis of the literature, the reader is referred to Nienhuis & Leuven (1998). In summary it can be said that river ecology has moved from a static view, describing spatial and temporal patterns and organisms, populations and communities in a zone or section of the river, through a still descriptive phase in which the occurring biota were correlated with physical and chemical environmental factors. In the consequent stage, structural linkages at various spatial and temporal scales, and biological processes became the focus, and explanations for patterns depended on carefully conceived causal pathways

According to Townsend (1996), most ecological concepts can be clustered into the broad spatio-temporal context of the catchment hierarchy of an entire river, the unit of study for the river ecologist. In this hierarchy spatial up-scaling is prominent, from the single habitat, via the landscape unit (or ecotope), to the spatial zonation pattern, to the tributary and finally to the entire river catchment. Temporal up-scaling may occur from local, short-term ecological processes to succession processes on ecosystem level, and finally to historical, or better, geological processes. Longitudinal linkages are shown prominently in the river continuum and nutrient spiralling concepts, while transversal linkages are demonstrated in the flood pulse concept. At present, we have much theory but too few rigorous tests at the catchment level to falsify Townsend's (1996) ideas.

Following Southwood (1977), ecological theory predicts that the medium, the physical habitat, rather than the organisms, determine the structure of ecosystems. The habitat provides the template on which evolution processes forge characteristic life history strategies. Thus through the effects of habitat conditions on the fitness of individual organisms in geological time, certain combinations of adaptations for survival and reproduction are assumed to be selected. According to Moss (2000), it is a consequence of Southwood's (1977) strictly ecological theory, that emphasis on individual, in policy terms often charismatic species, may be counter-productive for the conservation of freshwater ecosystems. For example, the euphoria about 'the salmon back in the Rhine' asks for comments. No wonder that the salmon is back in the Rhine after stocking the tributaries of the river with hundreds of thousands of young individuals. The salmon returns to the Rhine if the fish is able to migrate from the sea to its upstream spawning areas, finding the proper resting habitats, feeding items etc. on its way (de Groot, 1990, 2002). Weirs in the harnessed rivers are still preventing this. Maintenance of migratory salmonid stock may require major change in forestry policy, and operations around the spawning rivers; moreover, conditions in estuaries or on the open sea will affect the recruitment and survival of the stock. The chain of required habitats forms the template for the return of the salmon. Aquatic ecosystems provide many of the fundamental biogeochemical services upon which the continuity of live depends, and this provides a strong argument for conservation of functional values in aquatic systems through whole-system reconstruction rather than following the philosophy of terrestrial species-oriented conservation (Moss, 2000).

The 'habitat as a template' concept (Southwood, 1977) can be phrased in popular terms: 'create' the physical habitat, and than the species will follow. In a densely populated country like the Netherlands most natural ecosystems have been disturbed or were eradicated; almost all present-day landscapes and habitats have been shaped through cultural processes. Practical Dutch nature managers (Helmer et al., 1992; Overmars & Helmer, 1998) have, therefore, redefined the 'habitat as a template' concept into 'the genius of the place' concept, in which the dominant impact of man in shaping physical habitats is recognized. Natural physical (hydrological) and geo-morphological processes should be stimulated, and the subsequent development into full-grown ecosystems should be allowed to continue with a minimum of human interference. In this approach, nature development is seen as a cultural phenomenon, in which the existing landscape, including its historical and societal functions, should be revitalised commensurate with the present-day standards. The 'genius of the place' concept states that potentials of the landscape for a 'natural' development should be optimally exploited. Ideally, natural hydrological and geo-morphological processes should be allowed to proceed, instead of artificial designs from the drawing-table.

In the Netherlands, the habitat restoration in river basins starts from the 'stepping stone' concept (van Dijk et al., 1995; Cals et al., 1998), based on the assumption that local management projects should be set up along the longitudinal axis of the river Rhine at regular distances, in such a way that a connected chain of nature reserves comprising favourable conditions for the (potential) plant and animal species will be developed. One might argue that a drawback of the stepping stone concept is that a standard recipe for nature development will be applied to many localities, without accounting for the original spatio-temporal heterogeneity and ecological connectivity between the river course and the floodplains and integrity of the river and its floodplain. Basic insight is needed into the fluvial dynamics of the entire river, the original hydrodynamics and geomorphology, i.e. the ecologically acceptable flow regime (EAFR; Petts, 1996) as the template for habitat restoration. Once the template has been defined, a better understanding is needed of the reference ecosystems of the entire catchment (palaeo-references). A recent analysis of the Dutch water policy (Buijse, 2001) demonstrates that the Dutch water managers are fully aware of the theoretical limitations of the stepping stone concept. 'Water' should become a structuring principle in the Dutch spatial planning policy, and the attractive, but rather outdated concept of the national ecological network (Baerselman & Vera, 1989; LNV, 1990) should be redefined in the context of a natural and more sustainable water management.

River rehabilitation practice

Dutch river policy has recently embraced catchment scale management (V&W, 1998). Some non-conformist visions (a.o. 'the genius of the place') of Dutch non-governmental organizations (NGO's) for full ecological restoration of the Rhine and Meuse (De Bruin et al., 1987; Helmer et al., 1992) have been scrutinized by the Dutch Ministry of Transport and Public Works (Silva & Kok, 1994, 1996; V&W & LNV, 1994; V&W, 1996), and the paramount ideas have been incorporated in the Dutch national water policy (V&W, 1998).

The generally accepted governmental policy is now to allow the rivers to take more physical space, in order to avoid emergency scenarios during extremely high discharges, and consequent high water levels in the river ('Space for the river'; VROM & V&W, 1996;V&W, 1998). This change in policy had far reaching consequences for the pending river management. Each elevation of the water level caused by actions of man (building of houses; reclamation of wetlands; forestation; etc.) must be compensated for in a sustainable manner. The use of the winter bed of the river has severely been restricted: existing man-made constructions, such as farmhouses and brick-factories may be maintained, but no expansion of building sites is allowed. Ecological rehabilitation measures, comprising the natural development of floodplain forests, creating resistance and roughness for the water flow during maximum discharge, may not interfere with the unimpeded discharge of river water. All measures causing higher water levels than the calculated standard high water levels, connected to specific risk assessment levels, are not allowed, unless compensating measures will be taken. Compensating measures which are considered to reach this aim are (Fig. 4): (1) removal of obstacles in the river bed, such as ineffective groynes and summer dikes; (2) lateral compensation by land inwards rebuilding of the winter dike; (3) lowering (excavating) the river floodplains over very large surfaces. Presumably additional policy decisions have to be made for the period until 2015, guaranteeing that the legally agreed safety standard, 1/1250 per year, will hold at a discharge of 16 000 m^3 s^{-1}: (4) creating inland retention basins, inundation fields, to be used in case of extremely high river discharges; (5) creating 'calamity polders', spillways and overflows, inland abandoned land to be designed in the future to retain river water in case of an unprecedented river flood. These 'calamity polders' are heavily contested, owing to the fact that already centuries ago the now land inwards areas, the former floodplains, have been reclaimed, completely cultivated and densely populated and economically and legally fully occupied.

The theoretical concepts for river rehabilitation are predominantly catchment oriented, but the planning and implementation of rehabilitation schemes are mainly done at the level of the regional water boards, and participation of municipalities and local interest groups is important. The motives to start a river rehabilitation project are very diverse. Initiative is taken by a wide variety of stakeholders, either top-down government directed, or bottom-up, e.g. by a local group of nature conservationists. Harper et al. (1995) who reviewed numerous river rehabilitation projects, consider river restoration to deal in most cases with small scale, practical projects, without a 'great vision' (such as 'sustainable catchment management'). The targets of most European rehabilitation plans are set for small stretches of rivers (up to 10 km), and also for small rivers or streams ($Q < 10$ m^3 s^{-1}). The full time span for a larger rehabilitation project easily takes 20 years (Harper et al., 1995). As stated before, each country has its own specific physical planning and legislation procedures; these make border-crossing river rehabilitation projects (such as the Gelderse Poort along the Rhine, in the border area of the Netherlands and Germany) challenging but laborious (de Jong, 1999), unless the countries involved share the same objectives (cf. the Living Border Meuse; Toebat et al., 2000). In the section *'Quality assessment of rehabilitation projects'* we will dwell deeper upon this subject

De Waal et al. (1995) reviewed 66 river rehabilitation schemes recently carried out mainly in Western Europe. Man's most dramatic impact on fluvial systems is canalisation, which involves the direct modification and shortening of the main river channel in the past. This could explain the high percentage (75%) of rehabilitation projects dedicated to channel morphology. Nature conservation and restoration were by far the most common objectives of the rehabilitation projects (85%). A commonly cited constraint involves the inability to allow floodplain inundation because of the requirements to provide flood control. An other argument is the lack of control over floodplain land as a result of land ownership or because of urban stakeholders occupying the floodplain area.

The current practical objectives for ecological restoration of river basins in the European Union, based on Harper et al. (1995) and de Waal et al. (1995, 1998) can be categorised as follows:

(1) The first aim, of course, is the rehabilitation of nature and landscape, lost in the past. Important nature development measures are river re-meandering of the previously deepened, straightened and featureless channel, allowing overland flooding during high discharges, restoration of side-channels, excavation of new river courses, oxbow lakes, reinstatement of floodplain habitats, which serve rich wildlife and waterfowl. Nature development enhances the amenity and recreational potentials of the river.

(2) The second objective is flood protection which may lead to a 'win–win' situation in combination with ecological rehabilitation. Excavation and widening of the accreted floodplains enhance the water retaining volume of the floodplains. Combined with nature development motives, lowering the surface of the land enhances the frequency of flooding, while excavating the original, irregular geo-morphological pattern in the underground, creates a diversified pattern of habitats.

(3) The third motive deals with the (sport)fisheries interests. The spawning habitats for fish can be improved by restructuring river channels, creating fish passages, sand banks, gravel beds and boulder groups. Increase of the habitat diversity along fully canalised streams and restoration of connections with wetlands and oxbow lakes is likely to lead to improvement of riverine fish communities.

(4) The fourth motive is the opportunity for nature rehabilitation created by the European policy to set aside arable lands in the floodplain, that are no longer economically profitable. The fertilised arable fields can be converted to herb-rich floodplain grassland, particularly those on sandy-clay soils. Establishing non-fertilised wet meadows alongside rivers leads to a reduction of the nutrient leaching from cultivated farmland, and hence to improvement of the water quality of the stream.

Reference image and target image

Rehabilitating a river to its former or to a more semi-natural state is a challenging prospect. A common aim of rehabilitation projects is to make the river more natural, and 'naturalness' has become an important element of the enhancement ethology (de Waal et al., 1998). But what is considered 'natural' nowadays, might not have the same qualification centuries ago. Is a floodplain overgrown with hardwood or softwood floodplain forests more 'natural' than floodplain pastures, bordered by hawthorn hedges and studded with cows? If so, each rehabilitation process would need a reference image, i.e. a picture of a former state or condition of the river that will be used as a reference point during the rehabilitation process. The second and possibly the most important step in any rehabilitation scheme is to define a target image ('Leitbild' or vision) and associated project objectives. The target image sketches an ideal solution for the river and its floodplains, that is optimal under present conditions, opportunities and constraints. Successful river rehabilitation relies on integration of the skills and determination of specialists from a range of disciplines, from conception and planning through the implementation and appraisal (de Waal et al., 1995).

The historic reference situation necessary as a blueprint for ecological restoration, is a matter of much debate. Geological research in natural river levees of fluvial areas along the Dutch river Rhine revealed pollen diagrams from cores dated around 2700 years BP. The profiles showed an almost closed hardwood floodplain forest with small open spots in the vegetation, either caused by minor human activities or by natural grazing (de Klerk et al., 1997). These data are used as palaeo-reference ecosystems for nature development projects currently in progress on embanked floodplains in the Netherlands, where agricultural lands will be transformed into new nature areas. To re-establish these "near-natural wilderness areas" means allowing the natural process of vegetation succession in river floodplains for almost 200 years without human intervention. This is virtually impossible, and conflicts with several other functions of the rivers, viz. flood protection and shipping.

Postma et al. (1996) made an inventory of the present and past status of riverine landscapes in the Netherlands, classified as ecotopes (Fig. 5). In the present situation, roughly 75% of the surface area of the river floodplains of the Rhine is cultivated land, either dominating grassland (pastures), fields or built-up area. The current Dutch government policy is to decrease the acreage of cultivated land, to set aside, to abandon agricultural land and to stimulate the development of natural grassland and floodplain forests. The reference image dating back some centuries, shows dominance of hard- and softwood floodplain forests (roughly 60–70%), and this is in sharp contrast with the present situation, where only less than 5% of the floodplain forests is left. The target image (Leitbild) shows an intermediate position, a strong stimulus for natural grasslands, brushwood and floodplain forests, at the expense of classical agricultural land. The aim of the government is the doubling of the area of natural landscape along the large rivers to 30 000 ha in 2010.

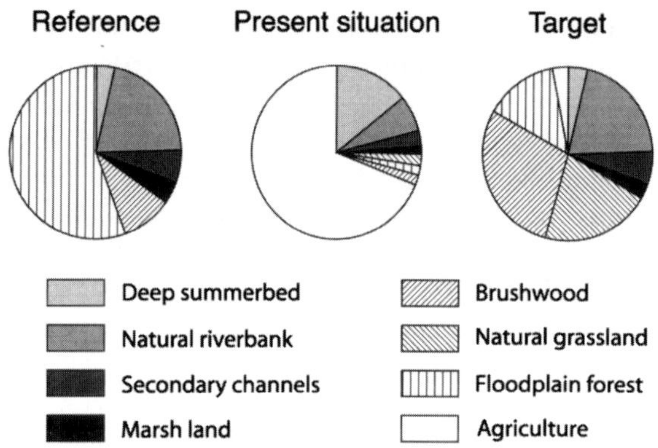

Figure 5. Relative share of the surface area of ecotopes in the Dutch lowland section of the river Rhine, showing the reference image, the present situation and the target image (Postma et al., 1996).

Quality assessment of rehabilitation plans

Dutch rehabilitation projects: successes and failures

Development of secondary channels

Van Dijk et al. (1995), Schulte-Wulwer-Leidig (1995), and Cals et al. (1998) gave a general overview of the progress of the ecological rehabilitation of the river Rhine. During the past twenty years water quality improved greatly owing to substantial international commitment (see section *Water quality and biodiversity*). A number of habitat restoration projects are carried out, as well in Switzerland, France, Germany as in the Netherlands. The river habitat improvements rely heavily on local and regional planning policies, and lack a catchment-oriented holistic approach. The ICPR (1998b) has made a start into this direction, carrying out an inventory of ecologically important areas as a first step towards an ecological network along the Rhine.

We will focus more specifically on the Dutch part of the Lower Rhine. Originally this lowland river was a meandering river with multiple channels, and the variety of flow velocities and water depths resulted in a large number of different river habitats. River regulation and intensive agricultural use of the floodplains resulted in massive loss of characteristic river communities and ecosystems, such as shallow flowing water habitats, unprotected river banks (without revetment), river dunes, floodplain forests, in particular hardwood floodplain forests, and floodplain grasslands.

The Rhine Action Programme, ratified by the Rhine countries in 1987, has accelerated existing programmes to enhance the quality of the water and the sediments in and along the river Rhine. A second important objective of the Rhine Action Programme was the restoration of lost and deteriorated river ecosystems. Successful projects are defined in this context as projects that function as substitute for lost habitats from the Lower Rhine basin, that bear a considerable potential for hydromorphological processes and the creation of riverine habitats, and that comprise characteristic river species, which will occupy these habitats. One of the possibilities to reach these aims is the digging of secondary channels and side channels, along natural, but silted up water courses in the floodplains. A secondary channel is a lateral channel, hydrologically connected with the main river channel, comprising flowing water, and natural (sandy) banks where erosion and sedimentation are the main driving forces in the formation of new habitats.

Since 1989 the possibilities to create man-made secondary channels in the floodplains of the Dutch rivers have been explored by means of literature studies and model experiments (Schropp, 1995; Schropp & Bakker, 1998). In 1994, the first two permanently flowing secondary channels along the river Waal were created at Opijnen (SC 1) and Beneden-Leeuwen (SC 2) (Fig. 3), and other projects followed during 1995–2001.Because of the experimental status of these secondary channels, the hydrological, morphological and ecological developments were intensively monitored. Post-project monitoring comprised a large number of abiotic and biotic variables. The ecological value of secondary channels will be illustrated by the monitoring results of selected groups of aquatic macro-invertebrates, fish and wading birds (Simons et al., 2001).

Aquatic macro-invertebrate species benefited almost immediately after the opening of the secondary channel (SC 1 in 1993, 122 species, and SC 2 in 1994 200 species; variation over the period 1993–1998 in SC 1 is 113–126 species, and over the period 1994–1997 in SC 2 181–257 species (Simons et al., 2001). The number of rheophilic species proper amounted to 5–10% of al the aquatic species. Compared with the aquatic macro-invertebrate data of the main channels of river Rhine branches in the Netherlands in 1995, 23% of the taxa were exclusively found in the secondary channels. The overall macro-invertebrate species spectrum, however, also counts a high number (15%) of exotic, non-native species, reflecting the image in the Rhine proper. In terms of density, the dominance of exotic species is even more prominent. The total number of individuals of *Corophium curvispinum* – an exotic species from the Danube – varied between 44 and 80% of the total number of all macro-invertebrate individuals over the period 1993–1998.

Both secondary channels were used as growing habitat for larval and juvenile eurytopic and rheophilic fish species (6–9 eurytopic species and 1–6 rheophilic species in SC 1 over 1994–1998; 2–5 eurytopic species and 3 rheophilic species in SC 2 over 1995–1998). Densities of young fish are higher in the secondary channels than in the groyne fields of the main river channel (Grift, 2001).

The shallow water and wet soils around SC1 and SC2 are used as foraging habitats by waders during autumn migration.. The steepness of the channel slope is decisive for the areas of feeding habitat available. Also this aspect should be worked out further.

A preliminary conclusion is that both restoration projects SC1 and SC2 are successful projects, although longer monitoring and a better analysis of the data should be done. In a more recent project, the Gamerensche Waard (Fig. 3), two smaller channels were excavated in 1996, and a larger one in 1999. All three have a larger discharge than SC1 and SC2. Especially the scale of dynamic processes, such as water flow, erosion, and sedimentation is obvious here. Sand martins (*Riparius riparius*) were triggered to nest in the steep eroded cliffs directly after the completion of the works, and rheophilic fish became abundant. (Grift, 2001; Jans et al., 2001). Other secondary channels are in the process of being excavated along the Waal (Schropp & Bakker, 1998). As yet, hydromorphological instability and the revealing of severely contaminated sediment deposits (an legacy from the 1960s) does not allow any conclusion about ecological developments in these projects.

Table 2. Assessment of a-biotic and biotic variables and management practice of rehabilitation projects (derived from Samborska ,2000). DW = Duursche Waarden; MiW = Millingerwaard; MeW = Meinerswijk; BK = Blauwe Kamer. > = increase, < = decrease; = no change of variable in target image, in relation to starting position. Objectives realised after execution phase: ++ very well; + considerably; ± partly; – not realised. X = no data available. Birds = breeding birds

	Variable	DW	MiW	MeW	BK
a-biotic	hydrodynamics >	+	+	+	+
	morphodynamics >	+	++	+	+
	water quality >	+	+	+	+
biotic	floodplain forests >	+	+	–	–
	wetlands, swamps >	–	–	–	+
	grassland <	+	+	±	±
	open water =	+	+	+	+
	typical plants >	+	+	+	+
	river grassland spp.>	+	+	X	X
	diversity vegetation >	+	+	X	+
	structure vegetation >	+	+	+	+
	diversity fish >	+	X	X	+
	abundancy fish >	+	X	X	+
	diversity birds>	+	X	+	+
	abundancy birds>	+	X	+	+
	diversity mammals>	+	+	+	X
	abundancy mammals>	+	+	+	X
management	nr. cattle ha^{-1}	0.2	0.3	0.3	0.3
	grazing practice	+	±	–	±
	plan/execution	1988/92	1992	1991	1991/92
	monitoring	++	±	–	++
	research	+	±	–	+

Entire floodplain rehabilitation

The restoration projects Opijnen and Beneden-Leeuwen (Simons et al., 2001) focus mainly on the rehabilitation of aquatic ecosystems proper. We evaluated the results of four ecological rehabilitation projects, comprising entire floodplains, in the Lower Rhine catchment, and checked whether the original aims have been realised. The four evaluated projects were: floodplains Duursche Waarden (IJssel – Rhine), Blauwe Kamer and Meinerswijk (Neder-Rijn – Rhine), Millingerwaard (Waal – Rhine) (Fig. 3; for details see Samborska, 2000). All four projects share roughly ten years of active ecological rehabilitation measures. Table 2 shows the a-biotic and biotic variables that have been assessed, together with the results reached. The main variables comprised increase/decrease of hydrodynamics, morphodynamics and water quality, diversity of vegetation units and diversity of plant species, number of species and abundance per species of mammals, breeding birds, and fish species. Additional assessment of the effects of grazing pressure on grasslands was done.

No general statement can be made regarding the 'success' or 'failure' of a specific project. The balance in all four projects is positive. All projects started from the dominance of cultivated, over-fertilised and levelled grassland or agricultural land (e.g. maize fields). Increasing the spatial heterogeneity by re-introducing more hydrodynamics and morphodynamics (excavation of sediment and dredging of silted up river branches), in combination with an extensive, near-natural grazing regime, enlarged within a few years the actual values of nature in all areas. The small size of the nature development areas in the floodplains has its drawbacks, hampering the return of critical species of the floodplain forests and the return of extensive swamps and wetlands. Many bird species and especially mammals need larger habitats as resting places, or to build up viable populations.

In general, the development of well structured brushwood canopies and hardwood floodplain forest takes more time than originally foreseen in the planning stage. Softwood floodplain forests on the lower parts of the floodplain, comprising willow species, black poplar and alder, show spontaneous colonisation, but strong wave wash and currents induced by heavy traffic on the river Waal may prevent settlement. Hardwood floodplain forests occupy the natural niche of the marshy and dry upper parts of the floodplains. Owing to the canalisation of the river and the construction of levees this habitat has disappeared almost everywhere. The input of viable seeds may be a limiting factor, and consumption and trampling by large herbivores may also hamper settlement. It will be very difficult to get the hardwood forests back under the present management conditions. The settlement of ash and elm is easier than that of the summer oak; it takes roughly 35–50 years for a hardwood floodplain forest to reach a semi-adult stage (van de Steeg et al., 1997; Siebel, 1998; van Splunder, 1998). The assessed ecological restoration projects are mainly dealing with terrestrial ecosystems, and less attention is paid to the rehabilitation of the open water bodies. Open pools and basins have still a rather rather poor water quality owing to seepage and leaching of nutrients from the over-fertilised former agricultural fields. Monitoring in the field of results reached, occurs irregularly and in an inconsistent way, related to the responsibilities of the land owner or nature management organisation engaged. These differences do hardly allow a comparison of the relative positions reached by the rehabilitation processes.

Interesting differences in rehabilitation successes are related to the hydro- and morphodynamic characteristics of the river branch under study. The IJssel is a low-dynamic river, and rehabilitation of the original physical processes, such as sand transport, scouring of the river bed, erosion, and sedimentation, hardly occurs. The Neder-Rijn has been, for 40 years, a weir-regulated river branch, with reduced and strictly controlled hydro- and morphodynamic processes. The Meinerswijk floodplain shows the lowest scores in terms of ecological rehabilitation success over the past ten years (Table 2). Meinerswijk is situated close to the city of Arnhem, on the Neder-Rijn, and buildings, such as an old brick factory, and archeological elements are conspicuous. Ruderal vegetation and grasslands are still dominating, and specific floodplain elements like marsh lands and reed swamps are hardly present. Floodplain forest development is retarded, and the grazing pressure by cattle is far too high.

The River Waal is the most promising one in terms of changes in hydro- and morphodynamic powers (Schoor et al., 1999). Illustrative is the development of the Millingerwaard in this context. High waters spread sand over the over-fertilised floodplain fields, leaving sandy trickles and small pools; new sand dunes and temporary sand ridges are formed, attracting plant and invertebrate species which were formerly absent. In 1992, an area of 14 ha of maize fields and grasslands in the floodplain of the Millingerwaard (Gelderse Poort) was set aside. Monocultures of *Brassica nigra, Cirsium arvense* and *Urtica dioica* dominated the scene in the first years after the fields were left fallow. In 1993 and 1995, the area experi-

Table 3. Ecological rehabilitation plans assessed by de Nooij (1999). Locations are indicated in Figure 3

Floodplain Waaloever Opijnen – Waal (Rhine)
Floodplain Loevestijn – Waal (Rhine)
Floodplain Duursche Waarden – IJssel (Rhine)
Floodplain Ossewaard – IJssel (Rhine)
Floodplain Blauwe Kamer – Nederrijn (Rhine)
Floodplain Amerongse Bovenpolder – Nederrijn (Rhine)
Floodplain Grote Bol Willige-Langerakse Waard – Lek (Rhine)
Floodplain Bosscherwaarden – Lek (Rhine)
Floodplain Liedensche Waard Batenburg – Maas (Meuse)
Floodplain Ossenkamp Gennep – Maas (Meuse)

enced extremely high river floods, creating sand dunes and eroding other parts of the fields. A layer of silt spread over the elevated fields, containing seeds and diaspores. This created numerous micro-habitats, and increased plant and animal diversity. In 1996, already 150 species of higher plants were counted. In 1997, the natural afforestation increased, controlled by semi-natural large herbivores, managed by the nature conservation agency (Bekhuis et al., 1995; Blankensteijn, 1997).

Quality assessment of rehabilitation projects

To answer one of the main questions posed in this paper, viz. concerning the use and application of ecological knowledge in rehabilitation projects, we performed a detailed quality assessment of ten ecological rehabilitation projects along Dutch stretches of the rivers Rhine and Meuse (Table 3; for details see de Nooij, 1999). The quality assessment was done via a best professional judgement procedure by river experts. The ecological criteria checked, were the potential of ecological rehabilitation in terms of (1) naturalness or physical integrity, (2) rarity of ecotopes, (3) diversity of ecotopes, (4) habitat functions based on the distribution of specific plants and animals, and (5) corridor function of the ecotopes in the river continuum. All rehabilitation plans assessed, originated in the period 1985–1995, implying that the boundary conditions of these projects appeared to be very stringent, mainly dictated by legal requirements regarding safety against flooding and facilities for transport by cargo ships. Boundary condition is that ecological rehabilitation may not lead to an increase of the hydraulic resistance of the riverbed, and hence to an increase of the design high water levels (River Act, 1908). Rehabilitation measures may not demand a relocation of the 'winter dike', the main levee between the inhabited polder and the river basin, or an adaptation of the depth and width of the main gully.

The conclusions from our study are:

1. Older plans for ecological rehabilitation are dominantly based on physical (hydrological and geomorphological) criteria, and far less on ecological criteria. More recent plans increasingly involve ecological criteria.
2. All plans are ecologically rather static, and based on the river ecotope classification (cf. Rademakers & Wolfert, 1994). The ecological potentials which are inherent in the natural succession, the spatial and temporal autonomous developments, such as the unpredictable appearance of dominant exotic species, and the moulding activities of extremely high river floods, have been ignored.
3. The quality of the historical reference image is decisive for the quality of the target image. This requires more detailed historical research, including the ecological interpretation of old maps.
4. The most stringent boundary conditions for ecological rehabilitation of the large rivers are stemming from the functions 'safety' and 'transport'. Protection against flooding and river navigation are major economic issues, determining to a large extent the potential ecological development (e.g. of the river Waal).
5. Ecological rehabilitation obviously is not so much an economic aim, but a cultural aim. One of the three arguments below, or a combination of these arguments is responsible for the fact that many plans for ecological rehabilitation of floodplains have not been executed, and will only be implemented in the long run: (a) the presence of severely contaminated sediments (class 4 polluted sediments, a legacy from the 'dirty' 1950s and 1960s), implying very expensive removal, (b) the laborious acquisition of agricultural land from private stakeholders, and (c) the lack of financial means for execution of the plans, notwithstanding the fact that the Dutch budgets are considerable, when viewed in an international context.

As said before, rehabilitation projects are local enterprises, lacking a catchment-oriented approach. The most comprehensive study in the context of the catchment hierarchy theory, done in the Netherlands, is the work of van den Brink (1994). He studied the impact of the hydrological regime along the lower Meuse and Rhine on the aquatic flora and fauna following transversal gradients, from the permanent water body of the central river through the elevated and almost completely isolated floodplain lakes, rarely connected to the river, to scour holes presently located on the landward side of the levee, and never connected to the river, except through the supply of groundwater. He investigated numerous localities and concluded that spatial isolation improves water quality and biodiversity. Along both rivers the biodiversity of the aquatic communities was found to decrease with increasing flooding frequency. Under the regime of a high flooding frequency a low diversity, turbid-water community exists, with dominance of phytoplankton biomass and filter-feeding invertebrates. Under a low flooding frequency a high diversity, clear-water community prevails with high phytoplankton and zooplankton diversity, macro-invertebrates and submerged water plants.

The work of van den Brink (1994) and van den Brink et al. (1996) bear consequences for ecological rehabilitation projects. It explains e.g. why at Dursche Waarden (IJssel – Rhine) the re-connection of isolated, infrequently flooded lakes with the main river channel resulted in a huge loss of macro-invertebrate species, an increase in phytoplankton biomass, and an increase in chironomid head capsule deformations, which might be attributed to an increased sediment pollution in these lakes due to increased influx of polluted river water. This image is not confirmed for the fish community where species richness in water bodies, permanently connected to the main river, is higher than in isolated waters (Grift, 2001). The observations on biodiversity patterns for various species groups are in line with the general patterns described along connectivity gradients: the highest diversity for fish is found in the river and in connected water bodies, while the highest diversity of macro-invertebrates and aquatic vegetation is found in temporarily disconnected floodplain lakes (Tockner et al., 1998).

Trans-boundary ecological rehabilitation of the rivers Rhine and Meuse

The Rhine Action Programme of 1987 was the first European programme considering the ecological rehabilitation of a large transboundary river, and substantial progress has been achieved in the course of the past decades (see section *Water quality and biodiversity*; Marteijn et al., 1999). That does not alter the fact that water management and connected ecological rehabilitation measures in the border area of Germany and the Netherlands are legally and politically separated items. The Germans are responsible for their part of the river Rhine, and the Dutch for their downstream part. In the German federation the separate states are almost autonomous with regard to their water policy. There exists a German Rhine-coordination platform, but this is an administrative body, and conflicts may easily arise between upstream and downstream interests concerning river management. In 1992, the Convention of Helsinki was agreed between most countries of the European Union. Sustainable use as part of integrated river basin management of trans-boundary rivers is covered by the Convention. This implies not only water quality as the oldest subject, but also the management of water quantity and the conservation and rehabilitation of ecosystems. In 1998 the treaty was ratified by the German and Dutch parliaments. Up to now, France did not sign the treaty, because of international conflicts concerning the river Meuse. The Convention of Helsinki, however, is of great importance for the Netherlands, the downstream country on the Rhine, enhancing the international responsibility and impact of the International Rhine Commission (Mostert, 1998)

Although the Rhine Action Programme of 1987 and the upcoming Rhine 2020 Programme (ICPR, 2000) agreed upon the execution of an international scheme for ecological rehabilitation of the river Rhine, in practice there are considerable differences between the German policy regarding ecological rehabilitation and the Dutch, especially when it comes to spatial planning and land use. This may be illustrated by

Table 4. Ecological rehabilitation policy in the Netherlands and in Germany, concerning the trans-boundary Gelderse Poort (GP); Millingerwaard (MiW) is part of GP; (derived from de Jong, 1999)

Netherlands	Germany
-intensive agriculture	-extensive agriculture
-offensive ecological rehabilitation	-defensive ecological rehabilitation
-changes in policy: GP is prestigious pilot project	-no changes in policy: GP is valuable cultural landscape
-integrated in national and regional policy plans	-not integrated in national and regional policy plans
-considerable financial means to execute plans	-restricted financial means to execute plans

the Gelderse Poort (Millingerwaard, Waal – Rhine), an international border-crossing nature reserve of 210 km^2 along the river Rhine, half in Germany, half in the Netherlands (de Jong, 1999). Table 4 shows the contrast between the strong offensive focus of the Dutch, and the conservative defensive attitude of the Germans regarding ecological rehabilitation of a large river floodplain area. The Dutch nature managers are in a process of changing intensively used agricultural land into 'new' spontaneous nature. The Germans regard their agricultural land as valuable cultural landscape, allowing only restrictive ecological rehabilitation measures. The Gelderse Poort offers a well documented example of the Dutch policy for nature development, where unexpected and unpredicted ecological developments created a new landscape (see section '*Dutch rehabilitation projects: successes and failures*'). In 2001, we still do not have an integrated trans-boundary policy plan for the Gelderse Poort. The recent lifting of the inner borders between countries in the European Union, and the growing political interest in regional, instead of national problems offer nevertheless a good basis for transboundary ecological rehabilitation schemes (Witsen, 1994).

The Netherlands and Belgium share the Meuse as joint border river over a stretch of 45 km, the Border Meuse. In 1994, the Belgian and Dutch authorities signed a declaration of cooperation, and meanwhile the project 'Living Border Meuse' is progressing slowly but successfully from the planning phase into the implementation phase: in 2000–2001 the first pilot project, a floodplain stretch of 2 km will be rehabilitated, by removal of the stone revetments along the banks, in order to enhance natural sedimentation and erosion processes. The cooperation between the Belgian and the Dutch water managers is an example of a win–win situation, because both partners have roughly the same objectives. The Dutch main priority is flood protection by creating more discharge capacity for the river Meuse. Derived objectives are gravel mining and nature development. In Belgium (Flanders) nature development is the main aim, since the desired safety is already guaranteed by reinforced levees. Unlike the Dutch project, however, the Belgium project will not comprise gravel extraction. The project includes lowering of the level of the river banks and re-constructing side channels, including the rehabilitation of natural habitats (Toebat et al., 2000).

An effective strategy to enforce the Dutch rehabilitation schemes to become economically feasible appears to be to agree an alliance with partners which play a important role in the physical planning and management of the rural areas. Important stakeholders are Ministries and Regional Agencies of the Environment, Nature and Water Management, companies for drinking water supply, brick factories, gravel mining companies and recreation boards. 'Win–win' agreements between the interests of nature rehabilitation and other societal interests may thus show considerable benefits for two or more parties. An example is the Dutch 'Gravel and sand for nature' project. Excavation of the floodplains and mining of gravel in the Meuse basin offers the sand- en gravel-miners a valuable resource, and at the same time it enhances the natural potentials of the river, 'rejuvenating' the silted-up floodplains by sediment removal. The main goal , of course, is that these excavation measures contribute to flood protection by facilitating the water flow in the river during extremely high discharges (van Leussen et al., 2000).

Conclusions

The river Lower Rhine comprises a man-dominated, strongly regulated catchment, polluted water and sediments, and annihilated and deteriorated ecosystems. The past 25 years the water quality and (to a lesser extent) the sediment quality improved considerably, leading to consequent improvement of biotic diversity. The rehabilitation of lost and destroyed ecosystems started recently. The non-refutable boundary conditions for restoration projects along the Rhine are safety against flooding and facilities for transport by cargo ships. Further recovery of the river biota depends on continued repression of pollution, an increase in morphological dynamics of the river catchment, and the development of the original natural habitats.

Holistic ecological theories to promote river catchment management are available, but rehabilitation projects rely heavily on local and regional planning policies. The use and application of ecological theory and empirical knowledge is increasingly playing a role in rehabilitation schemes, mainly focusing on nature conservation and restoration strategies, i.e. restoring habitat diversity and hence species diversity. Older plans for ecological rehabilitation are dominantly based on physical (hydrological and geomorphological) criteria, more recent plans fully involve ecological criteria.

The assessment of a number of projects along the Dutch part of the Lower Rhine leads to the conclusion that no general statement can be given concerning the 'success' or 'failure' of specific rehabilitation measures. The balance of 14 assessed projects is positive. Recent projects benefited from the wrong decisions made in earlier projects (viz. in Duursche Waarden the connection of isolated freshwater pools with the main river channel in 1989). All projects started with the dominance of cultivated, over-fertilised and levelled grasslands or maize fields. Increasing the spatial heterogeneity by re-introduction of larger hydrodynamical and morphodynamical forces i.e. excavation of sediment and dredging of silted up river branches and excavation of secondary channels, in combination with extensive, natural grazing by cattle on the fallow land, enlarged the actual values of nature in all areas within a few years. The quality of surface water bodies in the floodplains is rather negative, owing to retarded seepage of fertilisers via groundwater. The highly dynamic river Waal offers the best possibilities for the restoration of the natural sandy levees and back swamps. It will be very difficult to rehabilitate the hardwood floodplain forests under the present management conditions.

References

Admiraal, W., G. van der Velde, H. Smit & W. G. Cazemier, 1993. The rivers Rhine and Meuse in the Netherlands: present state and signs of ecological recovery. Hydrobiologia 265: 97–128.

Baerselman, F. & F. Vera, 1989. Nature development: an exploratory study for the construction of ecological networks. Ministry of Agric., Nature Manag. & Fisheries, Den Haag: 1–64.

Bekhuis, J., W. Bosman & H. Woesthuis, 1995. Millingerwaard. Jaarverslag 1993–1994. Stichting Ark, Laag Keppel (in Dutch).

Bernhart, H. H., 1990. De invloed van normalisering van de Bovenrijn op het riviersysteem en de maatregelen die daarvan het gevolg zijn. In Kamp, B. K., J. H. Mooij & J. Swart (eds), De Rijn, toestand en toekomst. Verslag van het internationale Rijncongres, 7–9 juni 1990, Arnhem (Gelderse Milieufederatie, Arnhem 1990): 97–131 (in Dutch).

Blankensteijn, H., 1998. Vanzelf natuur (11). Struiken en bomen bloeien in ooit maagdelijke wei. In: NRC Handelsblad 4-7-1998 (in Dutch).

Buijse, A. D., 2001. The role of wetland creation and restoration in Dutch water policy. In Jones, T. (ed.), WWF/EC Seminar Series implementing the EU Water Framework Directive. Proceed. The role of wetlands in river basin management. Seminar 2: 87–95.

Bij de Vaate, A., A. G. Klink, C. Bakker & M. J. R. Cals, 1996. Ecological river restoration in the Netherlands: effects on macroinvertebrate fauna in the Lower Rhine. Paper presented at 'River Restoration '96, Silkeborg, Denmark, September 9–13, 1996.

Cals, M. J. R., R. Postma, A. D. Buijse & E. C. C. Marteijn, 1998. Habitat restoration along the river Rhine in the Netherlands: putting ideas into practice. Aquatic Conserv.: Mar. Freshwat. Ecosystems 8:61–70.

CWB21, 2000. Waterbeleid voor de 21^{e} eeuw. Geef water de ruimte en aandacht die het verdient. Advies van de Commissie Waterbeheer 21^{e} eeuw aan staatssecretaris V& W en vz. Unie van Wateerschappen, Den Haag (in Dutch).

De Bruin, D., D. Hamhuis, L. van Nieuwenhuijze, W. Overmars, D. Sijmons & F. Vera, 1987. Ooievaar. De toekomst van het rivierengebied, Gelderse Milieufederatie, Arnhem (in Dutch).

De Groot, S. J., 1990. Herstel van riviertrekvissen in de Rijn een realiteit? 1. De Atlantische zalm (Salmo salar). De Levende Natuur 1990/3: 82–89 (in Dutch).

De Groot, S. J., 2002. A review of the past and present status of anadromous fish species in the Netherlands: is restocking the Rhine feasible? Hydrobiologia 478/Dev. Hydrobiol. 166: 205–218.

De Jong, D. M., 1999. Tussen natuurontwikkeling en Landschaftsschutz. Ph.D. Thesis University of Nijmegen, Eburon, Delft (in Dutch).

De Klerk, P., C. R. Janssen, J. H. J. Joosten & T. E. Tornqvist, 1997. Species composition of an alluvial hardwood forest in the Dutch fluvial area under natural conditions (2700 cal year BP). Acta Bot. Neerl. 46: 131–146.

Den Hartog, C., F. W. B. van den Brink & G. van der Velde, 1989. Brackish-water invaders in the river Rhine. Naturwissenchaften 76: 80–81.

De Nooij, R. J. W., 1999. Kwaliteitsbeoordeling van inrichtingsplannen voor natuurvriendelijke oevers langs grote rivieren. Verslagen Milieukunde nr. 174, Afdeling Milieukunde KUN, Nijmegen (in Dutch).

De Waal, L. C., A. R. G. Large, C. J. Gippel & P. M. Wade, 1995. River and floodplain rehabilitation in Western Europe: opportunities and constraints. Arch. Hydrobiol. Suppl. 101 – Large Rivers 9: 679–693.

De Waal, L. C., A. R. G. Large & P. M. Wade (eds), 1998. Rehabilitation of rivers. Principles and implementation. Landscape Ecology Series, John Wiley & Sons, Chichester.

Dister, E., 1990. Bescherming tegen overstromingen door revitalisering van ooigebieden aan de Bovenrijn'. In Kamp, B. K., J. H. Mooij & J. Swart (eds), De Rijn, toestand en toekomst. Verslag van het internationale Rijncongres, 7–9 juni 1990. Gelderse Milieufederatie, Arnhem 152–177 (in Dutch).

Dister, E., D. Gomer, P. Obrdlik, P. Petermann & E. Schneider, 1990. Water management and ecological perspectives of the Upper Rhine's floodplains. Regulated Rivers: Res. & Manag. 5:1–15.

Dynesius, M. & C. Nilsson, 1994. Fragmentation and flow regulation of river systems in the Northern third of the World. Science 266: 753–762.

Grift, P. E., 2001. How fish benefit from floodplain restoration along the Lower River Rhine, PhD Thesis University of Wageningen: 196 pp.

Helmer, W., G. Litjens, W. Overmars, H. Barneveld, A. Klink, A. Sterrenburg & B. Janssen, 1992. Levende Rivieren, Wereld Natuur Fonds, Zeist (in Dutch).

Harper, D., C. Smith, P. Barham & R. Howell, 1995. The ecological basis for the management of the natural river environment. In Harper, D. M. & A. J. D. Ferguson (eds), The Ecological Basis for River Management. Wiley, Chichester: 219–238.

Havinga, H. & A. J. M. Smits, 2000. River management along the Rhine: a retrospective view. In Smits, A. J. M., P. H. Nienhuis & R. S. E. W. Leuven (eds), New Approaches to River Management. Backhuys Publishers, Leiden: 15–32.

ICPR, 1987. Rhine Action Programme. International Commission for the Protection of the Rhine against Pollution, Technical Secretariat, Koblenz.

ICPR, 1998a. Action Plan on Flood Defence. International Commission for the Protection of the Rhine, Koblenz.

ICPR, 1998b. Betandsaufnahme de ökologischen wertvollen Gebiete am Rhein und erste Schritte auf dem Weg zum Biotopverbund. International Commission for the Protection of the Rhine, Koblenz: 71 pp.

ICPR, 2000. Rhine 2020 Programme for Sustainable Development and Action Plan 2005. International Commission for the Protection of the Rhine, Koblenz.

Jans, L., J. Backx, M. Greijdanus-Klaas, J. de Jonge, V. van der Meij, J. Oosterbaan, A. van der Scheer, M. Schropp & M. van Wijngaarden, 2001. Monitoring nevengeulen. Integrale jaarrapportage 1999/2000. RIZA Werkdocument 2001.062X. RIZA, Lelystad (in Dutch).

Junk, W. J., P. B. Bayley & R. E. Sparks, 1989. The floodpulse concept in river floodplain systems. Can. J. Fish. aquat. Sci. (SP) 106: 110–127.

Knaapen, J. P. & J. G. M. Rademakers, 1990. Rivierdynamiek en vegetatieontwikkeling. Rapport Staringcentrum, Wageningen: 83 pp.

Leuven, R. S. E. W., A. J. M. Smits & P. H. Nienhuis, 2000. From integrated approaches to sustainable river management. In Smits, A. J. M., P. H. Nienhuis & R. S. E. W. Leuven (eds), New Approaches to River Management. Backhuys Publishers, Leiden: 329–347.

LNV, 1990. Nature Policy Plan of the Netherlands in Outline. Ministry of Agricult., Nature Manag & Fisheries. Den Haag: 22 pp.

Middelkoop, H., 1997a. Embanked floodplains in the Netherlands. Geomorphological evaluation over various timescales. Ph.D. Thesis Utrecht University, Netherlands Geographical Studies 224: 1–341.

Middelkoop, H., 1997b. Growing plains: silting up of floodplains in the Netherlands under a changed climate. Change 38: 1–5 , RIVM, Bilthoven.

Moss, B., 2000. Biodiversity in fresh waters – an issue of species preservation or system functioning. Environm. Conservation 27: 1–4.

Mostert, E., 1998. River basin management in the European Union. How it is done and how it should be done. European Water Management 1 (3): 26–35.

Nienhuis, P. H., 1996. The North Sea Coasts of Denmark, Germany and the Netherlands. In Schramm, W. & P. H. Nienhuis (eds), Marine Benthic Vegetation. Ecological Studies 123: 1–470. Springer, Berlin.

Nienhuis, P. H. & R. S. E. W. Leuven, 1998. Ecological concepts for the sustainable management of lowland river basins: a review. In Nienhuis, P. H., R. S. E. W. Leuven & A. M. J. Ragas (eds), New Concepts for Sustainable Management of River Basins. Backhuys Publishers, Leiden: 7–33.

Nienhuis, P. H. & R. S. E. W. Leuven, 2001. River restoration and flood protection: controversy or synergism? Hydrobiologia 444: 85–99.

Nienhuis, P. H., R. S. E. W. Leuven & A. M. J. Ragas, 1998. New concepts for sustainable management of river basins. Backhuys Publishers, Leiden.

Overmars, W. & W. Helmer, 1998. Genius of the place. Aarde en mens 2 (2): 3–10 (in Dutch).

Petts, G. E., 1996. Water allocation to protect river ecosystems. Regulated Rivers: Res. Managem. 12: 353–365.

Postma, R., M. J. J. Kerkhofs, G. B. M. Pedroli & J. G. M. Rademakers, 1996. Een stroom natuur. Natuurstreefbeelden voor Rijn en Maas. Watersysteemverkenningen 1996: 1–102. RIZA-nota 95.060 (in Dutch).

Rademakers, J. G. M. & H. P. Wolfert, 1994. Het Rivier-Ecotopen-Stelsel. Ecological rehabilitation of the rivers Rhine and Meuse 61-1994, DLO/RIVM/RIZA, Lelystad: 1–77.

RIWA, 2000. Annual Report 1998. Part C: the Rhine and Meuse. International Association of River Waterworks, Amsterdam.

RWS, 1996. Rhein/Waal Fahrrinne/vaargeul Rotterdam-Duisburg. Rijkswaterstaat Directie Oost-Nederland, Arnhem/Wesel/Munster juni 1996.

Samborska, E. M., 2000. Evaluatie van de resultaten van het ecologisch herstel in het rivierengebied in Nederland. Verslagen Milieukunde nr. 183, Afdeling Milieukunde KUN, Nijmegen (in Dutch).

Schoor, M. M., H. P. Wolfert, G. J. Maas, H. Middelkoop & J. J. P. Lambeek, 1999. Potential for floodplain rehabilitation based on historical maps and present-day processes along the river Rhine, the Netherlands. In Mariott, S. B. & J. Alexander (eds), Floodplains: Interdisciplinary Approach. Special Public 163. Geological Society, London: 123–136.

Schropp, M. H. I., 1995. Principles of designing secondary channels along the river Rhine for the benefit of ecological restoration. Wat. Sci. Technol. 31: 53–58.

Schropp, M. H. I. & C. Bakker, 1998. Secondary channels as a basis for the ecological rehabilitation of Dutch rivers. Aquatic Conserv: Mar. Freshw. Ecosyst. 8: 53–59.

Schulte-Wulwer-Leidig, A., 1995. Ecological master plan for the Rhine catchment. In Harper, D. M. & A. J. D. Ferguson (eds), The Ecological Basis for River Management. John Wiley & Sons, Chichester: 505–514.

Siebel, H. H., 1998. Floodplain forest restoration. Tree seedling establishment and tall herb interference in relation to flooding and shading. Ph.D. Thesis, University of Nijmegen.

Simons, J., C. Bakker, M. H. I. Schropp, L. Jans, F. Kok & R. Grift, 2001. Man-made secondary channels in the river Rhine (The Netherlands); results of post-project monitoring. Regul. Rivers–Res. Manag. 17: 473–491.

Simons, J. & R. Boeters, 1998. A systematic approach to ecologically sound river bank management. In de Waal, L. C., A. R. G. Large & P. M. Wade (eds), Rehabilitation of Rivers: Principles and Implementation. Wiley, Chichester: 57–85.

Silva, W. & M. Kok, 1994. Natuur van de rivier. Toetsing WNF-plan Levende Rivieren. Eindrapport Ministerie Verkeer en Waterstaat, Den Haag (in Dutch).

Silva,W. & M. Kok, 1996. Integrale verkenning inrichting Rijntakken; IVR-rapport 1. Ministerie van Verkeer en Waterstaat, RIZA, Arnhem en Waterloopkundig Laboratorium, Marknesse (in Dutch).

Smit. H., R. Smits & H. Coops, 1994. The Rhine-Meuse Delta: ecological impacts of enclosure and prospects for estuary restoration. In Falconer, R. A. & P. Goodwin (eds), Wetland Management. Inst. Civil Engineers, London.

Smits, A. J. M., P. H. Nienhuis & R. S. E. W. Leuven, 2000. New approaches to river management. Backhuys Publishers, Leiden.

Southwood, T. R. E., 1977. Habitat, the templet for ecological strategies. J. anim. Ecol. 46: 337–365.

Tittizer, T. & F. Krebs (eds), 1996. Ökosystemforschung der Rhein und seine Auen – eine Bilanz. Springer Verlag, BerlIn 1–468.

Tockner, K., F. Schiemer & J. V. Ward, 1998. Conservation by restoration: the management concept for a river-floodplain system on the Danube river in Austria. Aquatic Conserv. mar. Freshw. Ecosyst. 8: 71–86.

Toebat, J., K. Lantmeesters, I. Hoet & H. Gielen, 2000. Het Vlaamse project 'Levende Grensmaas'. Natuurhistorisch Maandblad Natuurhistorisch Genootschap Limburg 89: 160–163 (in Dutch).

Townsend, C. R., 1996. Concepts in river ecology: pattern and process in the catchment hierarchy. Arch. Hydrobiol. Suppl. 113, Large Rivers 10: 3–21.

Townsend, C. R. & A. G. Hildrew, 1994. Species traits in relation to a habitat templet for river systems. Freshw. Biol. 31: 265–275.

Van den Brink, F. W. B., 1994. Impact of hydrology on floodplain lake ecosystems along the lower Rhine and Meuse. Ph.D. Thesis University of Nijmegen.

Van den Brink, F. W. B., G. van der Velde, A. D. Buijse & A. G. Klink, 1996. Biodiversity in the Lower Rhine and Meuse river floodplains: its significance for ecological management. Neth. J. aquat. Ecol 30: 129–149.

Van de Steeg, H. M., P. J. A. Vervuren & C. W. P. M. Blom, 1997. Ecologische waarden voor herstel van hardhoutooibossen. In Nieuwe Wildernis (Dieren, 1997) 3: 42–47 (in Dutch).

Van Dijk, G. M., E. C. L. Marteijn & A. Schulte-Wulwer-Leidig, 1995. Ecological rehabilitation of the river Rhine: plans, progress and perspectives. Regul. Rivers: Res. Mang. 11: 377–388.

Van Leussen, W., G. Kater & P. van Meel, 2000. Multi-level approach to flood control in the Dutch part of the river Meuse. In Smits, A. J. M., P. H. Nienhuis & R. S. E. W. Leuven (eds), New Approaches to River Management. Backhuys Publishers, Leiden: 287–306.

Vannote, R. L., G. W. Minshall, K. W. Cummins, J. R. Sedell & C. E. Cushing, 1980. The river continuum concept. Can. J. Fish. aquat. Sci. 37: 130–137.

Van Splunder, I., 1998. Floodplain forest recovery. Softwood forest development in relation to hydrology, riverbank morphology and management. Ph.D. Thesis, University of Nijmegen.

Van Urk, G. & H. Smit, 1989. The Lower Rhine geomorpological changes. In Petts, G. E., H. Moller & A. L. Roux (eds), Historical Change of Large Alluvial Rivers: Western Europe. John Wiley Sons, Chichester: 167–182.

VROM & V&W, 1996. Beleidslijn Ruimte voor de Rivier, Ministerie van Volkshuisvesting, Ruimtelijke Ordening en Milieubeheer & Ministerie van Verkeer en Waterstaat, Den Haag (in Dutch).

V&W, 1996. Achtergrondnota Toekomst voor Water. Project Watersysteemverkenningen. RIZA Nota 96.058. Rapport RIKZ 96.030. Rijkswaterstaat, Den Haag: 1–415 (in Dutch).

V&W, 1998. Fourth National Policy Document Water Management. Vierde Nota Waterhuishouding, Regeringsbeslissing, Ministerie van Verkeer en Waterstaat, Den Haag (in Dutch).

V&W, 2000. Europese Kaderrichtlijn Water, een tussenstand. Ministerie van verkeer en waterstaat, Den Haag (in Dutch).

V&W & LNV, 1994. Natuur aan het werk. Achtergronddocument – Ministerie Verkeer en Waterstaat (V&W) en Ministerie van Landbouw, Natuurbeheer en Visserij (LNV), Den Haag (in Dutch).

Ward, J. V. & J. A. Stanford, 1995. Ecological connectivity in alluvial river ecosystems and its disruption by flow regulation. Regul. Rivers-Res. Manag. 11: 105–119.

Witsen, J., 1994. Het Europa van de regio's. Afscheidsrede KUN. PIN-Nieuws, Nijmegen 1994: 17/7, 1–27 (in Dutch).

Zagwijn, W. H., 1986. Nederland in het Holoceen. Geologie van Nederland, deel 1. Rijks Geologische Dienst Haarlem / Staatsuitgeverij, Den Haag.

Hydrobiologia **478:** 73–106, 2002.
P.H. Nienhuis & R.D. Gulati (eds), Ecological Restoration of Aquatic and Semi-Aquatic Ecosystems in the Netherlands (NW Europe).

Lakes in the Netherlands, their origin, eutrophication and restoration: state-of-the-art review*

Ramesh D. Gulati & Ellen van Donk
Centre for Limnology, Netherlands Institute of Ecology, Rijksstraatweg 6, 3631 AC Nieuwersluis, The Netherlands
E-mail: gulati@cl.nioo.knaw.nl

Key words: external loading, lake management, biomanipulation, food web, Lake Loosdrecht, Wolderwijd, Zwemlust, phosphorus, *Cyanobacteria*, zooplankton, *Daphnia*, bream, fish, benthivores, planktivores, macrophytes, *Chara*, northern pike, Secchi-disc

Abstract

This article starts with a brief description of the origin and eutrophication of shallow Dutch lakes, followed by a review of the various lake restoration techniques in use and the results obtained. Most freshwater lakes in the Netherlands are very shallow (<2 m), and owe their origins to large-scale dredging and removal of peat during the early 17th century. They vary in area from a few hectares to a few thousand hectares, and are generally found in the northern and western part of the country. Most of them lie in the catchment areas of the major rivers: the Rhine, the Meuse and the Schelde. Because of their natural and aesthetic value, these lakes fulfil a recreational function. The lakes are important to the hydrology, water balance and agriculture in the surrounding polder country. The external input to the lakes of phosphorus (P) and nitrogen (N) and of polluted waters from the rivers and canals have been the major cause of eutrophication, which began during the 1950s. In addition, more recently climate changes, habitat fragmentation and biotic exploitation of many of these waters have probably led to loss of resilience and thus to accelerated eutrophication. Lake eutrophication is manifested essentially in the poor under-water light climate with high turbidity (Secchi-disc, 20–40 cm) caused usually by cyanobacterial blooms (e.g. *Oscillatoria* sp.), and loss of littoral vegetation. Despite recent perceptible reductions in external P inputs, non-point sources, especially of N from agriculture, still remain high and constitute a major challenge to the lake restorers. Lake recovery is also invariably afflicted by in-lake nutrient sources. These include P loading from the P-rich sediments, mineralization in the water and release by the foraging and metabolic activities of the abundant benthivorous and planktivorous fish, mainly bream (*Abramis brama*).

A variety of restoration techniques have been employed in the Dutch lakes: hydrological management, reduction of P in the external loads, in-lake reduction or immobilisation of P, and complementary ecological management. This last involves *biomanipulation*, or the top-down control of the food web. Hydrological management has resulted in an improvement in the lake water quality only in a few cases. The failure of lake restoration measures (e.g. in the Loosdrecht lakes, described as a case study) has led water managers to use biomanipulation in other lakes under restoration. Lake biomanipulation principally involves reducing the existing planktivore population, bream in most cases, and introducing piscivores such as northern pike (*Esox lucius*). Lake Zwemlust is discussed as a case study, with brief mention of some other small lakes which have been biomanipulated.

The restoration studies reveal that decrease of P to low levels is no guarantee that cyanobacterial populations will also follow suit. This is because cyanobacteria can withstand great variation in their P content and thus in their C:P ratios. Thus, for a unit weight of P, the Cyanobacteria can yield relatively more biomass and cause greater turbidity than, for example, green algae, which have relatively lower C:P ratios. This is possibly an explanation for the success of these filamentous Cyanobacteria in many Dutch lakes, and the failure of restoration endeavours. In addition, a falling trend in chlorophyll-*a* content in these shallow lakes does not set off an immediate increase in

* NIOO Publication no. 2991.

lake transparency because of resuspension of seston and inorganic suspended matter from the lake bottom by both wind-induced waves and fish foraging activity.

The zooplankton-grazing peak in spring, caused usually by large-bodied grazers, *Daphnia* spp., is invariably the first step in bringing about a clear-water phase. Subsequently, summer light conditions trigger optimal growth conditions for macrophytes, which then maintain the high water clarity by competing successfully with phytoplankton for nutrients, especially N. The 'return' of macrophytes, especially stoneworts(*Chara* spp.) in some lakes, has contributed to the sustaining of improved light conditions and success of the restoration measures. In addition to competing with phytoplankton for nutrients, the macrophytes exert their positive influence in manifold ways. They act as a major nutrient sink, provide refuges for zooplankton and young pike and reduce wind- and fish-induced bioturbation of sediment. Most restoration accomplishments in recent years have been attributed to the success of aquatic macrovegetation.

In general, the achievements of restoration work in the Dutch lakes, especially those using biomanipulation measures, are questionable: there are probably more examples of failures than of successes. The failures are generally linked not only to insufficient or no decrease at all in the autochthonous or in-lake nutrient loadings, but also to rapid increase of the planktivorous fish in the years following their reduction. A 75% reduction in the existing planktivore population has often been used as an arbitrary yardstick for effective reduction, but may not be sufficient. However, fish stock reductions to <50 kg FW ha^{-1} and maintenance at that level might have a greater chance of success, though maintaining the existing fish population at preconceived levels is difficult since for reasons not yet fully understood, piscivores, pike in particular, fail to develop sizeable populations. Studies so far have helped us recognise that for sustainability of the positive effects on water quality, 'natural development' should be central to future lake restoration programmes. Future restoration plans typically visualise lakes as integral parts of their landscape, and envisage their 'nature development'. Such thinking aims at reinforcing the lakes' shoreline vegetation to prevent erosion and improve the subtlety of the land–water transition (e.g. Volkerak Zoommeer lake system). Where in-lake P stocks have retarded the pace of lake recovery (e.g. Loosdrecht Lakes), excavation of 20–30 m deep pits in shallower lake areas to allow wind-induced shifting of the nutrient-rich upper sediment layers and burial in the pits in order to hinder P releases from the sediments is now under way. For some lakes the creation of artificial islands to reduce the wind fetch factor and erosion has been planned; in other cases, more natural development of the quasi-aquatic ecosystems by water-level management in order to encourage the shoreline macrovegetation to develop has been planned. Such plans also have the provision of extending the upper and lower limits for permissible annual water-level fluctuations and exploring the effects of transient drawdowns. Ideally, near-natural water levels, unlike the current levels, are under consideration as possibly being the best option, also bearing climate change in mind. However, the consequences of flooding and recessions on the ecosystems and other water uses by man still need to be thoroughly investigated. In short, the experiences acquired from the failures and some successes of the last two decades should pave the way to development of more enduring strategies for sustainable restoration of our lake ecosystems.

Introduction

In N and NW Europe there have been many reassessments and changes of strategies in respect of research into and management of inland waters. The emphasis of restoration studies on lakes has progressively shifted from one of seeking solitary solutions to specific water-quality problems, to dealing with environmental issues at a much broader level. The more sophisticated solutions are those in which the entire catchment area of the water body to be restored is taken into account, including the runoff. The present-day approach to lake eutrophication and pollution issues is to treat the water systems and their landscapes as one complex or entity. The remedial measures should ideally deal with the cause of ecosystem stress rather than with eradicating the undesirable symptoms alone. These measures need to be taken both outside and inside the water body to be restored (Vollenweider, 1987).

A challenging problem in industrialised W. Europe, especially in countries with intensive agriculture such as the Netherlands, is river and lake eutrophication caused by highly intensive agricultural practices as well as animal husbandry (poultry farming, piggeries and cattle farming). To achieve high productivity in agriculture, very high doses of fertilisers and manures containing both nitrogen (N) and phosphorus (P), but especially N, are applied to the fields. For example, the N applied to the fields in the Netherlands in 1990 amounted to about 450 kg ha^{-1} y^{-1} (GLOBE-EUROPE, 1992; review papers in Nienhuis & Gulati, 2002). This is about four times the average for W. Europe as a whole, about twice as high as in the European Union, and perhaps the highest dosage applied in any country in the world. The impact of such intensive fertilisation is far reaching: following seepage of nutrients to underground water and leaching, the nutrient-rich (particularly NO_3) water masses will re-emerge elsewhere to be transported via canals and streams to the lakes and rivers and to the sea. In eutrophied and polluted lakes, perennial algal blooms or their frequent recurrence and poor underwater light climate are well-known water-quality issues. It must be emphasised that like any complex hydrology problem, problems associated with eutrophication and pollution are too complex to solve within national boundaries, and need to be tackled at the international level and jointly.

In this paper, the origins, eutrophication and eutrophication control of lowland Dutch lakes will firstly be described. The various techniques of lake restoration, including hydrological, chemical and complementary approaches involving manipulation of the food web (biomanipulation or top-down control) will then be treated. Biomanipulation has received much attention in N. America and W. Europe during the last ten years, but especially in the Netherlands (see papers in Gulati et al., 1990b; van Liere & Gulati, 1992), Germany and the Nordic countries. The food chains of several shallow lakes in the Netherlands have been manipulated as a complementary remedial measure to ameliorate eutrophication and speed up recovery. State-of-the-art information on the following aspects will be discussed:

(a) The origin and eutrophication of lakes, and some long-term studies following restoration measures
(b) Restoration techniques involving hydrological and chemical manipulation aimed at P reductions in the inflows and in the lake itself, and manipulation of lake biota (biomanipulation) and case studies
(c) Modelling studies based on both empirical and theoretical information and the role of fish, macrophytes and zooplankton in lake restoration
(d) Salient aspects of restoration research, including achievements and failures, and future research directions

The lakes: their origin, functions and eutrophication

General

About 16% of the total area (41 864 km^2) of the Netherlands is covered by water, mostly classified as wetlands, includes riverine, estuarine and coastal ecosystems (Wadden Sea), freshwater lakes, of which Lake IJsselmeer is the biggest, and nutrient-poor fens (Fig. 1). The water-bodies vary considerably in area, depth, hydrology and physico-chemical and biological characteristics (see Best et al., 1993). During the last 2000 years, thousands of square kilometres of wetlands, including coastal salt marshes and shallow lakes, have been reclaimed for agriculture (Wolff, 1993). At present, the hydrology of these waters is being strongly influenced by transboundary rivers, especially the river Rhine.

Most freshwater lakes in the Netherlands are very shallow (<2 m), and owe their origin to large-scale dredging and removal of peat, which started in 1633 (see papers in van Liere & Gulati, 1992). They vary in surface area from a few hectares to a few thousand hectares, and are generally found in the north, north-west and west of the country, mostly in the catchments of the major rivers: the Rhine, the Meuse and the Schelde. The relatively older peaty lakes originated from erosion of peat and subsequent flooding. More recently, some of these lakes have become much deeper (10–50 m) due to excavation of sand (Gulati, 1972; Hosper, 1997).

The recreational function of these lakes, including navigation, is related to their natural and aesthetic value. De Haan et al. (1993) have discussed the impact of the reservoir function of the lakes on the limnology of the peaty lakes in the Netherlands with particular reference to the Friesian lake district (10 000 ha) in the north, and in the central part of the country. Low-lying as most of the lakes are, they play an important role in the hydrology and water balance and agriculture in the surrounding polder country. As part of the hydrological management of these lakes, in winter they receive water from the agricultural areas and polders, and in summer act as a source of water supply for various uses, including irrigation. Located in certain specific areas, the lakes make up 'lake districts'. Of the various lake districts, the focus here will restricted to those lakes that have been the subject of restoration research (Fig. 1)

rain water in ditches and ponds in urban areas to help replenish and improve retention of groundwater. The plans aim to restore the ecology of drainage ditches, give the rivers more room, and reduce emissions from various sources (agriculture, road traffic, atmospheric deposition).

Lake restoration is now among the major environmental issues relating to water management in general (papers in Nienhuis & Gulati, in press). Collaboration among scientists affiliated to Dutch universities and national research institutions and regional or local water management boards, but especially funding bodies, is playing an important role. The Institute for Inland Water Management and Waste Water Treatment (RIZA, Lelystad) and Institute for Coastal Waters (RIKZ, Middelburg) are state agencies falling under the *Rijkswaterstaat* (Directorate-General for Public Works and Water Management, Ministry of Traffic and Water Ways) whose task it is to monitor national surface waters. The two co-ordinate the monitoring and report on the quality of inland and coastal waters, respectively. RIZA has also carried out careful applied and experimental research on inland waters during the last two decades, and funded research projects at other research institutions. Such developments have helped to overcome some of the technical problems, financial snags and delays confronting lake restoration work in the Netherlands. Changes in philosophy and co-operation strategies have already led to increased understanding of ways of restoring water quality in lakes and their inflows. In some cases the results appear to be successful and sustainable. In addition, such co-operative works have paved the way to acquiring useful insights into the causal factors behind eutrophication and the resilience of the lakes to amelioration. Experience gained over the last two decades as well as international collaboration has thus been instrumental in reducing external nutrient loading rates, especially of P. This knowledge has certainly provided an impetus to lake restoration. Despite this, non-point nutrient discharges, especially N from agriculture, remain a major challenge to lake restorers.

Development of restoration research

Eutrophication control and lake restoration research

In the Netherlands, eutrophication control was one of the major environmental policy issues during the 1980s and 1990s (Hosper & Jagtman, 1990). Despite nutrient control policy and international programmes (Rhine Action Programme, North Sea Action Programme) and four National Policy Documents on Water, the P levels in lake inflows have remained higher than expected, as have in-lake concentrations. Persistent blooms of Cyanobacteria are still encountered in many shallow lakes since reduction in the external P loading to 0.3–0.4 g $m^{-2}y^{-1}$ or less has not produced the desired change in water quality. The Loosdrecht lakes study has served as model study to monitor the effects of nutrient reductions in other Dutch lakes under rehabilitation. The studies (1982–1990) were aimed at reduction of external nutrient loads and their effects on water quality, and were co-ordinated by the Centre of Limnology, Nieuwersluis (van Liere & Gulati, 1992). Of the border lakes in the Lake IJsselmeer area, since the mid 1970s RIZA has focussed attention mainly on the eutrophication of lakes Wolderwijd and Veluwemeer (Fig. 1). In the Reeuwijk Lakes, in-lake measures to reduce P have been used (van der Does et al., 1992; van der Vlugt et al., 1992; Boers et al., 1992). The Volkerak–Zoommeer lakes were created by damming a part of the North Sea estuary and flushing with freshwater in 1987. Physical and chemical measures for P reduction such as dredging and sediment removal (van der Does et al., 1992), P-inactivation (Boers et al., 1992) and flushing with P-poor, Ca-rich water (Hosper & Meijer, 1986; Jagtman et al., 1992) have not lead to stained improvements in water quality.

None of these options have succeeded in reducing the in-lake P loading effectively and promptly. Moreover, the in-lake P-reduction measures mentioned have had dramatic socio-economic consequences, the costs being exorbitant. They are estimated to amount to about €55 000 ha^{-1}, not including disposal of sediments rich in P or containing hazardous, toxic materials, for which there is no adequate solution. In addition, P release from the left-over and exposed sediment, from fish (see below and Fig. 3) and from resuspension of bottom materials caused by wind (Gulati & van Liere, 1992) is likely to annul the positive effects of the measures (papers in Gulati et al., 1990b).

Control measures and constraints

Most efforts to alleviate the detrimental and undesirable effects of eutrophication on aquatic systems address the problem of P reduction in the inflows (Edmondson & Lehman, 1981). This is also true for lake restoration works in the Netherlands, where control has focussed on external P loading from point sources (Hosper, 1998). The P levels in seepage water from deeper polders in the western part of the country may still exceed 1 mg l^{-1}. In addition, there are P inputs from precipitation, c. 0.1 mg P l^{-1}, and from water used to flush the lakes. There is no certainty that the water quality of the lake will improve even after nutrient load reductions (see van Liere & Gulati, 1992). Lake recovery is invariably afflicted by two factors: (1) internal P loading from the sediments (Sas et al., 1989; van der Molen & Boers, 1994); and (2) foraging and metabolic activities of the abundant benthivorous and planktivorous fish in these shallow lakes (Hosper & Jagtman, 1990). Fish hamper the pace of recovery by both stirring up the lake sediments, and their top-down negative feed backs. Control of nutrient loading from lake sediments has proven to be an even trickier task than controlling external nutrient loading.

Supplementary remedial measures are needed to overcome: (1) the augmented P release from the P-rich lake sediments, and (2) the abundance of planktivores, which prevent the larger zooplankters such as *Daphnia* spp. from developing and controlling phytoplankton. Deteriorating light conditions caused by algal blooms as well as continual sediment resuspension by benthivorous fish and wind prevent the submerged macrophytes from establishing, and thus competing for nutrients with phytoplankton. van der Molen and Boers (1999) have evaluated several restoration projects in the Netherlands and provided guidelines for defining targets and standards. They identify systems and measures best suited for restoration. For any amelioration to take place, the problems associated with reduction of nutrient inputs must be clearly identified, and targets should ideally relate to natural reference systems. Major lake functions such as recreation, agricultural water use and commercial fisheries impose constraints. For example, agricultural water use does not allow large water-level fluctuations. A detailed knowledge of the food-web structure and ecosystem functioning are among the essential prerequisites, and the extent of degradation that the water bodies have undergone needs to be quantified before corrective action can be taken. If feasible, the emphasis should be on re-establishing the lakes' ability to self-restore rather than 'intensive surgery' on the system (van der Molen & Boers, 1999). A brief survey of restoration work during the last 15 years or so follows below.

Early research to delineate the extent of the problem

Lake restoration efforts in the Netherlands started in the early 1980s and were aimed essentially at improving the under-water light climate of the lakes by controlling cyanobacterial blooms (Hosper, 1998), mainly of *Oscillatoria* spp. Control of external nutrient loadings was considered an option only for lakes where sewage was discharged and pollution control measures were operative. Research into the water quality of the Loosdrecht Lakes (the WOL project) started in 1982, and was one of the first major lake restoration projects (see below). Its main objective was to monitor ecosystem response to reduction of P loading in inlet water. The European Union (EU) funded the project and the Netherlands Institute of Ecology /Centre for Limnology, Nieuwersluis, co-ordinated it. About ten research institutions, including university departments, provincial authorities and RIZA (Lelystad) collaborated in carrying out the project, which ended in 1990 (van Liere & Gulati, 1992).

During the period 1987–1990, in conjunction with the Centre of Limnology, RIZA began a preparatory biomanipulation experiment in two small ponds (100 × 100 m; 1 m deep) dug out near Lake Wolderwijd (Province of Flevoland). RIZA also included several fishponds at Beesd, near Utrecht, maintained by the Organisation for Improvement of Inland Fisheries (OVB, Nieuwegein) in the experiment. The effects of fish exclusion on zooplankton were investigated in an enclosure in Lake Tjeukemeer (Richter, 1986). These pond and enclosure investigations were, however, frustrated by the lack of replication, inadequate controls, size-related sampling problems and high temporal and spatial variabilities.

The inter-institutional cooperative projects that commenced in 1982 at Lake Loosdrecht, in 1987 at Lake Zwemlust and Bleijswijkse Zoom, and in 1989 at Lake Wolderwijd, have provided useful insights into lake restoration problems (Gulati & van Donk, 1989; van Donk & Gulati, 1989; Gulati, 1990b; Gulati et al., 1990; van Donk et al., 1989, 1990a, b, c; van Liere & Gulati, 1992; Hosper, 1997; Meijer, 2000). The water quality concerns that triggered these basic studies led to a national initiative to hold the first international conference on lake biomanipulation entitled 'Biomanipulation, a tool for lake management' in Amsterdam in 1989 (Gulati et al., 1990b). At present, about ten regional and provincial water control boards, directorates for public works, water management authorities and RIZA are engaged in restoration, management and maintenance of over two dozen waterbodies. They have formed the national 'Ecological Platform for Lake Restoration', an informal, scientific discussion group that has virtually evolved into a national discussion forum with brainstorming sessions to discuss the projects now underway and future planning. There have been more than two dozen such projects since the late 1980s. The restoration techniques that have been used in the Netherlands will be briefly described below, followed by some case studies.

Techniques of lake restoration other than biomanipulation

Lake and reservoir management technologies in Europe, the US and Canada developed rapidly during the 1980s (Cooke et al., 1993), especially in the Netherlands, Denmark and other neighbouring countries. Since it has now been established that P and, to a lesser extent, N have the greatest impact on algal growth in freshwaters, most restoration techniques are directed towards reducing P in the inflows and in lakes. Lake managers are working in conjunction with researchers to unravel the P dynamics within the various lake compartments (sediment, lake littoral and pelagial) and biotic interactions (fish, zooplankton, algae and macrophytes). Techniques to reduce P by chemomanipulation (Cooke et al., 1993) have improved with time. Complementary restoration techniques such as biomanipulation have developed thanks to our improved knowledge of the food web in lakes during the last two decades.

Reduction of external nutrient loads

Wastewater treatment and diversion of nutrients from lake inflows are the foremost techniques used to reduce external nutrient loadings (Cooke, et al., 1993). The response of lakes to reductions in external loads has been reviewed by several workers (see e.g. Sas et al., 1989). Cullen & Forsberg (1988) reviewed the response of over 40 lakes to reduction in external P loads: almost half of these lakes showed only marginal or no clear reduction in lake P, and little reduction of chlorophyll-*a* content. However, Cooke et al. (1993) concluded that even though the in-lake P may not be lowered in response to a decrease in the external loading, an improvement in water quality in lakes is not precluded. The first well-recorded case study on external P reduction (in the Loosdrecht lakes during 1982–1990, van Liere & Gulati, 1992) did not bear this out in the first 10 years after start of nutrient reduction measures.

Loosdrecht lakes

The Loosdrecht lakes (Fig. 2: Lake Loosdrecht, Lake Breukeleveen and Lake Vuntus) in the Vecht River region between the cities of Amsterdam and Utrecht are among the best-investigated lakes in the Netherlands as far as the effects of reduction in external P loads are concerned. The regional Water Management and Sewerage Service (DWR) manages the lakes in the Amstel, Gooi and Vecht area. Until 1984, the lakes had been receiving P- and N-rich water from the nearby polluted Vecht River. P reduction in the inflow water began in 1984, and research on the lake ecosystem was carried out between 1982 and 1990.

The water budget of this lake system is regulated mainly by evapo-transporation, precipitation and complex drainage, including both seepage losses and infiltration gains. Between 1944 and 1984, the mean external P loading was about 1.0 g m^{-2} y^{-1} (Gulati et al., 1991b). The lakes became highly eutrophic between 1950 and the late 1970s due to both discharge of untreated household wastes and inlet of P-rich (up to 3 mg l^{-1}) river water. The light climate in the lake (Secchi-disc transparency depth = SD) had deteriorated to 0.30–0.40 m, preventing littoral macrovegetation from developing. Cyanobacteria (*Oscillatoria limnetica*) and other prokaryotes (*Prochlorothrix hollandica*) which dominated the lake seston, reached 200×10^3 filaments ml^{-1}. The zooplankton in the lake was dominated by smaller-bodied rotifers, four species of cyclopoid copepods and *Bosmina* spp. (Gulati et al., 1985, 1992), with low numbers of *Daphnia cucullata*. The bream, a planktivorous fish, dominated the existing fish population (range, 200–400 kg FW ha^{-1}).

The restoration measures were aimed mainly at increasing the SD to c.1 m depth, the water quality norm for recreational lakes in the Netherlands. In 1984, the P-rich input water from the Vecht River was replaced by water from the Amsterdam Rhine Canal, which was first treated with ferric chloride in a sedimentation basin to reduce its P concentration. Changes in the lake ecosystem were monitored for several years (van Liere & Gulati, 1992). The P dynamics of the lake water and the sediments showed notable changes (Keizer & Sinke, 1992). There were simultaneous routine measurements of the phytoplankton (chlorophyll-*a*, P and

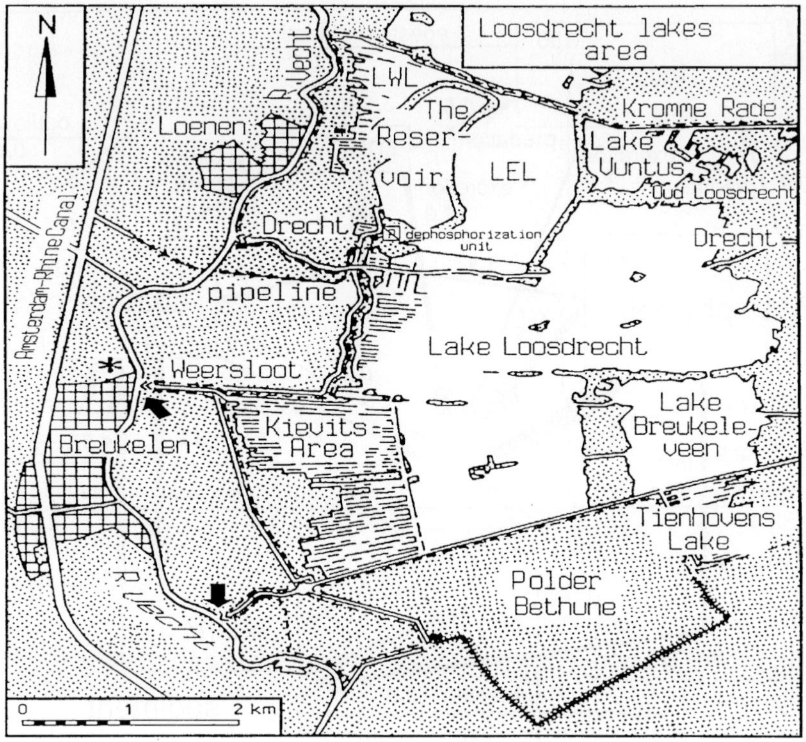

Figure 2. Loosdrecht lakes (Loosdrecht, Breukeleveen and Vuntus), with the Vecht River running north-south on the western side of the lakes. Arrows indicate the position of sluices which supplied water from the Vecht River to the lakes before the source of water supply was changed in 1984 (Source: Fig. 1 in van Liere & Janse, 1992).

N contents) and zooplankton (biomass, composition, densities, grazing and P and N) (Gulati et al., 1992).

No change was observed in either phytoplankton or chlorophyll-*a* levels, or underwater light climate. Cyanobacteria and detritus continued to dominate. However, the C: P ratios of seston increased (Gulati et al., 1991a, 1992) because of a decrease in particulate P. The seston C:P ratios and *Daphnia* numbers were significantly correlated on the basis of 9-year data, not including those relating to Lake Breukeleveen. In the Loosdrecht and Vuntus lakes, the annual mean abundance of *D. cucullata* ranged from 104 to 0.7 ind. l^{-1} and mean seston C:P ratios varied from 250 to 500 (molar). In Lake Breukeleveen, the daphnid densities were relatively higher for a given seston C:P ratio, especially in the 2 years after the lake was subjected to biomanipulation (van Donk et al., 1990c). That the high seston C:P ratios in the lakes constrained *Daphnia* abundance was confirmed later in laboratory studies (DeMott & Gulati, 1999; DeMott et al., 2001). In contrast, the P requirements of *Bosmina* spp. and cyclopoid copepods appeared to be lower and there was no relationship with seston C:P ratios. The abundance of native *Daphnia* sp., *D. cucullata*, in the Loosdrecht lakes was limited by the dietary P, in addition to predation by bream, the dominant planktivore (Lammens et al., 1992).

The P content and fluxes at various trophic levels in the water column and lake sediment, including the inflows and outflows (Fig. 3), show that roughly fifty per cent of particulate P in the lake, c. 150 mg P m^{-2}, was present in fish in 1987 (van Liere et al., 1992; Janse et al., 1992). The decrease in external P load or gross inflow from 3.3 to 1.0 mg m^{-2} d^{-1} reduced the diffusive release rates of SRP (soluble reactive phosphorus) from the aerobic sediment (measured at 20 °C) from 1 mg $m^{-2}d^{-1}$in 1984 to c. 0.40 mg $m^{-2}d^{-1}$ in 1990. The mineralization and excretion rates of inorganic P in lake sediment and from benthos during summer were c. 3 mg m^{-2} d^{-1}, i.e. an order of magnitude higher than the diffusive release rate. The downward seepage of P was estimated at 0.6 mg m^{-2} d^{-1}, which agrees well with the average retention rate of P, i.e. surplus based on difference between inflow and outflow. Sediment burial and diagenesis would thus appear to be ineffective mechanisms for with-

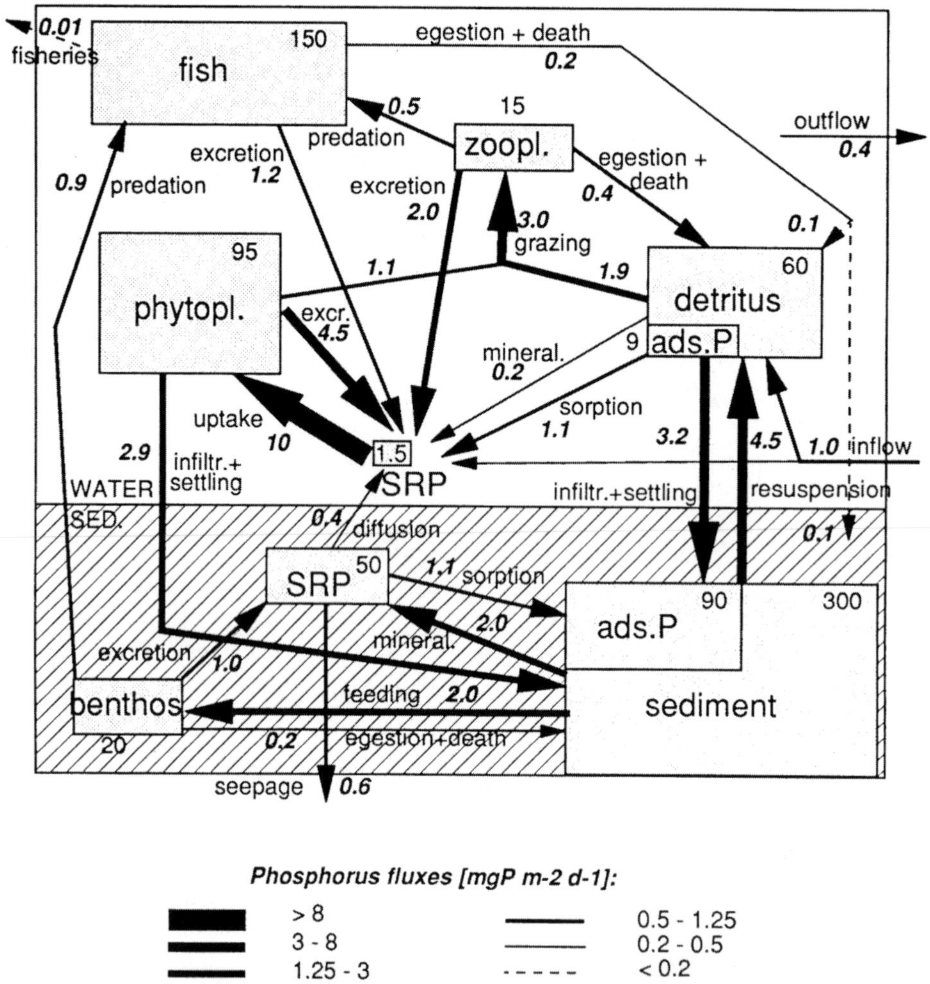

Figure 3. Flow diagram of phosphorus content (mg P m^{-2}) and fluxes (mg P m^{-2} d^{-1}) between trophic levels, water and sediment of Lake Loosdrecht for the period April–September, 1987, based on mathematical model PC-LOOS (Janse et al., 1992) (Source: Fig. 3 in van Liere & Janse, 1992).

drawing P from the lake. Interestingly, between 1982 and 1991, the TP content of the upper 2-cm sediment decreased from 0.94 to 0.60 g kg^{-1} DW. About one-fifth of this was incorporated in easily degradable organic matter and thus potentially bio-available.

Janse et al. (1992) simulated the P cycle by employing their dynamic mathematical model PC Lake, and using data on the hydrology and mass and rate processes at all trophic levels. The model predicted that: (1) a further reduction of the external load would lead to a gradual decrease of the TP level in the lake, (2) irrespective of the external loads, P removal by dredging and iron addition may result in a rapid but reversible recovery, and (3) increased P retention in the sediments would retard improvements in water quality. The model outcome was validated by an observed reduction in TP from 130 to 80 μg l^{-1}. In addition, flushing with nutrient-poor water, chemomanipulation or even biomanipulation were considered ineffective options because the P loading level was still high (van Liere & Janse, 1992). Model evaluations were aimed at a further reduction of P loading from 0.35 to 0.10 mg m^{-2} y^{-1}.

Other examples of nutrient reduction

Verstraelen et al. (1992) described eutrophication problems in lakes in the Amstel, Gooi and Vecht River areas threatened both by discharge of wastewater and the continuing ingress of nutrient-loaded water from the Vecht River to compensate for a water deficit due to other causes. The restoration programme of the Ankeveen lakes and Kortenhoef lakes, outlined by DWR, was aimed at preventing discharges of both wastewater and supply from the Vecht River entering the lake. The long-term goal was to partially restore the original groundwater flows, and increase the supply of seepage water from the Horstermeer polder. The Kortenhoef lakes have insufficient alternative ways of replacing supplies from the Vecht River. The available additional supply options are either very expensive or less acceptable from a social viewpoint. Removal by dredging of lake mud and c. 75% reduction of fish biomass during 1992–1994 in Lake Hollands Ankeveen led to an improvement in water quality.

Bootsma et al. (1999) describe the eutrophication abatement programme for the Lake Naardermeer nature reserve a few kilometres north of Ankeveen lakes (Fig. 1), one of the very few natural lakes in the Netherlands. The restoration project, started in 1985, was aimed at reducing the external P load using P-poor inlet water. Dredging of the lake-bed sediment was confined to certain areas. Water quality and aquatic vegetation were monitored and a modelling approach was used for the lake's management. Even though sediment P release appeared to retard lake recovery, by the mid 1990s, turbidity had decreased and the vegetation comprising *Najas marina* and Characean species typical of the lake had re-established itself over large areas. The model correctly predicted significant changes in aquatic vegetation.

Hydrological management: flushing with nutrient-poor water

Hydrological management involves replenishing the lake with water from an extraneous source or from another lake with lower nutrient levels and preferably rich in Ca and HCO_3. Dilution as a restoration tool, therefore, implies necessarily reducing the concentration of nutrients in lake water to limiting concentrations (Cooke et al., 1993). The use of nutrient-poor water for this purpose also reduces the suspended seston. To achieve the best dilution effects, the timing of the flushing is important. In addition, the quality and quantity of the water to be used needs to be worked out beforehand.

Only in a few cases in the Netherlands has improvement in the lake water quality been achieved by hydrological management. Although the technique has great potential, it depends greatly on the sustained availability of good quality water. Hosper (1998) suggested that the winter period, when algal growth is generally at a minimum and more water is available, is to be preferred for flushing. According to Hosper (1997, 1998), cyanobacterial blooms can be virtually wiped out from well-mixed water bodies by using a flushing rate of 0.75 month^{-1} during November–February. He cites the case of the Friesian lakes and lakes in the western provinces in which water quality can potentially be improved using water from the IJsselmeer-Markermeer lakes (Fig. 1) for flushing.

Lake Veluwe (3240 ha; mean depth, 1.3 m), one of the border lakes (Fig. 1) in the Lake IJsselmeer area (Hosper, 1984, 1997; Jagtman et al., 1992: see also Cooke et al., 1993) was flushed using nutrient-poor polder water during 1978–1979, concurrently in 1979 with P-reduction in the sewage water entering the lake. The main objective of the measure was to reduce seston mass, comprised mainly of *Oscillatoria*, and thus reduce pH and high internal loading of P. Polder water (TP, 80–100 μg l^{-1}) was used for flushing the lake water (TP, 400–600 μg l^{-1}) in winter (November–March) at a rate of 3.6% d^{-1}. The buffering capacity of Ca and HCO_3-rich polder water, through its effects on pH, had already facilitated a reduction of TP by the end of the first winter. The TP level in the lake dropped 3–4-fold after the first winter flushing, ranging between 100 and 200 μg l^{-1}. The *Oscillatoria* bloom (SD, 0.15–0.25 cm) virtually disappeared, aided by successive cold winters. During 1982–1983, the maximum internal P loading had decreased to 0.8 mg m^2 d^{-1}, about 15% of what it had been 1979. By 1985, diatoms and green algae had become dominant for the first time in two decades. Despite a 2–3-fold decline in chlorophyll, the SD did not increase beyond 20 cm. Further recovery of the lake was, however, impeded by the sediment P, and wind and fish-induced resuspension of large amounts of detritus and inorganic matter from the lake bottom.

In addition, Lake Wolderwijd (Fig. 1), the site of a large biomanipulation experiment during 1990–1991 (Meijer & Hosper, 1997), was flushed with polder water without success (see below). Hosper (1984) and Hosper & Meijer (1986) questioned the efficacy of flushing for further improvement in water quality, mainly because the nutrient concentrations in the lake and the 'dilution' water from the polder became very similar. Because good quality 'dilution water' is scarce and the logistics of transporting this water to lakes are neither easy nor cheap, the technique has not become popular in the Netherlands.

In-lake measures

Cooke et al. (1993) discuss in detail the techniques of reducing in-lake nutrient concentrations that are available. Such reductions are indispensable because, as noted, reduction of the external P loading will fail to improve water quality if the loadings from the sediments are high. This is especially true for shallow Dutch lakes like the Loosdrecht lakes. Despite external load reductions, there is no guarantee that the water quality in such lakes will improve (van Liere & Gulati, 1992). Internal P loading from the sediments invariably afflicts lake recovery. Careful removal in the upper, often loose, sediment layers, is quite expensive, and although immobilisation of P in the sediments by chemical fixation is an alternative to sediment dredging, these techniques have been attempted in only a few bodies of water in the Netherlands.

Dredging combined with other measures

Does et al. (1992) describe sediment dredging as an additional restorative measure to improve water quality in the peaty Lake Geerplas (28 ha; 1.9 m), one of the Nieuwkoopse lakes (Fig.1), where the external P loading was reduced substantially, from 0.9 to 0.2 g m^{-2} y^{-1}(Does et al., 1992). Geerplas had previously become eutrophied as a result of drainage from market gardens and inlet of nutrient-rich water from a bird colony. The high turbidity of the lake was caused by both high chlorophyll levels and TP content (0.45 mg l^{-1}). The restoration, which started in 1980, included treatment of point sources, hydrological isolation, chemical P removal in the inlet water, post treatment in a helophyte filter, and dredging in 1990 in which the top loose peaty layer, 0–25 cm thick, was removed. The sediment extraction did not lead to the desired reduction in TP. Even though chlorophyll concentration decreased by <50%, the proportion of Cyanobacteria in phytoplankton did not decline, and the transparency increase from 0.3 to 0.4 m was marginal. After dredging, the loose top mud layer that had formed on the exposed hard sediment apparently had the same P release characteristics as the sediments had before the dredging (see Hosper, 1998).

Phosphorus fixation in sediments using iron (III) chloride

Boers et al. (1992) used ferric chloride solution as a P-binding chemical agent to inactivate P in the sediments, iron being a natural binder of phosphate. A daily dose of 100 g m^{-2} of Fe^{3+} was added to the sediments of eutrophic Lake Groot Vogelenzang (area 18 ha; average depth, 1.75 m) in October and November 1989. The $FeCl_3$ solution was diluted about 100 times with lake water and mixed with the surface sediments using a water jet. Three weeks later, the concentrations of TP, chlorophyll-*a* and suspended solids decreased. However, about 3 months later, the TP level rose again. P release from intact sediment cores, which had decreased from 4.0 to 1.2 mg m^{-2} d^{-1} ($n = 5$), had risen to c. 3 mg m^{-2} d^{-1} ($n = 5$) a year later. The P retention time, based on the disappearance rate of chloride, was only one-tenth of the estimated 1 year. Failure to procure a protracted improvement in water quality was ascribed to the high external loading due to rapid flushing with P-rich water, or to the loss of binding capacity of the $FeCl_3$ due to reduction or binding with carbonate or sulphide, or both these causes. The results obtained were tentative, and long-term availability of the iron to bind P needs substantiation.

Boers et al. (1992) considered the operational costs of c. €7500 ha^{-1} cheap and the method effective compared with 'fisheries management' (food web manipulation). However, such a cost-benefit comparison should also include the long-term sustainability of the techniques. If applied in conjunction with biological control measures, this method may yield more lasting results than if applied alone. The results of the chemical manipulation were disappointing, mainly because the effects of P sedimentation by flocculation were offset by the high flushing rate with water rich in P.

Artificial mixing

Artificial mixing has been carried out only in a few Dutch lakes, mainly because the lakes are well mixed, being shallow. However, in stratifying lakes with recurrent blooms of *Microcystis,* artificial mixing has some restoration potential. Visser et al. (1996) applied this technique in Lake Nieuwe Meer (area 1.32 km^2; mean depth 18 m), near Amsterdam. This hypertrophic lake, which is stratified in summer, suffered from perennial *Microcystis* blooms. Mixing was done with success during spring through summer 1992, and repeated intermittently during spring 1994 in order to save energy. The plume of compressed air released in the water layers just 1 m above the sediment mixed the water column above, so that the *Microcystis* was entrained in the turbulent flows and was prevented from accumulating in surface layers. The mixing, however, prevented sedimentation of the non-cyanobacterial forms (flagellates, green algae and diatoms). The Cyanobacteria could thus no longer exploit the optimal light conditions in the upper mixed layers. The non-cyanobacterial forms, on the other hand, benefited from the mixing, which also enhanced silicate availability to diatoms in the upper layers. A decrease in the pH on mixing helped to bring about a shift in the species composition to non-cyanobacterial forms.

In a model study in the Biesbosch reservoirs, Oskam (1978) ascribed a decrease in phytoplankton biomass maximum to light limitation caused by an increase in the mixing depth. In Lake Nieuwe Meer, however, Visser et al. (1996) observed that mixing led to an increase in phytoplankton biomass because of reduced sedimentation losses rather than to an increase in the growth rates. Unlike reduction in external loading, mixing produces instantaneous results without decreasing the nutrient concentrations. Moreover, for deeper lakes, mixing is more effective than reduction of external nutrient loading.

Liming

Most experience relating to liming comes from Sweden and Norway, where liming has been chosen as a national strategy to preserve species threatened by acidification (Henrikson & Brodin, 1995). The larger Dutch lakes are strongly buffered due to high calcium and bicarbonates. The liming has, however, been attempted in fens, moorland pools and shallow soft-water wetlands in the south and south-east of the country, which are acidic (van Dam & Buskens, 1993). Atmospheric depositions of sulphates and ammonium in the pools are estimated at 44–50 mmol m^{-2} yr^{-1} and 84–103 mmol m^{-2} yr^{-1}, respectively. Between 20 and 70% of the sulphate inputs is removed and due to nitrification, 40–70% of ammonium escapes to the air or sediments, and the nitrate virtually disappears. The alkalinity thus produced ranges from 12 to 52 meq m^{-2} yr^{-1}, with pH-values rising to 4.1–5.4. The reduced sulphur compounds tend to accumulate in the sediments and their oxidation in dry summers causes pH values to decrease to c. 3.7.

Paleolimnological studies reveal that both acidification and eutrophication are major threats to the pool biota (macrophytes, desmids, diatoms, macrofauna, fishes and amphibians). Afforestation exacerbates the acidification effects further and reduces wind dynamics. Reductions of acid atmospheric deposition to <40 mmol m^{-2} yr^{-1} and of ammonia to <30 mmol m^{-2} yr^{-1} are needed for recovery. Lamers et al. (2002) and Roelofs et al. (2002) have reviewed liming and its ecological effects in the Netherlands in detail.

Biomanipulation: a lake restoration strategy

Introduction

Lake biomanipulation is synonymous with so-called biological control in lake restoration. Relatively small changes in the biological relationships between organisms produce favourable changes in lakes (see e.g. Edmondson, 1991: pp. 281–282). Two main criteria for good water quality are a decrease in phytoplankton biomass and improvement in the under-water light climate. Biomanipulation complements nutrient reduction in lake restoration: if applied in conjunction with other measures, it speeds up the processes of lake rehabilitation.

Although Shapiro et al. (1975) first used the term 'biomanipulation' in 1975, the history of biomanipulation is much older. Cooke et al. (1993) cite Caird (1945) as probably the first to observe that phytoplankton growth reduced greatly after stocking of largemouth bass (piscivorous fish) to a 15-ha lake in Connecticut (USA). However, the pioneering study of Hrbácek et al. (1961) on the role of fish stock in ponds in influencing zooplankton species composition is better known: it drew our attention for the first time to the role of fish in the size-structuring of zooplankton. The paper by Brooks & Dodson (1965) and the so-called 'size-efficiency hypothesis' proposed by Hall et al. (1976) increased our insight into food web relationships, including the role of predation by planktivorous fish on the body size, composition and densities of zooplankton prey, and of piscivorous fish in controlling planktivorous fish (Brooks & Dodson, 1965; Zaret, 1980; Lazzaro, 1987). The zooplankton community can thus shift to predominantly larger zooplankton, which in turn cause heavy mortality of phytoplankton leading to a short- or long-term clear-water phase in lakes. The clear-water phase in lakes is a manifestation of reversal of some of the undesirable top-down food web effects triggered by reducing the existing planktivore population. Such biomanipulation studies are now well documented (review papers in Gulati et al., 1990: Gophen, 1990; Lammens et al., 1990; Benndorf, 1990).

The first international biomanipulation conference, *Biomanipulation, a tool in lake management*, held in the Netherlands in 1989 (Gulati et al., 1990; Lammens

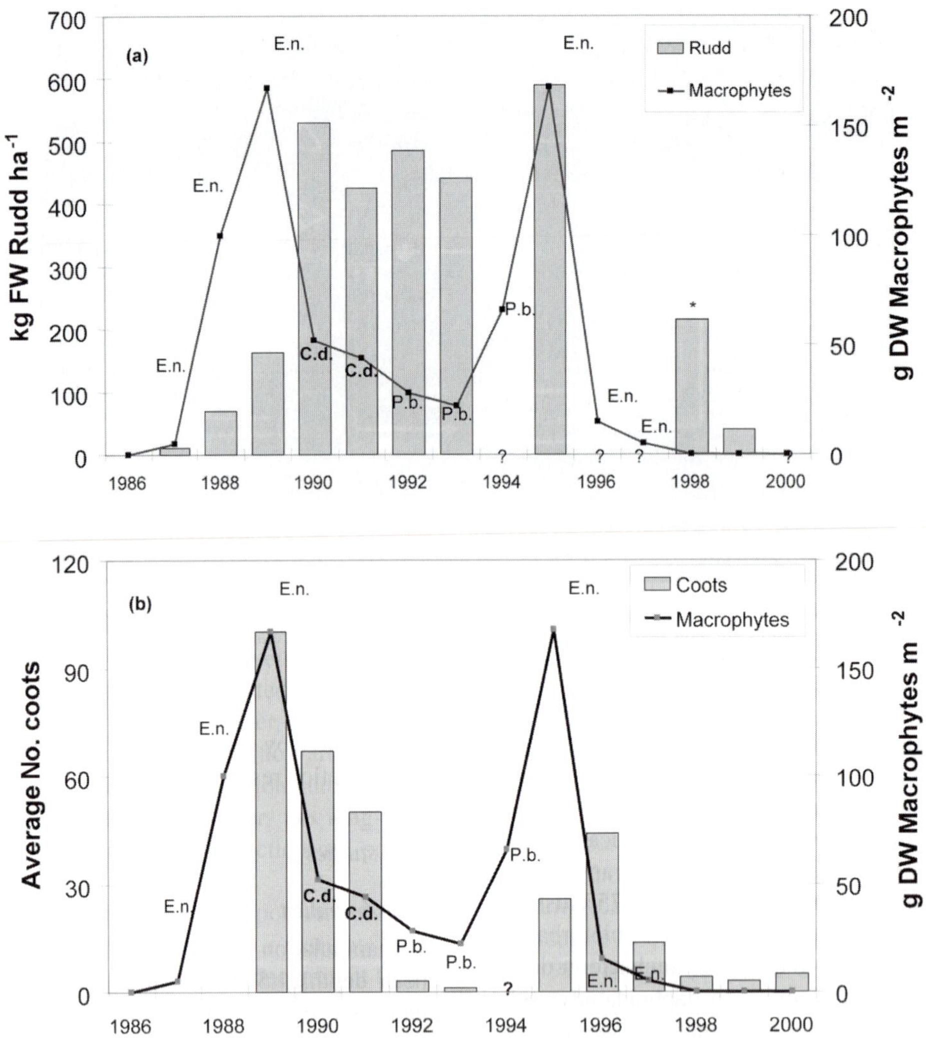

Figure 5. Mean annual biomass of macrophytes and rudd (*Scardinius erythrophthalmus*) (upper panel), and of macrophytes and coot numbers (lower panel) in Lake Zwemlust from 1986 to 2000. The lake has been biomanipulated twice: in March 1987 and in April 1999. Abbreviations: E.n., *Elodea nuttallii*; C.d. *Ceratophyllum demersum*; P.b. *Potamogeton berchtholdii* (Source: van de Bund & van Donk, 2002).

and a refuge for zooplankton. About 1500 pike (*Esox lucius*) fingerlings (4 cm), 140 adult rudd (*Scardinius erythrophthalmus*), and about 1 kg (wet weight) of daphnids (*D. magna, D. hyalina*) were introduced.

Within a few weeks of the biomanipulation, phytoplankton showed an explosive growth, followed by rotifers (>4000 ind. l^{-1}) in late April. Cyclopoid copepods, and their nauplii were the first crustaceans observed, followed by *Bosmina* sp. and *Daphnia* spp. *Daphnia galeata* appeared first and was followed by *D. cucullata, D. magna* and *D. pulicaria*. By early July phytoplankton had decreased dramatically. In the subsequent 6–8 weeks, though not dense in number, calanoid copepod, *Eudiaptomus gracilis*, and *D. pulicaria* and *D. magna*, were the major grazers (Gulati, 1989). The chlorphyll-*a* remained <5 μg l^{-1} and the water clarity increased to >1.5 m. The rudd spawned in July, and about 20% of the transplanted pike survived. The existing fish population was c. 20 kg FW ha^{-1}. Chironomids, mainly *Chironomus plumosus*, developed remarkably (8000 ind. m^{-2}).

In 1988, phytoplankton (chlorophyll, 140 μg l^{-1}) increased in mid March. The rotifer densities were low. Although *D. pulicaria* (130 ind. l^{-1}) was the main grazer, its fecundity was very low. Typical littoral forms (*Simocephalus vetulus* and *Chydorus sphaericus*) and ostracods became abundant in open water. *Elodea nuttallii* and *Chara globularis* were the main macrophytes and *Mougeotia* the main filamentous alga (Ozimek et al., 1990). The high water transparency followed by a marked decrease in phytoplankton apparently out-competed for N by the macrophytes (van Donk et al., 1993). In 1989, the water remained clear until early summer. However, *Volvox aureus* colonies appeared in large numbers in mid summer. The phytoplankton productivity in 1989 was, however, only 50% of that in 1988. *Keratella cochlearis* (3000 ind. l^{-1}) was the dominant rotifer. In midsummer, *D. galeata* (140 ind. l^{-1}) was the main grazer. The community grazing rates, which ranged from 125 to 340% d^{-1}, caused the SD to reach the lake bottom. In 1989, the existing macrophyte population consisted mainly of *Elodea* sp. (c. 170 g DW m^{-2}), which covered nearly 80% of the lake area (Ozimek et al., 1990) (Fig. 5a). Nearly three-quarters of both TN and TP in the lake were stored in the submerged macrovegetation (Table 3: van Donk et al., 1990b). The macrofauna inhabiting the lake bottom and hydrophytes was quite diverse (Kornijow et al., 1990; Kornijow & Gulati, 1992a,b). The snail *Lymnaea peregra*, which used *Elodea* as a substrate, numbered 102 ind. m^{-2}. The rudd spawned three times in 1989 and the existing fish population was 106 kg ha^{-1} (van Donk et al., 1990). The condition of both 1^+ and 2^+ pike was poor. During the summer of 1989, there were on average some 100 coots in the lake vicinity, feeding mainly on *Elodea* (van Donk et al., 1994a) (Fig 5b), but their numbers decreased in subsequent years when *Elodea* was replaced by *Ceratophyllum* sp.

During 1990–1991, the positive effects of the preceding years were reversed: *Bosmina longirostris* replaced the *Daphnia* spp., the copepod and rotifers increased markedly. The zooplankton grazing rates declined to <50% d^{-1}, and in 1991 *Bosmina* was the main grazer. The existing macrophyte population decreased, with a shift in dominance from *E. nuttallii* in 1988 and 1989 to *Ceratophyllum demersum* in 1990 and 1991 (van de Bund & van Donk, 2002) (Fig. 5a). The dominance of *Potamogeton berchtholdii* during 1992–1994 signalled a serious deterioration in water quality. Rudd comprised c. 90% of the existing fish population of 400–500 kg ha^{-1}, and the 0^+ pike and older fish (2^+ and 3^+) the remaining 10%. However, the condition of the rudd deteriorated in 1991, with fish standing crop decreasing to 110 kg ha^{-1}. Both the rudd and 0^+ pike ate daphnids (E.H.H.R. Lammens: unpublished data), so that *D. pulex* disappeared. Summing up, the rudd appeared to play a crucial role, both for the changes in zooplankton and for macrophytes: they selectively consumed *Elodea*, thereby causing a shift in 1990 to *Ceratophyllum* (van Donk, 1998).

During 1992–1995, alternating periods of clear water and turbid water occurred within the same year, with recurring cyanobacterial abundance (van de Bund & van Donk, in press). In 1995, *Elodea* returned but decreased sharply in 1996. The existing rudd population was highest in 1995 (Fig. 5a). The changes in the plankton composition are reported in Romo et al. (1996). Although the large cladocerans were almost absent during 1995–1999, the rotifers became abundant. In short, the lake had reverted to a situation similar to that prior to the restoration measures in 1987.

The biomanipulation, which was repeated in late April 1999, involved a reduction of fish, chiefly rudd, to <40 kg ha^{-1} (Fig. 5a). As in 1987, this measure also led to a marked increase in water transparency in 1999. The short-term responses to the measures in 1987 and 1999 were similar as regards the effects on water transparency. However, in contrast with 1987, the fish removal in 1999 was not 100% and the Cyanobacteria (mainly *Microcystis*) persisted. As well as this, despite the return of clear water conditions in 2000, the macrophytes did not directly develop in response to the 1999 measures; instead, a layer of filamentous algae covered the lake's sediment in 1999. In 2001, however, both *Ceratophyllum* and *Potamogeton* reappeared (not shown in Fig. 5).

Lastly, it may be remarked that repeated fish stock reductions appear to be a reasonable management strategy to keep Lake Zwemlust clear; to sustain this improvement in the water quality on long-term basis, however, reduction of nutrient loading via seepage water to the lake is indispensable

Lake Wolderwijd

The lake (2650 ha, mean depth 1.5 m) had suffered from cyanobacterial blooms (*Oscillatoria agardhii*), turbid water and virtual lack of submerged vegetation since the early 1970s (Meijer & Hosper, 1997; Meijer et al., 1994b). As mentioned above, from 1980 onwards, the lake was flushed with polder water low in P and high in Ca. The flushing was increased from 1 to 2.10^6 m^3 $month^{-1}$ in 1988–1989 to 4 to 7. 10^6 m^3 $month^{-1}$ during 1991–1992, which reduced the water residence time by half to c. six months. A 0.3–0.5 cm iron grid was installed in the sluice to preclude fish immigration to the lake. Despite a 50% decrease in the concentrations of TP and chlorophyll-*a*, SD increased from 0.20 to only 0.30 m.

To achieve the objective of 1 m SD depth, a large biomanipulation experiment was carried out in 1990–1991 (Meijer & Hosper, 1997). The fish stock, mainly bream and roach (*Rutilus rutilus*), was reduced during the early spring period in 1991 by about 75%, from 205 to 45 kg ha^{-1}. In May 1991, the lake was stocked with 217 pike fingerlings ha^{-1}. The SD of the lake increased to 1.8 m, mainly because of the grazing by *Daphnia galeata*. Nevertheless, the clear-water phase lasted only for about 6 weeks. *Daphnia* disappeared in July due to food limitation, and though algal biomass increased, the daphnids did not recover, due to predation by the mysid shrimp *Neomysis integer*. The reduced biomass of the predator perch caused the shrimp to develop abundantly. Macrophytes failed to establish in the lake, probably because of the cold weather. Most of the young pike died, probably due to lack of shelter. The pike were therefore not able to control the production of 0+ fish. The submerged vegetation, especially Characeae, expanded remarkably from c. 28 ha in 1991 to 438 ha in 1993. The water over the *Chara* meadows was clear, helped by the decreased wave action and increased sedimentation of the suspended particulates in these areas. Meijer & Hosper (1997) hypothesised that expansion of stoneworts meadows might ultimately result in a longer-lasting clear-water state. They advocated fish stock reduction aimed at a spring clear water phase for further expansion of *Chara* vegetation. In line with this, commercial fishing is being continued on a limited scale. Lammens et al. (2002) endorsed this approach, citing their long-term data analyses on fishing in various lakes including the nearby Lake Veluwemeer, which revealed positive effects of moderate fishing on the expansion of *Chara* vegetation and maintenance of high water clarity (see the 'Discussion section').

Reeders et al. (1989) and Reeders & bij de Vaate (1990) tested the feasibility of using zebra mussel (*Dreisseina polymorpha)* as a bio-processor for water quality management. They compared the filtering rates (F) of the mussel in Wolderwijd and two other lakes. The F was inversely related to the suspended matter content, irrespective of temperatures in the range 5–20 °C. The animals (L, 18 mm) cleared c. 30 and 60 ml h^{-1} mussel^{-1} at seston levels between 20 and 40 mg DW l^{-1}. The sigmoidal relationship between L and F indicates that the weight-specific F declines as the animals grow older. Pseudofaeces production increased with increase in seston levels. About 675 mussels m^{-2} are required to consume the daily phytoplankton production in Lake Wolderwijd (Noordhuis et al., 1992). Research in ponds shows that in the presence of mussels, the Cyanobacteria (*Oscillatoria* and *Aphanizomenon*) disappear (Noordhuis et al., 1992). In laboratory bioassays using mixtures of food types, the F on *Microcystis* did not differ from green algae (Dionisio Pires & van Donk, in press).

Biomanipulation of other lakes

Lake Breukeleveen (area 180 ha; mean depth 1.5m) is one of the five Loosdrecht lakes (Fig. 2), which as noted, did not respond to the nutrient reduction measures. In 1989, almost two-thirds of the existing fish population of 157 kg ha^{-1}, comprising c. 90% fish in poor condition, including pikeperch, was removed using stop nets and seine netting (van Donk et al., 1990c). The lake was restocked (400 fish ha^{-1}) with 0^+ (2–3 cm) pike and large-bodied *Daphnia* spp. (12 ind. m^{-3}: *D. pulex* and *D. hyalina*). The experiment failed due to rapid re-growth of the bream population and an increase in predatory Cladocera (*Leptodora kindtii*), in addition to the suppression of the *Daphnia* grazing by filamentous Cyanobacteria, mainly *Oscillatoria* sp. (van Donk et al., 1990b). It is, however, not clear if the reduction in fish biomass in the lake was inadequate. A recruitment of fish from the adjacent lake is among the conceivable causes of the failure of the measure, as fish barriers in the lake inlets did not prevent immigration to the lake. A mesocosm study some years later also showed that wind-induced turbulent mixing adversely affected the SD (van Donk et al., 1994b). To reduce stirring up of the sediment by fish and wind, the water authorities started excavating in May–June 2002 several deep pits (20–40 m) in different lake parts to allow wind and wave-induced shifting and burial of the loose, nutrient-rich lake sediments in these pits. It is expected that this measure will also retard the in-lake nutrient releases from the sediments.

As in other Dutch lakes, macrophyte vegetation in Lake Bleiswijkse Zoom (area, 14.4 ha; mean depth, 1.1 m; length, 2 km; width, 50–200 m) disappeared due to the dominance of Cyanobacteria (*Aphanizomenon* and *Anabaena* spp.). An attempt in 1981 failed to reduce the existing fish population and water quality did not improve (Meijer et al., 1994a, 1995; Meijer, 2000). Before biomanipulation in April 1987, bream and white bream dominated the fish stock (about 750 kg ha^{-1}). The experimental lake part, the Galgje (3.1 ha), was separated from the main lake (Zeeltje), by a metal screen at the narrower lake part to prevent the fish from entering the Galgje. About 85% of the total fish biomass was removed from the Galgje by seine netting and electro-fishing, and 800 individuals (3 cm) of pikeperch (*Stizostedion lucioperca*) were introduced during the summer, mainly to control the 0+ bream. In 1988, 3500 0+ pike and 90 1+ pike were introduced. The seston mass (<33 μm) in Galgje decreased to about one-fourth of that in the control part, with a discernible increase in water transparency due to reduced sediment resuspension, as has been observed in bream reductions in other Dutch lakes (Lammens, 1989). *Chara*, rather than zooplankton grazing (Gulati, 1990b), was responsible for the increase in water transparency in autumn. Bioassay experiments confirmed that *Chara* successfully competed for nutrients with the phytoplankton (Meijer, 2000). Food limitation in the experimental part adversely affected the survival of pikeperch and the pikeperch failed to regulate O+ cyprinid production. The experiment failed to produce long-term positive effects.

The Volkerak-Zoommeer lake system (area 6150 ha; mean depth 5.2 m) was created in 1987 by isolating a part of the Eastern Scheldt, a tidal estuary in the Rhine Delta in the south-west Netherlands (Fig. 1), and constantly flushing it with river water. The lake was quite clear during the first few years, despite high P loading. The first freshwater species were encountered within the four months of the lake's isolation from the estuary and flushing with freshwater. The water clarity increased in March and early April 1990 (SD, 3 m) due to the high grazing pressure exerted by *Eurytemora affinis*, and in the subsequent few months by *Daphnia pulex* (Gulati & Doornekamp, 1991). The light conditions deteriorated during 1991–1994 due to blooms of Cyanobacteria. Both Cyanobacteria and predation by fish depressed the zooplankton and their grazing. In 1992, the recruitment of fish, especially roach, was high (c. 40 kg ha^{-1}) causing the daphnids to disappear. In the subsequent 2 years, the existing fish population increased three-fold. To obtain the SD target of 2 m, reductions of P and fish biomass (bream and ruff) were considered (Breukers et al., 1997).

Although macrophyte coverage increased to 22% in 1991 (Schutten et al., 1994), the improvement in water clarity was minor because the nutrient loading continued to be high. In addition, consumption by waterfowl of the plants retarded the improvements. The bottom materials resuspended by wind action contributed more to the turbidity than matter transported by the river (Tamminga, 1992). The key factors for integrated water management of the Volkerak–Zoommeer include creating optimal spawning conditions for northern pike and water level regulation (Ligtvoet & de Jonge, 1995). To achieve these objectives, reed and reed grass were planted during 1990–1991, and over 40 islands were constructed in the lake and the shoreline and banks reinforced to curb wind-induced erosion. In addition, grazing by livestock and geese of the littoral vegetation had to be checked in order to improve the propensity of the land-water transition to natural development of biodiversity.

In 1987, the Binnenschelde (area 178; mean depth 1.5 m) was created within the Eastern Scheldt estuary in the SW Province of Zeeland for recreational purposes (Fig. 1). The lake became a freshwater lake, though the chloride content (600 mg l^{-1}) remained high. In 1988, dredging of part of the lake resulted in only a transient reduction of the internal P loading. Between 1989 and 1992, aquatic plants, northern pike and perch were introduced. The lake water became clear in the autumn of 1990, despite an increase in internal P loading, coinciding with the appearance of larger-bodied daphnids. The water remained clear for several subsequent years. The aquatic vegetation (fennel pondweed, stoneworts & water milfoil) increased in 1991. The light conditions worsened in 1993, however, and nutrient and chlorophyll concentrations increased markedly in 1995. *Daphnia* populations appeared to be controlled strongly by *Neomysis*. In 1997, the transparency norm of 1 m for recreation was not achieved due to an increase in cyanobacterial populations, and by 1999 the transparency dropped to just 0.30 m. Thus, despite the restoration measures, the Binnenschelde ecosystem has shown no signs of stability or self-sustaining ecosystem resilience. It has recently been proposed to revert to a saline situation ($>16\,000$ mg Cl l^{-1}) in the lake by letting in sea water from the Eastern Scheldt estuary. The proponents of this alternative believe that a saline situation is a compromise guaranteeing sustainable clear water for the Binnenschelde.

Lake Noorddiep (31 ha; depth, 0.9–2.6 m) is a long and narrow water body which was formerly a branch of the IJssel River, close to its mouth in the IJssel Lake (Fig. 1). It is hypertrophic with TP concentration of 0.25 mg l^{-1}. The lake was divided in 1987 into three parts, demarcated by road bridges. In the manipulated part (4.5 ha; depth, 1.5 m), 75% of the existing fish population (545 kg ha^{-1}), mainly bream and other cyprinid-like fish, were removed in the winter of 1987–88, and pike was introduced (Meijer et al., 1990, 1994ab, 1995). The control part (16 ha) contained about 800 kg ha^{-1} of fish, three-quarters of which was large bream. There was a sustained improvement in the underwater climate of the manipulated part for 8 years, caused by *Daphnia* grazing in the spring periods though not in the summers. Consequently, mats of filamentous macro-algae developed on the lake bottom, and the littoral macrophytic vegetation flourished. The fish biomass remained roughly at about half the level it had been before the measures, and the pike population seemed to thrive well in the littoral region. The bream seems to have been largely replaced by YOY (young-of-the-year) populations of roach and perch. This lake is probably unique in that the piscivorous fish stock has exceeded 25% of the total fish stock and because the measures taken have produced long-term positive effects.

Several other small lakes (Fig. 1) have either been biomanipulated or biomanipulation measures have followed nutrient reduction measures (Meijer et al., 1999). The main objective in most of the cases has been to reduce algal blooms and improve the light climate. Massive fish reductions have generally led to long- or short-term improvements initiated by *Daphnia* grazing and sustained by *Chara* development, as has happened in Lake Duiningmeer. Similarly, the introduction of pike and perch and zebra mussel (in Lake Waay) has had a positive effect on the light conditions and vegetation. Predation on daphnids by *Neomysis*, as has happened in Lake Sondelerleijen, however, can nullify such positive effects (see Hosper, 1997). Lastly, slight deepening of lake levels, together with fish stock reductions by emptying, have improved the light climate dramatically in some lakes, the IJzeren Man being one example. However, most of the studies relating to these lakes appear to have an informal character, being geared to local recreational purposes, and only selected ecosystem components have been monitored.

Discussion

Model studies and their contribution to lake restoration

From their analysis of data relating to 49 shallow lakes, van der Molen & Boers (1994) concluded that the parameters that apply to empirical models for P loading and P concentration are different for lakes prior to and after nutrient reductions. If the external P loading has not been reduced, the summer P levels will depend on the external and hydraulic loads. Thus, for such lakes the 'classic' models (Vollenweider type of relations) would be adequate. In contrast, where the external loading has been reduced, the measured internal loading will account for most of the variations in P concentration during summer. In some 230 Dutch lakes and ponds, including those that had been biomanipulated (van der Molen & de Boer, 1994; Portielje & van der Molen, 1999), levels of both TP and TN have generally tended to decline since the early 1980s. Any increase in the N: P ratios is caused by a greater reduction in the P emissions rather than an increase in N emissions.

The ratios for maxima of chlorophyll-*a* and TP and chlorophyll-*a*: TN were discernibly higher in the systems dominated by Cyanobacteria than other algae. Cyanobacteria possess the capacity to withstand greater changes in P concentration, which is also reflected in the greater variation in the C:P ratios. This is perhaps also the key to the success of these filamentous Cyanobacteria in many Dutch lakes, and the snag in lake restoration. A decrease in P to very low levels is thus no guarantee that the cyanobacterial populations will follow suit. As well as this, a decreasing trend in chlorophyll-*a* in such wind-exposed, shallow lakes will not ensure an immediate increase in lake transparency because of the wind-induced resuspension of detritus from the lake bottom. The time needed to improve the underwater light climate will thus be longer than expected, the average depth being the crucial factor. Interestingly, the model showed that a severe winter spell followed by a dry spring period will lead to a greater decrease of chlorophyll-*a* levels in the ensuing summer. This has been attributed to the collapse of *Oscillatoria*, though this generalisation does not hold for some *Oscillatoria* lakes such as the Loosdrecht, Reeuwijk and Nieuwkoop lakes (see also Hosper, 1998).

Portieltje & van der Molen (1999) state that a submerged macrophyte areal cover of $>5\%$ is often related to lower levels of chlorophyll-*a* and dissolved nutrients. At an areal cover $>10\%$, the decrease in the ratios of C (chlorophyll-*a*): P implies a greater decrease in C (Cyanobacteria biomass) than P. The authors attribute this to control of algae by both bottom-up and top-down effects in the presence of macrophytes. Thus, a concurrent action of the nutrient and biological control is implicit here. The latter, that is, zooplankton and its grazing pressure on algae, would increase concomitantly with the increase in macrophyte levels, as has been confirmed by field data from biomanipulated lakes reviewed here. System-specific linear relationships between the TP and TN concentrations measured in the incoming water have enabled us to assess the permissible loads for individual lakes more confidently than with combined data from many lakes (Portieltje & van der Molen, 1999). Consequently, the chain of events initiated by a decrease in nutrient loading, complemented by biomanipulation measures and culminating in increased water transparency, is a complex one and depends on biological variables as well as other system characteristics.

Scheffer et al. (1997) demonstrate that analyses of patterns of cyanobacterial dominance in the field show that the algal community is a hysteretic system with two alternative equilibria or stable states: one dominated by phytoplankton and the other one by submerged macrophytes. Their simple competition model shows that hysteresis should in fact be expected from differences in physiology between Cyanobacteria and other algae. The Cyanobacteria are not only superior competitors under conditions of low light, but promote such conditions because they can cause a higher turbidity per unit of P than other algae. This mechanism of hysteresis helps us to understand why and how in shallow lakes, Cyanobacteria, when dominant, resist and thwart restoration efforts by nutrient reduction. Janse (1997) used his integrated lake model PCLake to investigate at what level of nutrient loading the transition between the two alternate states (Scheffer et al., 1997) would occur. To describe the competition between phytoplankton and macrophytes he used the nutrient contents in the lake, including the upper sediment, and the top-down food-web effects. The model was run for a hypothetical, representative shallow lake in the Netherlands, using a realistic range of nutrient loadings to simulate the various initial conditions. The model response was highly non-linear and showed hysteresis: the loading level at which a transition occurred depended on the initial conditions. Empirical data on chlorphyll-*a* to P relations were compared with the model results. The 'critical nutrient level' appears to be influenced by lake dimensions and the net sedimentation rate. The reduction in nutrient loading will thus have to be accompanied by some additional measures. The PCLake model (Janse, 1997) was also applied specifically to Lake Zwemlust in order to examine the impact of grazing by fish and birds on the macrophytes and on the transition from a clear to a turbid state, using ten variables for the sensitivity analysis (Janse et al., 1998). The model predicts a transition from an algal-dominated state before the biomanipulation measures to dominance of rooted perennial vegetation thereafter. The grazing on macrophytes by fish and birds causes the lake to return to an algal-dominated state. The model results were quite sensitive to the zooplankton-grazing rate, and in the presence of herbivory only, to macrophyte growth rate parameters. In his discussion on alternative stable states, Moss (Fig. 2: 1999) illustrates the mechanisms by which a lake switches from one stable state to the other. Biomanipulation is a 'reverse switch' and leads to a clear-water state dominated by macrophytes, while 'forward switches', which can cause the lake to return to an algal-dominated state, involve mechanical damage, vertebrate grazing (fish, birds) and other loss factors. While Scheffer et al. (1993) and Jeppesen et al. (1994) emphasise the importance of 'critical nutrient thresholds', attributing change to algal-dominated states to nutrient increases, Moss (personal communication, Prof. B. Moss; Moss, 1999) cites a host of additional factors (from grazing of macrophytes by birds, mechanical plant damage, use of herbicides and pesticides and heavy metal toxicity to *Daphnia*, piscivore kills, and even rising salinity), other than nutrient increases, which will act as forward switches to trigger an algal-dominated state in lakes. Moreover, lake biomanipulation, which is essentially a reverse switch, does not necessarily mean that nutrient levels should be low while the clear-water phase lasts (Moss, 1999). For example, works on experimental ponds (Irvine et al., 1989) and Zwemlust (present study) demonstrate reverse switches operating even at high nutrient levels. Nevertheless, 'nutrient availability still has a role in the alternative state model in that the threshold for operation of forward switches seem to be depressed at increasing nutrient concentrations, and biomanipulation is more effective if nutrient levels are reduced (Jeppesen et al., 1994; Moss, 1999).

A multivariate analysis of phytoplankton and food-web changes in Lake Zwemlust (Romo et al., 1996) revealed nutrient limitation for phytoplankton during the clear-water periods caused by macrophytes and zooplankton (size-selective) grazing. The small-sized, fast-growing phytoplankton forms with high surface area to volume ratios favour such a limitation. The macrophytes compete successfully for nutrients with the phytoplankton, especially for N. Although phytoplankton and fish serve as the major sinks of nutrients during the turbid-water period, and macrophytes during the clear-water period, the two appear to play disparate roles in the nutrient turnover and availability for phytoplankton during the two stable states.

Lake restoration by biomanipulation: models, hypotheses, and concerns

Biomanipulation studies have greatly stimulated theory and hypothesis-forming for food web interactions. The field and theoretical studies have developed virtually side-by-side. These studies include: (1) alternative stable states (Moss, 1990; 1998) and models based on these (Scheffer, et al., 1997; Scheffer, 1998) and nutrient level changes (Moss, 1998); (2) top-down 'cascading effects' (Carpenter & Kitchell, 1988, 1992, 1993; Carpenter et al., 1985); and (3) bottom-up effects (McQueen et al., 1986). Evidence for the existence of clear-cut stable state lake categories in the Netherlands and elsewhere is somewhat weak. This is true even for the Lake Zwemlust study, which is considered a relatively successful long-term case study of lake biomanipulation in the Netherlands based on well-documented data (Gulati, 1995a,b; van Donk, 1998; van de Bund & van Donk, 2002). In this lake, clear water throughout the year was recorded only in the first few years after biomanipulation in 1987. In the years after 1990, the clear water was limited to spring and early summer and Cyanobacteria blooms tended to occur in autumn. Moreover, for testing the validity of the stable state model over longer periods, the lakes need to be left alone after the restoration measures. However, this did not strictly happen in Lake Zwemlust or in any other lakes in the Netherlands. It would be interesting to compare these extreme stable states for their nutrient contents and nutrient turnover rates. We know that the nutrients in turbid lakes are locked up mainly in the existing populations of phytoplankton and fish (Fig. 3), and in clear-water lakes mainly in the macrophytes. We can thus hypothesise that 'though the overall nutrient contents may be quite comparable during the two stable states, water clarity is manifested in extremely differing states because the rates of turn over (recycling) and availability patterns of nutrients differ'.

To sum up, the results of the Dutch biomanipulation studies (see Meijer, 2000) are ambiguous, with more long-term failures than successes recorded. The failures are, by and large, related to insufficient or no decrease in the in-lake nutrient loading but also to the rapid increase in planktivorous fish in the years following the fish reduction measures. It is pertinent to recall the concern of de Melo et al. (1992) that biomanipulation is at the stage of becoming entrenched as a lake management tool and is accepted unquestionably in the general literature. Based on an analyses of data relating to 118 response cells of the manipulated waters, the authors have cast doubts on the 'trophic cascade theory' of Carpenter & Kitchell (1992, 1993) because the top-down response weakened at the zooplankton–phytoplankton level. In 80% of the cases analysed, such a response was either absent or was unclear. In view of this revelation, the staunch proponents of biomanipulation for lake restoration would be well advised to temper their enthusiasm. de Melo et al. (1992) suggest that far from being convincing, the biomanipulation/trophic-cascade/top-down theory may be unsoundly based, with many half-truths and much 'hand-waving and overexploitation' of the data. On the other hand, Hansson et al. (1998) are optimistic: they conclude that 'biomanipulation is not only possible, but also a relatively inexpensive and attractive method for management of eutrophic lakes', particularly as a follow-up measure to the reduced nutrient loadings. The criticism expressed by de Melo would appear to be somewhat ill-founded, since it seems that relatively more deeper water bodies were included in their analysis than the shallow ones of the Dutch and Danish studies. Recently, Benndorf et al. (2002) have attributed the failure of most biomanipulation studies in deeper lakes to extremely high P loading so that the phytoplankton did not decrease. However, Drenner & Hambright (2002) did not find any support for this in their analysis. On the other hand, as nutrient dynamics differs markedly between the shallow and deep lakes (Moss, 1998) partly because of differences in sediment–water interactions, this difference is likely to affect the efficacy of the restoration measures as well as predispose the shallow lakes to be colonised by macrophytes.

Drenner & Hambright (2002) have critically reviewed the recent literature relating to the 'concept of cascading trophic interactions' – the trophic cascade hypothesis (Carpenter et al., 1985; Carpenter & Kitchell, 1988) which fishery biologists and water managers trying to improve the water quality have come to accept as gospel. According to this hypothesis, a rise in piscivore biomass causes a decrease in planktivore biomass, an increase in biomass of herbivore zooplankton, and a decrease in biomass of phytoplankton. Drenner & Hambright (2002) find limited evidence for the hypothesis, and consider it unwise to accept a hypothesis without adequately proving it. They observe that in 22 of the 39 published studies, the piscivore top-down effects on phytoplankton biomass were confounded by the simultaneous reductions of nutrients and planktivores; that is, the effects cannot be attributed solely to piscivore manipulation. In most of the remaining 17 non-confounded studies, piscivore effects on phytoplankton biomass were absent. The regression slope for chlorophyll: TP was lower in lakes with planktivores plus piscivores than in lakes with planktivores alone. Using chlorophyll-*a* as a surrogate for phytoplankton, this implies a greater phytoplankton decrease in the presence of piscivores. There is thus support for the 'trophic cascading interactions' hypothesis; that is, the lakes with piscivores contained relatively less phytoplankton biomass, regardless of the TP level. Drenner & Hambright have thus suggested using the regression slope of chlorophyll: TP as an indicator of 'functional piscivory'.

The regression relationships for chlorophyll: TP for many Dutch lakes, both those under restoration (Hosper, 1997; Meijer, 2000) and others, indicate two to three times more chlorophyll-*a* per unit weight TP than the ratios discussed by Drenner & Hambright (2002). Such lakes are dominated by filamentous Cyanobacteria, which have high and variable C:P ratios (DeMott & Gulati, 1999; DeMott et al., 2001). An important implication here is that for a unit weight of P, the Cyanobacteria can cause greater turbidity than, for example, green algae because of their relatively lower C:P ratios. Studies of over 250 Dutch lakes (CUWVO, 1987) have shown that the slope of chlorophyll: P regression relationships has exhibited a decrease in recent years. As Drenner & Hambright (2002) have observed, this decrease is, however, not related to changes in fish composition (increase in piscivores) in unconfounded lakes with both piscivores and planktivores. Moreover, the recent decreases in both chlorphyll-*a* and TP levels in some individual groups of Dutch lakes (e.g. the Loosdrecht lakes) (authors' unpublished data) confirm trends relating to changes in the regression slope for chlorphyll-*a* and P for eutrophic lakes in general in the Netherlands. As such, the phytoplankton decrease is not related to the cascading top-down effect, but to the delayed bottom-up effect, which causes a relatively greater decrease in chlorophyll-*a* than in P, in response to the reduction in the external P loads that began 15 years previously. The regression slope between TP (X axis) and chlorophyll (Y axis) has not decreased because of piscivore additions or increases but because of a decrease in P loading in general, which causes the cyanobacterial densities to decrease.

Central role of fish in restoration

Our knowledge that fish play a major role in the ecosystem (Carpenter & Kitchell, 1992; see e.g. papers cited in Moss, 1998) has contributed to the concept of biomanipulation, which Moss (1998) has even called 'the lynchpin of shallow lake restoration'. There are good reasons for assigning a key role to fish in lake restoration: the fish are a relatively easy instrument or supplementary instrument for restoration and management (Lammens, 1999). The effects of top-down manipulation are virtually instantaneous, and more tangible in shallow lakes (Jeppesen, 1998) than in deep lakes, except if the macrophytes are abundant. This is in contrast to the effects of hydrological and nutrient control measures, which need time to produce the desired results. Fish management strategies essentially involve reducing and regulating the impact of planktivorous fish by reducing their numbers or removing them entirely and restocking with piscivores. The literature relating to the European lakes (Hansson et al., 1998) shows that fish removal rates vary from 25 to 100%, although we know that most effective biomanipulation measures have involved attempting to remove 75 to 100% of the entire fish community (Hansson, 1998; Moss, 1998; Meijer, 2000). Although continual fish management would result in more sustainable and mutually acceptable changes for both fishermen and water-quality managers, the cost–benefit aspects may not make this a practical solution.

In terms of existing standing stocks of fish, bream was perhaps the most important planktivore/benthivore in most shallow Dutch lakes until lake restoration studies were initiated in the late 1980s. Lammens et al. (in press) have compared the development of bream populations in Dutch lakes and the long-term, indirect effects on water quality parameters (SD, chlorophyll-*a*, macro-vegetation and macrofauna), in relation to fishery exploitation in lakes in various areas. In Lake Veluwe, the c. 80% reduction of the bream population, from c. 100 to 20 kg ha^{-1} after 5 years of fishing led to a striking improvement in water clarity, and accelerated expansion of the *Chara* beds in shallower parts (Hosper, 1997). The densities of zebra mussels in the shallower areas have also increased and chlorophyll levels have declined, with perceptible increases in SD in open water. In the Friesian lakes, routine seine fishery has not affected the bream population despite high catches of 40–50 kg ha^{-1}. Losses caused by fishing have generally been offset by good

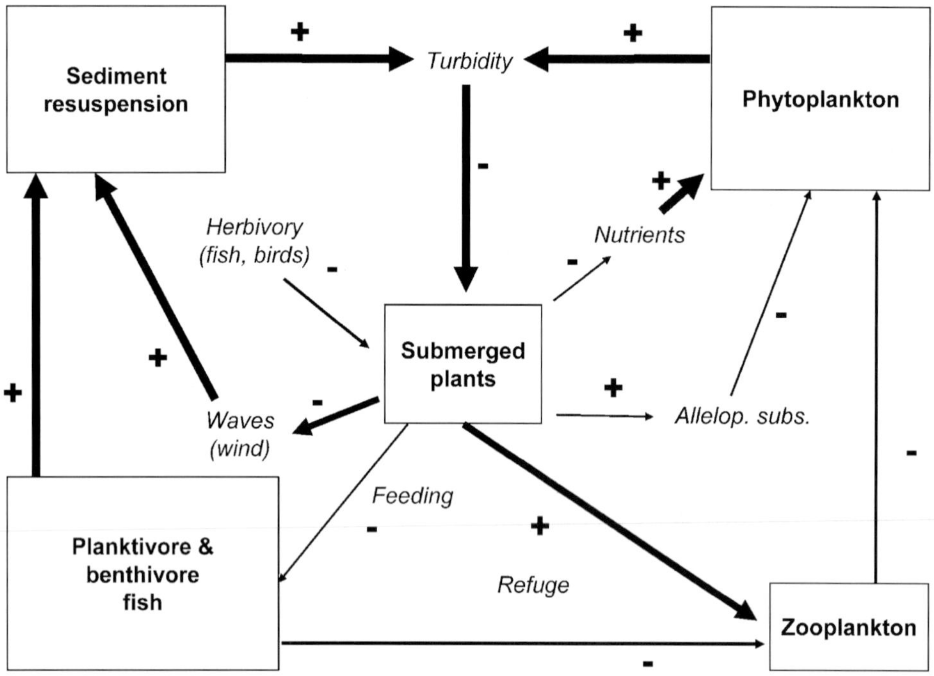

Figure 6. Diagrammatic illustration of the mechanisms and factors causing sediment resuspension and turbidity in shallow eutrophic lakes, especially in the Netherlands. After restoration measures, submerged plants, which are adversely affected by turbidity, start to contribute to improved light climate through both their direct and indirect feed-back effects. The thickness of arrows indicates the relative importance of the feed-backs.

recruitment and higher growth rates due to higher temperatures. The decrease in the chlorophyll-*a* level and increase of transparency have only been marginal. In Lake Volkerak, bream appeared first in 1988, within a year of the lake becoming a freshwater lake. By 1998, the bream had reached c. 140 kg ha^{-1}, and chlorophyll-*a* level had increased from 5 to 45 μg l^{-1}. However, in the same period, the vegetation cover decreased from 20 to 10% and SD declined from its maximum of c. 3 m in 1988 to c. 1 m in 1998. The indirect effects of the unexploited bream population in Lake Volkerak until 1998 differed from those observed in the Friesian lakes.

In Dutch lakes, the role of bream and planktivores in general is crucial in many ways (Fig. 6). In the first place, because the fish cause a regeneration of P via both digestive activity and excretion (Lazzaro, 1987), their feeding activities contribute importantly to in-lake recycling of nutrients and eutrophication process (Fig. 3). In addition, in view of the high body P, fish excretion and mortality and decomposition are important sources of P recycling within the Lake Loosdrecht water column (Fig. 3; van Liere & Gulati, 1992). Secondly, as the fish feed size-selectively on the larger zooplankters (especially the daphnids), they adversely affect the zooplankton grazing, and the subsequent increase in phytoplankton and detritus causes high turbidity. Thirdly, the benthivores (especially bream) negatively influence the under-water light climate by resuspending the bottom sediments during their foraging activities (Fig. 6). Sediment resuspension can promote aerobic mineralization of P in open water as well as recycling of P (Fig. 3) and fixation of P in Fe complexes if the redox potential is high. The planktivorous and benthivorous fish thus retard the pace of restoration by contributing to P-flux directly through their metabolic processes and to bioturbation in the upper sediment layer. Model studies (Meijer et al., 1990) show that >50% of the turbidity in shallow Dutch lakes can be ascribed to sediment resuspension by benthivores. The reciprocal SD values and resuspended inorganic solids were directly related to existing benthivore populations of up to about 650 kg FW ha^{-1} (Fig. 7a,b). Moreover, at such high levels the benthivores can potentially reduce SD to 0.4 m, irrespective of the algal contribution to turbidity.

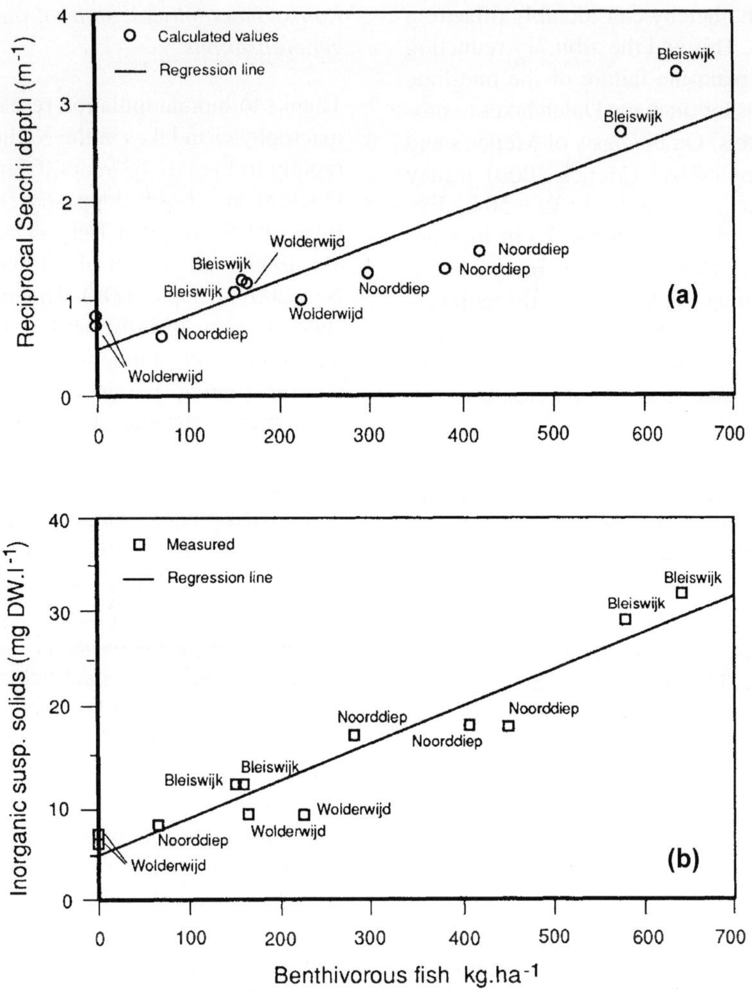

Figure 7. (a) (upper panel). Regression line between biomass of benthivorous fish and reciprocal Secchi-disc depth based on Secchi depth (SD) model restricted to 1 m depth. Calculated values of SD (July–September) are for three biomanipulated lakes (Lake Bleiswijk and Lake Noorddiep and a control and experimental pond in Lake Wolderwijd), for both experimental and control parts of the three water bodies. Model values for SD are computed from seston excluding the algae, i.e. comprising inorganic suspended solids (0.75) and detritus (0.25). Note: model calculations reveal that at their biomass level of 600 kg ha^{-1} benthivorous fish alone can reduce SD to 0.4m (1/SD = 2.5), i.e. excluding the effects of algae on SD; (b) (lower panel). Regression line between the biomass of benthivorous fish and inorganic suspended solids (ISS) in three lakes based on SD model as in upper panel, (a) Measured values (July–September) of ISS in lakes are also shown. (Source of Fig. 7: Meijer et al., 1990.)

It would thus seem that there is no other option than for lake rehabilitation measures aimed mainly at minimising the nutrient levels and other negative effects of fish on lake transparency to be directed at reducing the planktivores and benthivores. The latter, although the predominant fish in the shallow Dutch lakes (Meijer, 2000), have not received adequate attention. However, there are no ready-made ways of determining the amount of fish to be eliminated or restocked. In many Dutch lake studies, about 75% reduction of the existing population is advocated so as to produce the desired effects (see Meijer, 2000). This hypothetical reduction percentage does not, however, take account of the existing population in the lake concerned, or the standing stock to be achieved after such a reduction. Based on a realistic estimation that existing fish stocks in the Dutch lakes will vary between 200 and 1000 kg FW ha^{-1}, a 75% reduction of the stock will result in a remaining fish stock of between 50 and 200 kg FW ha^{-1}, a factor four variation. Moreover, we need to keep in mind that the fish reductions will often stimulate recruitment of the YOY fish, thereby considerably offsetting the reduction effects. This and the arbitrary reduction percentage might explain the failure of the one-time biomanipulation measures in many Dutch lakes to produce the desired effects. On the basis of Meijer's and other research studies at RIZA (Meijer, 2000), it may be surmised that reducing planktivores to <50 kg FW ha^{-1} and maintaining them at that level will increase the chances of success. However, maintaining a hypothetical existing fish stock is likely to be a difficult task if piscivores such as the northern pike (*Esox lucius*) fail to develop even moderate population numbers, for reasons not yet well understood, and a situation that applies to the Dutch lakes.

The trophic-level response or cascading effect within the foodchain (Carpenter et al., 1985), should ideally trickle down through the zooplankton to phytoplankton. However, the top-down effects are known to gradually lessen in the lower trophic levels (McQueen et al., 1986; Drenner & Hambright, 2002). Moreover, changes to fish foraging strategies are likely to weaken the cascading effects. This would at least seem to be the case with bream in many Dutch lakes: because their size- structure varies according to age, bream switch from planktivory to benthivory as they grow (Lammens et al., 1990).

The Lake Loosdrecht studies (Gulati & van Liere, 1992; van Liere & Gulati, 1992; van Liere & Janse, 1992) demonstrate that not including lake sediment, almost half the particulate-P in the lake (≈300 mg P m^{-2}) is 'locked into' the fish (Fig. 3). The body P regenerated (excretion, egestion and mortality death) by fish was about 1.4 mg P m^{-2} d^{-1}, which is equivalent to about 17 per cent of total P mineralised in water column, and about 140% of the external loading. Meijer et al. (1994a,b) reported that fish reduction of 150 kg ha^{-1} in Lake Wolderwijd led to a decrease in P-loading via fish equivalent to 60% of the external loading. Although P release from the sediment by fish foraging has not been quantified, it could be substantial. Moreover, since the levels of TP and planktivorous fish in lakes are directly related (Meijer et al. 1990), fish stock reductions are imperative for lake restoration measures. The inclusion of directly measured P data to P flux in the model studies will help us predict more accurately the extent of fish reductions required to produce both tangible and sustainable restoration effects.

Role of macrophytes: state-of-the-art and some generalisations

Thanks to biomanipulation research, the literature on macrophytes in lakes in the Netherlands has increased rapidly in the last 15 years (Ozimek et al., 1990; van Donk et al., 1993; Jeppesen et al., 1998; van den Berg, 1999; van den Berg et al., 1999; Scheffer et al., 1999; van Nes et al., 1999, Meijer, 2000; van Nes, 2002; Coops, 2002). The more recent papers excellently sum up information on the structuring role and impacts of submerged macrophytes on ecosystem functioning. We know that because of their bulk, macrophytes are much less efficient than microalgae at taking up nutrients, and can build up huge biomass and reduce availability of nutrients for algae. That macrophytes compete successfully with algae for nutrients, especially N, has become quite evident from biomanipulation studies (Kufel & Ozimek, 1994). Many of them have access to nutrients from both the sediment (Barko & James, 1998) and the water. Secondly, the macrophytes provide a refuge for larger-bodied zooplankters against fish predation (Moss, 1990; 1998), thereby promoting zooplankton grazing (Timms & Moss, 1994). Thirdly, the macrophytes considerably reduce fish-induced bioturbation, and prevent wind-induced resuspension (Fig. 6) of the essentially non-algal component (Gons et al., 1991) in the sediments (Fig. 7b); they also prevent shoreline erosion, thus allowing increased sedimentation. Consequently, the bioavailability of P will decrease and the light climate will improve (see Barko & James, 1998). Increased sediment stability appears to contribute to the seasonally persistent clear water patches associated with the *Chara* meadows (e.g. in Lakes Botshol, Veluwe and Wolderwijd). Fourthly, because of their huge biomass, which includes that of the colonising periphyton, and their long generation time, the macrophytes act as a major nutrient sink throughout most of the vegetative period. In addition, denitrification in the macrophyte beds limits algal growth further (Meijer et al., 1994b). Lastly, allelopathic substances released by macrophytes can have a negative impact on phytoplankton (Fig. 6), though the ecological significance of this mechanism is still unclear (van Donk & de Bund, 2002).

The development of macrophytes, which reinforces the process of lake clearing (Hosper, 1997), has in some lakes (Lake Veluwe, for instance) reached nuisance proportions, affecting mid-summer recreation activities in particular. The enhanced ability of the plants to invest in over-wintering structures (van Nes, 2002) leads to the prolongation of the macrophyte-dominated state. If the external nutrient loads continue to be high, once established, aquatic vegetation might reach nuisance proportions and adversely affect recreation. The costs and benefits of the vegetation need to be weighed up and straightforward management strategies devised for some lakes if these lake areas are to remain partly or wholly free of aquatic plants and not impede recreation (van Nes et al., 1999). This could facilitate a compromise between recreational needs and preventing dense algal blooms.

Role of zooplankton

Biomanipulation essentially involves alteration of the fish community to facilitate development of *Daphnia* in order to increase grazing pressure on algae (Moss, 1998). Not many studies have adequately addressed the role of zooplankton in relation to lake restoration, including biomanipulation (Timms & Moss, 1984; Gulati 1990a,b; Gulati et al., 1992; 1995b). Both empirical studies (Gulati, 1990a,b), and models (Janse et al., 1998; Jeppesen et al., 1999) point out the importance of zooplankton grazing for initiating clear water. Based on multiple regression analyses of data from 37 Danish lakes, Jeppesen et al. attribute the clear-water conditions in eutrophic, macrophyte-rich lakes, especially in summer, to zooplankton grazing on phytoplankton. They hypothesise that the role of zooplankton grazing in water clarity in macrophyte-rich lakes may be greater in eutrophic than in mesotrophic to lakes. A close monitoring of the chain of events leading to the formation of a clear-water phase in the spring period would undoubtedly show that such a phase is invariably triggered by the herbivorous zooplankton. We need, however, to analyse how the effects on phytoplankton cascade via zooplankton. The clear-water phase during spring (Sommer et al., 1986) has been documented frequently in lakes undergoing biomanipulation. Invariable sharp increases in grazing by larger-bodied daphnids were noted in several case studies done in the Netherlands (Gulati, 1989, 1990b; Hosper, 1997; Meijer, 2000) and other countries (e.g. Timms & Moss, 1984: Sondergaard et al., 1990).

The availability of data on length to filtering rate and on densities of the major grazers in Dutch lakes, and of regressions of specific clearance rates of zooplankton community and seston concentration in lakes allow us to make model calculations to predict the prospects of a clear-water phase (Gulati, 1990b; Gulati et al., 1992). Here we have compared biomanipulated lakes with non-biomanipulated ones, including those in which only nutrient control measures were taken to compare (1) mean individual weight of the crustacean communities (W) and mean seston levels, and (2) individual crustacean weight (W) and rotifer densities (Fig. 8a,b). Interestingly, the pooled data of the two lake types show significant negative correlations ($P < 00.1$). W in the biomanipulated lakes is up to an order of magnitude greater but seston concentration by same magnitude lower than in the non-biomanipulated lakes. Likewise, the rotifer numbers in the biomanipulated lakes do not exceed 1000 ind. l^{-1} but in the non-biomanipulated lakes reach up to 5000 ind. l^{-1}. The much higher values for W are due to much lower size-selective predation by fish in the biomanipulated lakes, and these larger-bodied zooplankters, usually belonging to *Daphnia* spp., in turn exert high grazing pressure, reducing the seston. The much lower rotifer densities in biomanipulated lakes indicate that when fish predation is low, large-bodied Cladocera are superior competitors for food than rotifers. An important biological water management strategy is thus stimulation of the conditions that promote the growth and development of larger-bodied zooplankton in order to trigger other food-web changes that will improve the under-water light climate (Gulati, 1990b).

Other means of lake manipulation

In addition to the zooplankton, fish and macrophytes, the zebra mussel, *Dreissena polymorpha*, is a potential candidate for biological lake management in Dutch lakes. Exploratory work (Reeders et al., 1989; Reeders & bij de Vaate, 1990, 1992; Noordhuis et al., 1992) shows that, being an efficient filter feeder, the mussel can be used as a biofilter to reduce the suspended matter in the lake water (Reeders & bij de Vaate, 1990), and to concentrate toxic waste materials such as heavy metals in certain lake parts colonised by the mussels. In Lake IJsselmeer, the areas colonised by zebra mussels have a relatively high SD, as confirmed by remote sensing studies carried out at RIZA. Since 1994, when the zebra mussels returned, this has also been true for certain areas of Lake Veluwe, and more recently for other border lakes (personal communication, Dr S.H. Hosper). These observations, plus the fact that zebra mussels are good food for diving ducks (bij de Vaate, 1991; Noordhuis et al., 1992; van Eerden, 1998), suggest that the zebra mussel may be effective in lake

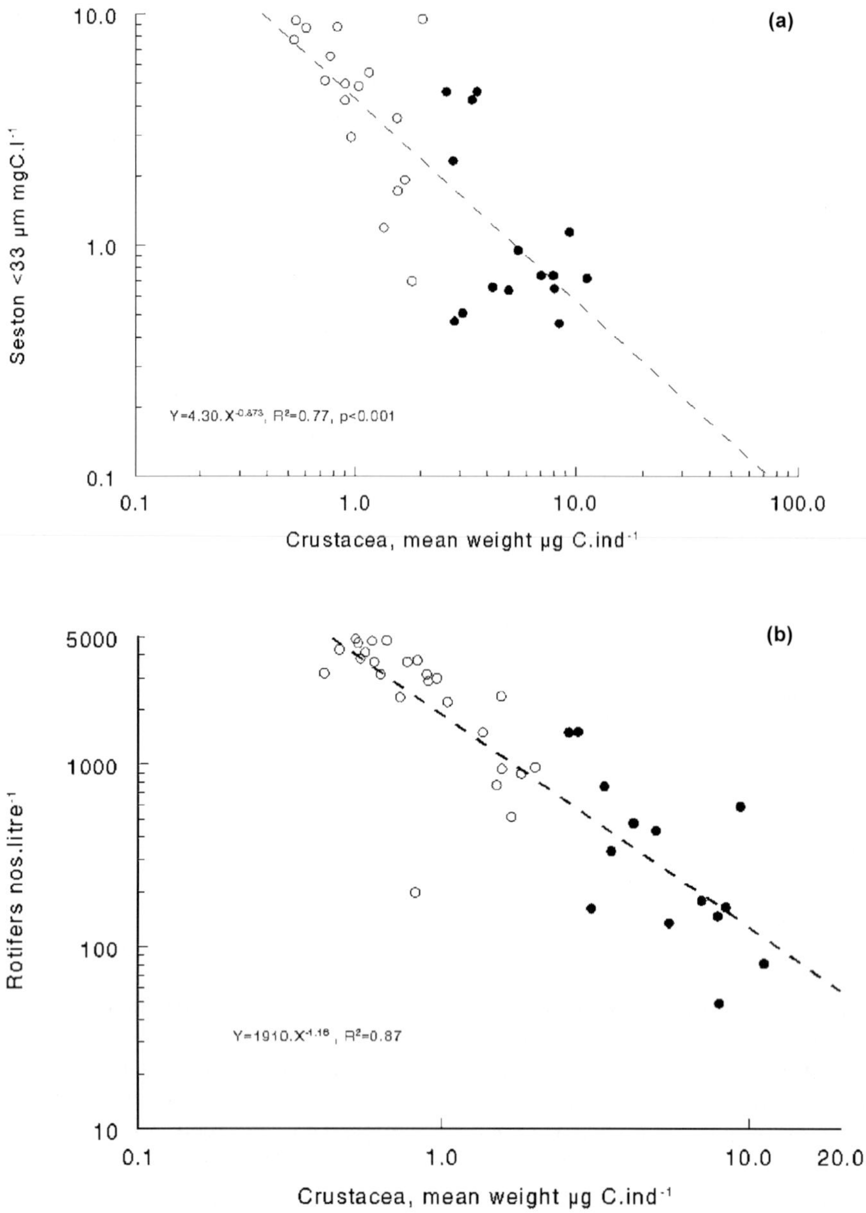

Figure 8. (a) (upper panel). Power curves for regression relationships between mean individual weights (W) of the crustacean communities and mean seston levels using pooled data of biomanipulated lakes (shaded circles) and non-biomanipulated ones(open circles), showing that the animals are heavier and seston mass lower in biomanipulated lakes. Data are means of several measurements in each of about seven lakes during 6-month period (April–October). (b) (lower panel). Individual crustacean weight (W), as in panel above and rotifer densities in the two groups of lakes as in upper panel, showing that in the presence of larger bodied crustaceans in biomanipulated lakes, the rotifer densities decrease significantly. Coefficient of determination R^2 is highly significant for both regression lines. (Based on unpublished data of first author.)

restoration. However, we know little about the optimal substratum or sediment needed for the zebra mussel to establish successfully, nor do we know much about their sudden and *en masse* disappearance from Dutch lakes. The lack of a proper substratum in lakes due to eutrophication is a plausible cause for the inability of the mussels to establish in these lakes. That the macrophytes form a suitable, natural substratum for the mussels (Reeders & bij de Vaate, 1990) augurs well for those waters where macrophytes have returned and are well established.

The other major change in the lake ecosystems of border lakes, especially since 1995, is the strong increase in the piscivorous, benthivorous and herbivorous water bird numbers. The bird increases had started by 1990, after the recolonisation of border lakes by *Chara* sp. and the subsequent improvements in water clarity. Other accompanying increases (macrophytes, zebra mussels, and smaller fish, both roach and perch) have apparently led to overexploitation of the available food resources, and the situation has yet to stabilise (data from unpublished RIZA Reports).

General conclusions: lessons learnt & future approaches

Some generalisations can be drawn from experiences relating to the use of various techniques in restoration works on Dutch lakes. The pooled monitoring data of 231 lakes in the period 1980–1996 indicate that the summer median concentrations of both TP and chlorophyll decreased by 56%, and those of TN by only 22%. The SD, however, improved by just 17% (Table 2: van der Molen & Portielje, 1999). Only in a few lakes was there more than marginal SD improvement in response to the reduction of nutrients in the inflows or in-lake reductions. The long-term studies (Loosdrecht lakes) show that even after about a decade of nutrient control measures, the lakes failed to exhibit any improvements in water quality. The causes of the ongoing eutrophication symptoms are invariably to be found in the in-lake stockpiles of P, both biotic (fish) and abiotic (sediment), and their slow dynamics. However, more recent studies in the Loosdrecht lakes using stable-isotope ($\delta^{13}C$) tracking of carbon transfer (Pel et al., submitted) show that the cyanobacterial community is losing resilience because of ongoing reduction in the external P loading during the last two decades. The adequacy of such methods to probe the trophic links in detail may provide sensitive means to detect and come to grips with the impact of anthropogenic activities and climate mediated stresses on pelagic food webs (Pel et al., submitted). The study shows that in the Loosdrecht lakes, the relative contribution of *Prochlorothrix*-like filaments to the total cyanobacterial filaments, including those of *Oscillatoria*, decreased from some 50% during 1988–1992 to 20% during 1997–2001. This decrease is probably related to the reduction in the pulsed releases of P from the lake sediments, which *Prochlorothrix* sp. can exploit better than *Oscillatoria*. Densities of *Oscillatoria* sp. have also decreased in the past decade, and a clear-water phase was observed in the lakes during the spring of 2001 (personal communication Dr R. Pel), almost two decades after the P reduction measures were first started.

Restoration measures complemented by biomanipulation have generally elicited a better response and a greater degree of lake recovery. In the case of the successful biomanipulation experiment in Lake Zwemlust, despite the unabated inputs of N and P via seepage water there was timely development of macrophytes crucial to limiting phytoplankton growth. In contrast, in the larger lakes such as Breukeleveen (van Donk et al., 1990c) where the nutrient inputs from external sources were reduced prior to the lake's biomanipulation, the measures failed. Ineffective reduction of fish stock combined with lake size, hydrology and wind-induced resuspension of the sediments prevented improvement in the lake's light climate. Flushing and sediment removal are promising techniques but have not gained popularity in the Netherlands because of the scarcity of good quality water and the unfavourable cost/benefit ratio. Overall, there has been as much valuable experience gained from the failures as from the transient successes. The studies have helped us understand that sustainability of the positive effects on water quality is central to the remedial measures. Lake restoration plans for the future typically envisage 'nature development', emphasising that a lake is an integral part of landscapes which include other aquatic systems and semi-aquatic and terrestrial ecosystems (see several papers in Nienhuis & Gulati, 2002). Such proposed measures include reinforcing the lakes' shoreline vegetation to prevent erosion and improve the propensity of the land-water transition to develop a natural biodiversity (e.g. Lake Volkerak–Zoommeer). In other cases (e.g. Lake Breukeleveen), the water authorities have started excavating several deep pits (20–40 m) within the shallow lake parts to allow wind-induced shifting and burial of the loose, nutrient-rich lake sediments in these pits and thus retard in-lake nutrient releases from the sediments. In some other waters, creation of artificial islands to reduce the wind fetch factor and erosion is planned. RIZA is investigating the feasibility of deploying water-level management to encourage the shoreline macrovegetation to develop and for greater natural development of the aquatic and semi-aquatic ecosystems. The plans envisage extending the upper and lower limits of the permissible annual fluctuations and exploring the ef-

Jeppesen, E., Ma. Sondergaard, M. Sondergaard, & K. Christoffersen, 1998. The structuring role of submerged macrophytes in lakes. Ecological Studies. Springer Verlag Vol. 131: 423 pp.

Jeppesen, E., J. P. Jensen, M. Sondergaard & T. Lauridsen, 1999. Trophic dynamics in turbid and clear-water lakes with special emphasis on the role of zooplankton for water clarity. Hydrobiologia 408/409: 217–231.

Keizer, P. & A. J. C. Sinke, 1992. Phosphorus in the sediment of the Loosdrecht Lakes and its implications for lake restoration perspectives. Hydrobiologia 233: 39–50.

Kornijów, R. & R. D. Gulati, 1992a. Macrofauna and its ecology in Lake Zwemlust, after biomanipulation I. Bottom fauna. Arch. Hydrobiol. 123: 337–347.

Kornijów, R. & R. D. Gulati, 1992b. Macrofauna and its ecology in Lake Zwemlust, after biomanipulation II. Fauna inhabiting hydrophytes. Arch. Hydrobiol. 123: 349–359.

Kornijow, R., R. D. Gulati & E. van Donk, 1990. Hydrophyte-macroinvertebrate interactions in Zwemlust, a lake undergoing biomanipulation. Hydrobiologia 200/201: 467–474.

Kufel, L. & T. Ozimek, 1994. Can *Chara* control phosphorus cycling in Lake Luknajno (Poland)? Hydrobiologia 276: 277–283.

Kufel, L., A. Prejs & J. I. Rybak (eds), 1997. Shallow Lakes '95 – Trophic Cascades in Shallow Freshwater and Brackish Lakes. Developments in Hydrobiologia 119. Kluwer Academic Publishers, Dordrecht: 416 pp. Reprinted from Hydrobiologia 342/343.

Lamers, L. P. M., A. J. P. Smolders & J. G. M. Roelofs, (2002.) The restoration of fens in the Netherlands. Hydrobiologia 478/Dev. Hydrobiol. 166: 107–130.

Lammens, E. H. R. R., 1989, Causes, consequences and success of bream in Dutch eutrophic lakes. Hydrobiol. Bull. 23: 11–18.

Lammens, E. H. R. R., 1999. The central role of fish in lake restoration and management. Hydrobiologia 396: 191–198.

Lammens, E. H. R. R., R. D. Gulati, M. L. Meijer & E. van Donk, 1990. The first biomanipulation conference: a synthesis. Hydrobiologia 200/201: 619–627.

Lammens, E. H. R. R., N. Boesewinkel-de Bruyn, H. Hoogveld & E. van Donk, 1992. P-Load, phytoplankton, zooplankton and fish stock in Loosdrecht Lake and Tjeukemeer – Confounding effects of predation and food availability. Hydrobiologia 233: 87–94.

Lammens, E. H. R. R., E. H. van Nes & W. M. Mooij, 2002. Differences in exploitation of the bream population in three Dutch lakes and their relation to water quality. Freshwat. Biol. 47.

Lazzaro, X., 1987. A review of planktivorous fishes: their evolution, feeding behaviours, selectivities, and impacts. Hydrobiologia 146: 97–167.

Ligtvoet, W. & S. A. de Jong, 1995. Ecosystem development in Lake Volkerak: Concept and strategy. In: Hosper S. H., Gulati L. van Liere & R. M. M. Rooijackers (eds), Integrated Water Resources Management. 31: 239–243.

McQueen, D. J., J. R. Post & E. L. Mills, 1986. Trophic relationships in freshwater pelagic ecosystems. Can. J. Fish. aquat. Sci. 43: 1571–1581.

Meijer, M. L., 2000. Biomanipulation in the Netherlands. 15 years of experience. Ph.D. Thesis University of Wageningen: 206 pp.

Meijer, M. L., M. W. de Haan, A. W. Breukelaar & H. Buitebveld, 1990. Is reduction of the benthivorous fish an important cause of high transparency following biomanipulation in shallow lakes? Hydrobiologia 200/201: 303–315.

Meijer, M. L. & H. Hosper, 1997. Effects of biomanipulation in the large and shallow Lake Wolderwijd, The Netherlands. Hydrobiologia 342/343: 335–349.

Meijer, M. L., E. Jeppesen, E. van Donk, B. Moss, M. Scheffer, E. Lammens, E. van Nes, J. A. van Berkum, G. J. de Jong, B.A. Faafeng, J.P. Jensen, 1994a. Long-term responses to fish-stock reduction in small shallow lakes: interpretation of five-year results of four biomanipulation cases in the Netherlands and Denmark. Hydrobiologia 275/276: 457–466.

Meijer, M. L., E. H. van Nes, E. H. R. R Lammens, R. D. Gulati, M. P. Grimm, J. Backx, P. Hollebeek, E. M. Blaauw & A. W. Breukelaar, 1994b. The effects of fish stock reduction: the consequences of a drastic fish stock reduction in the large and shallow lake Wolderwijd, The Netherlands – can we understand what happened? Hydrobiologia 275/276: 31–42.

Meijer, M. L., E. H. R. R. Lammens, A. J. P. Raat, J. G. P. Klein Breteler & M. P. Grimm, 1995. Development of fish communities in lakes after biomanipulation. Neth. J. aquat. Ecol 29: 91–101.

Meijer, M. L., I. de Bois, M. Scheffer, R. Portielje & H. Hosper, 1999. Biomanipulation in shallow lakes in The Netherlands an evaluation of 18 case studies. Hydrobiologia 408/409: 13–30.

Moss, B., 1990. Engineering and biological approaches to the restoration from eutrophication of shallow lakes in which aquatic plant communities are important components. Hydrobiologia 200/201: 367–377.

Moss, B., 1998. Shallow lakes, biomanipulation and eutrophication. Scope News Letter 29: 1–45.

Moss, B., 1999. Britsh Phycological Society Presidential address 1999: from algal culture to ecosystem; from information to culture. Eur. J. Phycol. 34: 193–203.

Nienhuis, P. H. & R. D. Gulati (eds), 2002. Ecological Restoration of Aquatic and Semi-Aquatic Ecosystems in the Netherlands. Developments in Hydrobiology 166. Kluwer Academic Publishers Dordrecht: 234 pp. Reprinted from Hydrobiologia 478.

Nienhuis, P. H., J. P. Bakker, A. P. Grootjans, R. D. Gulati & V. N. de Jonge, 2002. The state of the art of aquatic and semi-aquatic ecological restoration projects in the Netherlands. Hydrobiologia Hydrobiologia 478/Dev. Hydrobiol. 166: 219–233.

Noordhuis, R, H. H. Reeders & A. bij de Vaate, 1992. Filtering rate and pseudofaeces production in zebra mussels and their application in water quality management. Limnol. Aktuell. 4: 101–114.

NW4, 1997. Fourth National Policy Document on Water Management Government Proposals Summary (English summary): in National Presentation on Freshwater: Netherlands. 24 pp.

Oskam, G., 1978. Light and zooplankton as algae regulating factors in Eutrophic Biesbosch reservoirs. Ver. int. Ver. Limnol. 20: 1612–1618.

Ozimek, T., R. D. Gulati & E. van Donk, 1990. Can macrophytes be useful in biomanipulation of lakes – the Lake Zwemlust Example. Hydrobiologia 200/201: 399–407.

Pel, R., V. Floris & H. Hoogveld (submitted). Cryptic production fuels lake food webs: using the hidden isotopic heterogeneity in phyto- and zooplankton to unmask disparity in trophic C-transfer. Limnol. Oceanogr.

Perrow, M. R., M. L. Meijer, P. Dawidowicz & H. Coops, 1997. Biomanipulation in shallow lakes: state of the art. Hydrobiologia 342/343: 355–365.

Portielje, R. & D. T. van der Molen, 1999. Relationships between eutrophication variables: from nutrient loading to transparency. Hydrobiologia 409: 375–387.

Reeders, H. H., & A. bij de Vaate & F. J. Slim, 1989. The filtration rate of *Dreissena polymorpha* (Bivalvia) in three Dutch lakes with reference to biological water quality management. Freshwat. Biol. 22: 133–141.

Reeders, H. H. & A. bij de Vaate, 1990. Zebra mussels (*Dreissena polymorpha*): a new perspective for water quality management. Hydrobiologia 200/201: 437–450.

Reeders, H. H. & A. bij de Vaate, 1992. Bioprocessing of polluted suspended matter from the water column by the zebra mussel (*Dreissena polymorpha* Pall.). Hydrobiologia 239L 53–63.

Richter, A. F., 1986. Biomanipulation and its feasibility for water quality management in shallow eutrophic water bodies in the Netherlands. Hydrobiol. Bull. 20: 165–172.

Roelofs, J. G. M., E. Brouwer & R. Bobbink, 2002. Restoration of aquatic macrophyte vegetation in acidified and eutrophicated shallow soft water wetlands in the Netherlands. Hydrobiologia. 478/Dev. Hydrobiol 166: 171–180.

Romo, S., E. van Donk, R. Gylstra & R. D. Gulati, 1996. A multivariate analysis of phytoplankton and food web changes in a shallow biomanipulated lake. Freshwat. Biol. 36: 683–696.

Sas, H., I. Ahlgren, H. Bernardt, B. Boström, J. Clasen, C. Forsberg, D. Imboden, L. Kamp Nielsen, L. Mur, N. De Oude, C. Reynolds, H. Schreurs, K. Seip, U. Sommer & S. Vermij, 1989. Lake restoration by reduction of nutrient loading: Expectations, experiences, extrapolations. St. Augustin 1: Akademia Verl., Richarz. St. Augustine, Germany.

Scheffer, M., 1998. Ecology of Shallow Lakes. Chapman & Hall. London. 357 pp.

Scheffer, M., S. H. Hosper, M. L. Meijer, B. Moss & E. Jeppesen, 1993. Alternative equilibria in shallow lakes. Trend. Ecol. Evol. 8: 275–279.

Scheffer, M. R. S., A. Gragnani, L. R. Mur & E. H. van Nes, 1997. On the dominance of filamentous Cyanobacteria in shallow, turbid lakes. Ecology 78: 272–282.

Scheffer, M. E. van Nes & H. Coops, 1999. Dynamics and stability of *Chara* sp. and *Potamogeton pectinatus* in a shallow lake changing in eutrophication level. Hydrobiologia 409: 335–342.

Scheffer, M., S. Carpenter, J. A. Foley, C. Folke & P. Walker, 2001. Catastrophic shifts in ecosystems. Nature 413: 591–596.

Schutten, J., J. A. van der Velden & H. Smit, 1994. Submerged macrophytes in the recently freshened lake system Volkerak-Zoom (the Netherlands), 1987–1991. Hydrobiologia 276: 207–218.

Shapiro, J., V. Lamarra & M. Lynch, 1975. Biomanipulation: An ecological approach to lake restoration. In Brezonik, P. L. & J. L. Fox (eds), Water Quality Management through Biological Control. University of Florida, Gainesville: 85–96.

Sommer, E. U., Z. M. Gliwicz, W. Lampert & A. Duncan, 1986. The PEG-model of seasonal succession of planktonic events in freshwaters. Arch. Hydrobiol. 106: 433–471.

Sondergaard, M., E. Jeppesen, E. Mortensen, E. Dall, P. Kristensen & O. Sortkjaer, 1990. Phytoplankton biomass reduction after planktivorous fish reduction in a shallow, eutrophic lake; a combined effect of reduced internal P loading and increased zooplankton grazing. Hydrobiologia 200/201/: 229–240.

Tamminga, G. H., 1992. Interpretation of sedimentation data measured in a former tidal channel Lake Volkerak. Hydrobiologia 235: 107–118.

Timms R. M. & B. Moss, 1984. Prevention of growth of potentially dense phytoplankton populations by zooplankton grazing in the presence of zooplanktivorous fish in a shallow wetland ecosystem. Limnol. Oceanogr. 29: 472–486.

Van Dam, H. & R. F. M. Buskens, 1993. Ecology and management of moorland pools – balancing acidification and eutrophication. Hydrobiologia 265: 225–263.

Van de Bund, Wouter J. & Ellen van Donk, 2002. Short- and long-term effects of zooplanktivorous fish removal in Lake Zwemlust: 15 years of data. Freshwat. Biol. 47.

Van den Berg, M. S., 1999. Charyophyte colonisation in shallow lakes: processes, ecological effects and implications for lake management. Ph.D. Thesis. Free University, Amsterdam: 137 pp.

Van den Berg, M. S., M. Scheffer, E. van Nes & H. Coops, 1999. Dynamics and stability of *Chara* sp and *Potamogeton pectinatus* in a shallow lake changing in eutrophication level. Hydrobiologia 409: 335–342.

Van der Does, J., P. Verstraelen, P. Boers, J. van Roestel, R. Roijackers & G. Moser, 1992. Lake restoration with and without dredging of phosphorus- enriched upper sediment layers. Hydrobiologia 233: 197–210.

Van der Molen, D. T. & P. C. M. Boers, 1994. Influence of internal loading on phosphorus concentration in shallow lakes before and after reduction of the external loading. Hydrobiologia 276: 379–389.

Van der Molen, D. T. & P. C. M. Boers, 1999. Eutrophication control in the Netherlands. Hydrobiologia 396: 403–409.

Van der Molen, D. T. & R. Portielje, 1999. Multi-lake studies in The Netherlands: trends in eutrophication. Hydrobiologia 409: 359–365.

Van der Vlugt, J. C., P. A. Walker, J. van der Does & A. J. P. Raat, 1992. Fisheries management as an additional lake restoration measure – biomanipulation scaling-up problems. Hydrobiologia 233: 213–224

Van Donk, E., 1998. Switches between clear and turbid water states in a biomanipulated lake (1986–1996): the role of herbivory on macrophytes. In: Jeppesen E., M. Sondergaard & Christoffersen (eds). The Structuring Role of Submerged Macrophytes in Lakes, Ecological Studies, Spring Verlag, New York: 290–297.

Van Donk, E. & R. D. Gulati (eds), 1989. Biomanipulation in the Netherlands: applications in freshwater ecosystems and estuarine water. Proceedings of a congress organised by the Netherlands Hydrobiological Society, Amsterdam, 1988. Hydrobiol. Bull. 23: 99 pp.

Van Donk, E. & R. D. Gulati, 1991. Ecological management of aquatic ecosystems: a complementary technique to reduce eutrophication-related perturbations, In: Ravera, O. (ed) Terrestrial and Aquatic Ecosystems: Perturbation and Recovery. Ellis Horwood Chichester: 566–575.

Van Donk, E. & A. Otte, 1996. Effects of grazing by fish and waterfowl on the biomass and species composition of submerged macrophytes. Hydrobiologia 340: 285–290.

Van Donk, E. & W. J. van de Bund, 2002. Impact of submerged macrophytes including Charophytes on phyto- and zooplankton communities: allelopathy versus other mechanisms. Aquat. Bot. 72: 261–274.

Van Donk, E., R. D. Gulati & M. P. Grimm, 1989. Food web manipulation in Lake Zwemlust: positive and negative effects during the first two years. Hydrobiol. Bull. 23: 19–34.

Van Donk E., R. D. Gulati, & M. P. Grimm, 1990a. Restoration by biomanipulation in a small hypertrophic lake: First-year results. Hydrobiologia 191: 285–295.

Van Donk E., M. P. Grimm, R. D. Gulati & J. P. C. Klein Breteler, 1990b. Whole-lake food-web manipulation as a means to study community interactions in a small ecosystem. Hydrobiologia 200/201: 275–289.

Van Donk, E., M. P. Grimm, R. D. Gulati, P. G. M. Heuts, W. A. de Kloet & L. van Liere, 1990c. First attempt to apply whole-lake food-web manipulation on a large-scale in the Netherlands. Hydrobiologia 200/201: 291–301.

Van Donk E., R. D. Gulati, A. Iedema & J. T. Meulemans, 1993. Macrophyte-related shifts in the nitrogen and phosphorus contents of the different trophic levels in a biomanipulated shallow lake. Hydrobiologia 251: 19–26.

Van Donk, E., E. De Deckere, J. G. P. Klein Breteler & J. T. Meulemans, 1994a. Herbivory by waterfowl and fish on macrophytes in a biomanipulated lake: effects on long-term recovery. Verh. int. Vere. Limnol. 25: 2139–2143.

Van Donk, E., M. P. Grimm, R. D. Gulati, P. G. M. Heuts, G. Blom, K. Everards & O.F.R. Tongeren, 1994b. Use of mesocosms in a shallow lake to study the effects of different restoration measures. Arch Hydrobiol. Beih. 40: 283–294.

Van Eerden, M. R., 1998. Patchwork – patch use, habitat exploitation and carrying capacity for water birds in Dutch freshwater wetlands. Ph.D. Thesis State University of Groningen, Groningen: 448 pp.

Van Liere, L. & R. D. Gulati (eds), 1992. Restoration and Recovery of Shallow Eutrophic Lake Ecosystems in the Netherlands. Developments in Hydrobiologia 74. Kluwer Academic Publishers, Dordrecht: 305 pp. Reprinted from Hydrobiologia 233.

Van Liere, L. & J. H. Janse, 1992. Restoration and resilience to recovery of the Lake Loosdrecht ecosystem in relation to its phosphorus flow. Hydrobiologia 233: 95–104.

Van Nes, E. H., 2002. Controlling complexity in individual-based models of aquatic vegetation and fish communities. Ph.D. Thesis, University of Wageningen: 199 pp.

Van Nes, E. H., S. M. van den Berg, J. S. Clayton, H. Coops, Marten Scheffer & E. van Ireland, 1999. A simple model for evaluating the costs and benefits of macrophytes. Hydrobiologia 415: 335–339.

Verstraelen, P. J. T. J., Wisserhof, L. J. Rodic & R. Eijsink, 1992. Eutrophication control strategies for three shallow Vecht lakes in the province of North Holland. Hydrobiologia 233: 235–245.

Visser, P. M., B. W. Ibelings, B. van der Veer, J. Koedood & L. R. Mur, 1996. Artificial mixing prevents nuisance blooms of the cyanobacterium *Microcystis* in Lake Nieuwe Meer, the Netherlands. Freshwat. Biol. 36: 436–450.

Vollenweider, R. A., 1987. Scientific concepts and methodologies pertinent to lake research and lake restoration. Schweiz Z. Hydrol. 49: 129–147.

Walz, N. & B. Nixdorf (eds), 1999. Shallow Lakes, 1998: Trophic Interactions in Shallow Water and Brackish Water Bodies. Hydrobiologia 408/409: 394 pp.

Wilson, H., 1999. Legislative challenges for eutrophication control in Europe. Hydrobiologia 395/396: 389–401.

Wolff, W. J., 1993. Netherlands-wetlands. Hydrobiologia 265: 1–3

Zaret, T. M., 1980. Predation and Freshwater Communities. Yale University Press, New Haven & London 187: pp.

Hydrobiologia **478:** 107–130, 2002.
P.H. Nienhuis & R.D. Gulati (eds), Ecological Restoration of Aquatic and Semi-Aquatic Ecosystems in the Netherlands (NW Europe).

The restoration of fens in the Netherlands

Leon P.M. Lamers, Alfons J.P. Smolders & Jan G.M. Roelofs
Department of Aquatic Ecology & Environmental Biology, University of Nijmegen, 6525 ED Nijmegen, The Netherlands
E-mail: leon.lammers@sci.kun.nl

Key words: desiccation, eutrophication, fen, peatland, restoration, review

Abstract

The present paper reviews the major environmental problems in Dutch aquatic and semi-terrestrial fens: desiccation, (internal and external) eutrophication, acidification, habitat fragmentation and intoxication. It discusses both the positive and the negative consequences of the restoration measures taken in Dutch fens, and strongly emphasises the biogeochemical and biological processes and factors responsible for the deterioration of plant and animal communities in fens. Only with the knowledge of these key processes and factors are optimal restoration and management measures possible. Finally, important gaps in knowledge are pointed out and a call for new research is made.

Introduction

This paper discusses the restoration of Dutch mineral-rich peatlands with either surface or submerged peat accumulation. Although some fens are brackish (8.5–85 mmol Cl l^{-1}; mainly in the western part of the Netherlands), the majority are freshwater systems. We focus mainly on aquatic and semi-terrestrial fens (for terrestrial fen meadows, see Grootjans et al., 2002). Fens differ from ombrotrophic mires (bogs) in that the biogeochemical processes in their top layer are strongly influenced by the influx of mineral-rich groundwater or surface water (telluric water) (Gore, 1983; Wheeler & Proctor, 2000). Minerals may be supplied either by vertical flow (groundwater discharge, upward seepage), by horizontal flow (supply of surface water or groundwater) or a combination of both. The assumption that this automatically implies meso- or eutrophic conditions is absolutely incorrect. Studies have shown that mineralisation rates and nutrient availability are not necessarily lower in bog plant communities comparing with fen plant communities. The uptake of nutrients, rather than their availability, appears to form the principal constraint for plant growth in bogs (Waughman, 1980; Verhoeven et al., 1988; Koerselman & Verhoeven, 1992; Aerts et al., 1999). The term 'minerotrophic' strictly refers to the mineral richness (in particular of calcium (Ca) and magnesium (Mg)), whereas 'eutrophic' refers to the high availability of primary nutrients (in particular phosphorus (P), nitrogen (N), and potassium (K), and also carbon (C) for some aquatic systems). The key variables for the distinction between both mire types are, therefore, mineral concentrations, acid neutralising capacity (ANC), and often also pH, all variables being generally higher in fens. The ANC in fens is provided by bicarbonate buffering and by the exchange of base cations (Ca and Mg) against hydrogen ions on cation exchange sites in the peat. For an assessment of the ANC, bicarbonate concentrations, total cation exchange capacity (CEC) and base saturation (BS) are, therefore, important. Buffering by bicarbonate in surface and sediment pore water provides the first and most important mechanism against acidification. When almost all bicarbonate has been consumed, e.g. during desiccation, the pH drops to <5 and acids are buffered by cation exchange at binding sites in the soil or sediment. The power of the latter mechanism is determined by the total concentration of cation binding sites (CEC), and the BS, which is the percentage of the CEC occupied by base cations (in equivalents of charge). CEC and BS differ considerably between fen types, and BS is periodically lowered in fens where the water table drops below surface level during summer.

Figure 1. This old school print shows a peat-digger piling up peat blocks on the baulk. Note the long and rectilinear fen ditches, from which peat has been dredged from both sides. (Bos & Kwast, Wolters Publish., Groningen, Den Haag.)

Fens in the Netherlands

In the Netherlands, the nature of minerotrophic peatland areas has been strongly shaped by the extraction of peat, creating a distinctive pattern of turf ponds (called '*petgaten*' in Dutch), and baulks ('*legakkers*') onto which the extracted peat was deposited to dry (Figs 1 and 2). Many dams were blown away by storms, thus creating large shallow lakes called broads ('*wieden*'). Many of the former smaller turbaries got terrestrialised, and the present vegetation is strongly dependent on the type of management (Van Wirdum, 1991; Den Held et al., 1992; Van Wirdum et al., 1992; Wheeler & Shaw, 1995). The typical Dutch peatland type, a semi-natural landscape (Westhoff et al., 1970) created by anthropogenic activity, is comprised of many species-rich plant communities and offers an important habitat to waterfowl. Dutch fens contain many internationally threatened plant communities and species (Vermeer & Joosten, 1992). Like other wetland types, most fen peatlands are under the protection of the National Nature Policy Plan. Due to their unique character, the remaining large fen areas have been recognised as Ramsar Sites, i.e. Wetlands of International Importance. Some sites, such as the National Park *De Weerribben* (Overijssel), were already recognised in 1980 (Frazier, 1999). This stresses the need for conservation and restoration of Dutch fens with high priority, both from a national and international point of view.

Most of the remaining fens in the Netherlands are peatland remnants of the Holocene floodplain, which was once covered by extended mire areas (Pons, 1992). Different successional stages were present, from open water dominated by hydrophytes and helophytes, to floating fens and carr woods. In areas where water was less rich in minerals, like in the contact zones with elevated Pleistocene parts of the country, various types of transitions from fen to bog were present. The characteristic continuum between

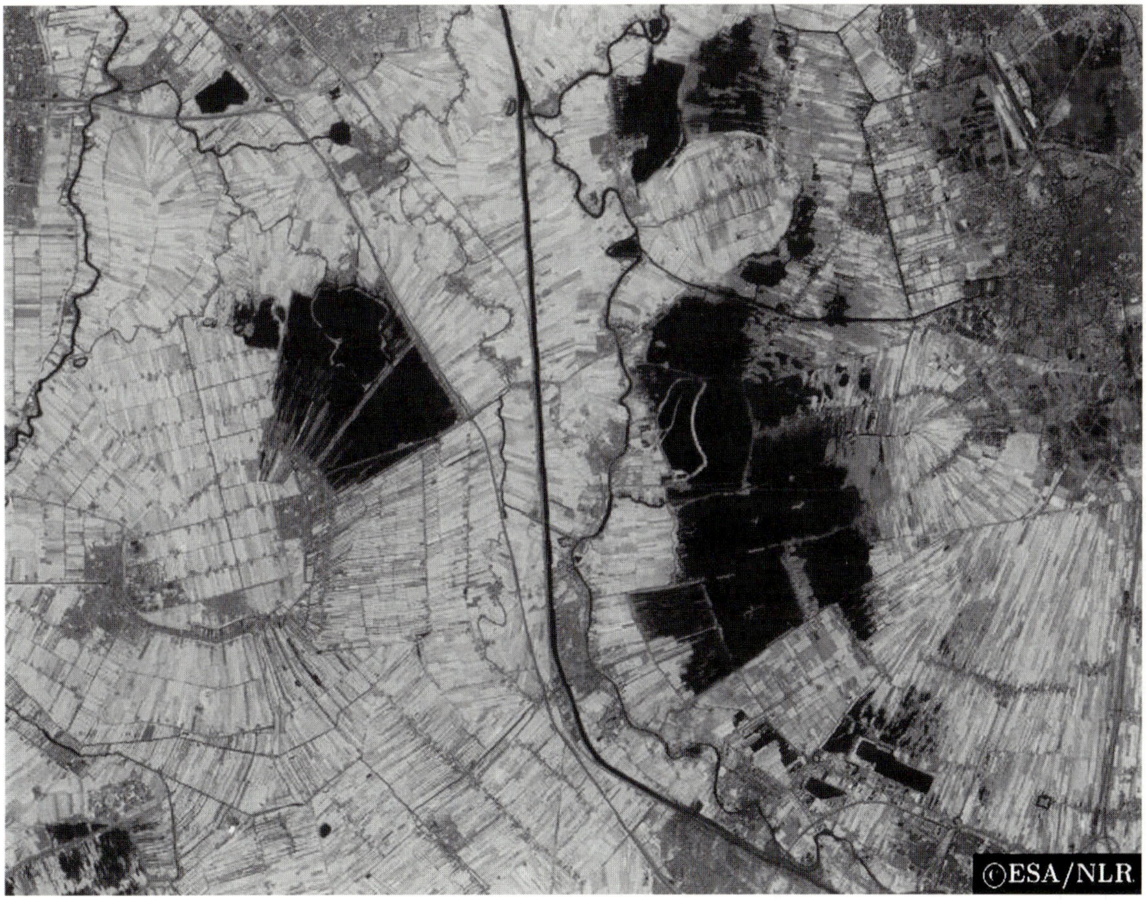

Figure 2. A satellite image of a fen area near Utrecht clearly shows the rectilinear method of peat extraction, which is characteristic for most Dutch fens. The destruction of peat baulks, mostly by storms, resulted in shallow lakes (dark) like the Vinkeveensche Plassen (left) and Loosdrechtse Plassen (right). European Space Agency (ESA) / National Aerospace Laboratory (NLR). Scale 1:155 000.

fens and bogs has now almost entirely vanished in the Netherlands, largely because of anthropogenic activities (see also Roelofs et al., 2002). Floodplain bogs developed in Holocene areas on locations where the accumulation of peat became elevated above the water level and telluric water could no longer reach the top layer (Pons, 1992; Schultz, 1992). Between 0 AD and 1200–1500 AD, however, these bogs were inundated by the rising sea level and the initiation of agricultural land use (Pons, 1992; van de Ven, 1994; van 't Veer et al., 2000). Most of the present fens developed on this drowned bog peat. In areas (formerly) influenced by the sea, like the Northwestern part of the Netherlands, fens developed under brackish conditions. In many Pleistocene areas, small fens developed due to the seepage of calcareous water from higher plateaus, or because of flooding or infiltration of riverine water. At present, small fragments of these peatlands still exist, scattered over the Pleistocene regions. Both in the Holocene lowlands and in the Pleistocene brook valleys, fens also developed on reservoir basins ('boezems' in Dutch), used to store excess water during winter. After they lost their storage function, valuable plant communities (fen meadows) developed in these areas flooded in winter.

Almost all recognised fen reserves are in the Holocene part of the Netherlands. Their area ranges between 200 ha and 600 ha, a few reserves being larger 1300–4500 ha (e.g. De Wieden and De Weerribben in Overijssel, Alde Feanen in Friesland). In addition, there are many small fens in both Holocene and Pleistocene areas, generally being up to 10 or 20 ha but still containing an important portion of the plant communities threatened in the Netherlands.

Due to the wide diversity in habitats, both terrestrial and aquatic, and the high availability of food,

Figure 4. The use of allochthonous water to maintain high water tables during summer caused eutrophication in this peat extraction pond. Filamentous algae dominate, while some *Stratiotes aloides* plants can barely persist (inlay). Photo L. Lamers.

1997). Moreover, desiccation itself generates acidification through the chemical oxidation of Fe and sulphide, generating sulphuric acid (Lamers et al., 1998a). The drop of the pH to values around 4 leads to an enormous increase in the mobility of potentially toxic metals, including heavy metals in the soil. The aluminium concentrations measured in acidified areas are shown to be toxic to rare species in litter fens (e.g. *Cirsium dissectum*). This toxicity can be counteracted by simultaneously raised Ca concentrations (de Graaf et al., 1997). The effect of desiccation or isolation is therefore twofold: a decrease of the pH (and concomitant increased 'free' metal concentrations) and a decrease of the concentration of detoxifying Ca. A hydrological problem specific to brackish fens is desalinisation, caused by altered agricultural water management and by the damming of the former Zuiderzee by the Aflsluitdijk, which has converted the inland sea into a freshwater lake. As a result of this, the chloride concentrations have dropped below 28 mmol Cl^{-} l^{-1}, a value considered to be the minimum for brackish fen development (R. van 't Veer, pers. com.).

In freshwater fens, however, acidification is not the only reason for undesired species replacement. *Scorpidium scorpioides* is, for instance, quite capable of surviving and growing under mineral-poor conditions. Nutrient availability is also an important determinant of the direction of vegetation changes. *Sphagnum squarrosum*, for instance, can efficiently utilise nutrients in eutrophic environments, making high growth rates possible. In species replacement, therefore, both water quality changes and interspecific competition seem to be involved (Kooijman & Bakker, 1995). Similar changes, generally regarded as problems, are also caused by the isolation from mineral-rich surface water. This isolation may, however, be caused by both anthropogenic and natural processes. As with desiccation, the minerotrophic vegetation loses the supply of vital minerals and, concomitantly, soil ANC decreases.

Desiccation may stimulate decomposition and mineralisation. Grootjans et al.(1986) have shown that the N availability increased due to long-term desiccation. In contrast, the P availability decreases because the capacity of the soil or sediment to bind P strongly increases during desiccation due to the oxidation of Fe. These opposing effects of drought on the availability of both nutrients, in addition to direct drought effects, cause adverse shifts in the vegetation composition. Characteristic sedges disappear and grasses like

Holcus lanatus and *Agrostis canina* take over (Grootjans et al., 1986). Acidification during desiccation, e.g. by the oxidation of iron sulphides can, however, tone down or even nullify this stimulation of decomposition and mineralisation (Lamers, 2001).

The eutrophication of surface and/or groundwater poses another severe threat to fens. Nutrient influx from agricultural areas and sewage has led to a strong increase of PO_4^{3-} and NO_3^- availability. To compensate for the shortage of water in nature reserves and agricultural areas, water from the rivers Rhine and Meuse is (directly or indirectly) used on a large scale (Roelofs, 1991). Aquatic communities lose their submerged plants and become dominated by fast growing, lemnid species and macro-algae (Fig. 4). Thus, vast fen areas (ditches, pools, lakes) show a strong decline of their biodiversity and change into species-poor plant communities. Likewise, (semi-) terrestrial fens loose many species. Highly competitive species like *Glyceria maxima, G. fluitans, Phragmites australis* and *Schoenoplectus (Scirpus) lacustris* become dominant. This type of eutrophication, in which nutrients are imported, is termed 'external eutrophication' to distinguish it from 'internal eutrophication' caused by internal mobilisation of nutrients. The latter has also been reported to lead to fen deterioration. Water shortage in Dutch fen peatlands is generally offset by the use of riverine water, having a higher net alkalinity. In fens containing moderately buffered peat, the influx of alkaline surface water has been shown to increase the decomposition of organic matter, by neutralising the organic acids in organic particles (McKinley & Vestal, 1982; Brock et al., 1985; Kok & van de Laar, 1991; Smolders, 1995; Brouwer et al., 1999). Thus, overall mineralisation is stimulated, leading to eutrophication. In most fens, the stimulating effect of Ca (as bicarbonates and carbonates) on decomposition and consequently on P-mineralisation appears to completely nullify any possible effects of Ca on P binding, at the pH prevailing (Fig. 5; Roelofs, 1991; Smolders, 1995; Brouwer et al., 1999).

Moreover, river waters are characterised by relatively high concentrations of sulphate because of natural weathering of sulphate containing rocks, anthropogenic dumping and sulphur runoff from agricultural areas. Groundwater and surface water are sulphate-enriched by desiccation and by NO_3^- pollution. In the first process, sulphate is mobilised from iron sulphide (FeS_x) deposits by oxidation and, in the second, NO_3^- is used by denitrifiers to oxidise sulphides to sulphate (Lamers et al., 1998a, 1999a). In freshwater

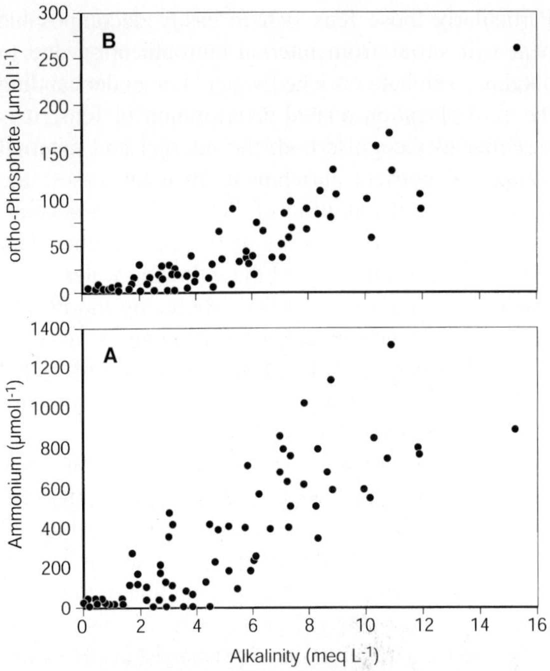

Figure 5. Relationship between alkalinity and (A) NH_4^+ or (B) PO_4^{3-} concentrations in sediment pore water for 70 randomly selected fen sediments in the Netherlands. From: Smolders 1995.

systems, microbial sulphate reduction is generally limited by the availability of sulphate. Increased sulphate concentrations in fens will, therefore, stimulate sulphate reduction. As the consumption of organic acids is accelerated, overall decomposition will also be stimulated. In addition, the sulphide formation resulting from sulphate reduction generates alkalinity, stimulating the decomposition further (as explained above). The sulphide formed binds to iron (hydr)oxides in the sediment forming iron sulphides such as FeS_2 (pyrite) and FeS. As a result, PO_4^{3-}, the major limiting nutrient, is released from Fe~P compounds and diffuses into pore and surface water (Boström et al., 1982; Caraco et al., 1989; Roelofs, 1991; Smolders & Roelofs, 1993, 1995; Koerselman & Verhoeven, 1995; Lamers et al., 1998b; Beltman et al., 2000) (Fig. 6). This sulphate-driven internal eutrophication is a general process in systems that are limited in energetically more favourable electron acceptors, like oxygen and NO_3^-. Therefore, sulphate pollution of groundwater also leads to eutrophication in waterlogged or flooded fen meadows (Jansen & Roelofs, 1996; Lamers et al., 1998b). The response appears to be regulated by the availability of readily decomposable organic matter, indicated by the peat profile (Kemmers, 1996).

Particularly those fens rich in easily decomposable peat will suffer from internal eutrophication due to alkaline, sulphate-enriched water. For understanding the eutrophication-related deterioration of fens, it is essential to recognise both the internal and external sources of nutrient enrichment. In many cases, the increase in concentration of PO_4^{3-} due to accelerated mineralisation appears to be much higher than the PO_4^{3-} concentration of the inflowing waters (Koerselman & Verhoeven, 1995). Reducing the PO_4^{3-} influx, for instance, by chemical stripping of the nutrient (dephosphatising) or by flow through a constructed wetland or an extended supply route, is in this case insufficient to prevent eutrophication.

Nutrient availability in fens is in several ways linked to pH, both biologically and chemically. Increasing the alkalinity and the pH will, as explained above, stimulate decomposition and mineralisation in slightly acidic fens. The chemical binding of PO_4^{3-} is closely related to the pH of the peat. The binding of PO_4^{3-} to Fe and aluminium is strongest at pH values around 6. Binding to calcium as $CaHPO_4$ (optimum at pH 7–8) is of lesser importance, but the binding of PO_4^{3-} as apatite ($Ca_{10}(PO_4)_6(OH)_2$) strongly increases above pH 6.5–7 (Stumm & Morgan, 1981). When pH decreases to about 4.5, PO_4^{3-}-binding stays more or less equal, but a decrease in the pH to 4 or less will dissolve metal phosphates (in the Dutch fens mainly Fe∼P) and increase P availability. At the same time, free NH_4^+ concentrations increase due to exchange with protons at cation exchange sites in the peat. Therefore, the chemical nutrient binding capacity in fens will change significantly only at pH <4 or at pH >7. Indeed, Beltman et al. (1996a) found that both N and P-availabilities in fens abruptly increased below pH 4.1.

A much less known causal factor for the severe deterioration of fens is the enforcement of highly stable water levels. In more natural situations, as in the past, water levels fluctuated throughout the year, being lower in the summer and higher during winter. Current water level regimes in the Netherlands tend to be the opposite: lower winter levels to enable rapid runoff of access water from agricultural land, and relatively high and stable summer levels provide water for growth and evapo-transpiration. In most cases, this unnaturally high summer level can only be maintained when allochthonous river water is used. As stated earlier, this may often lead to internal eutrophication because of enrichment with bicarbonate and sulphate. Another problem that is caused during the high water level is

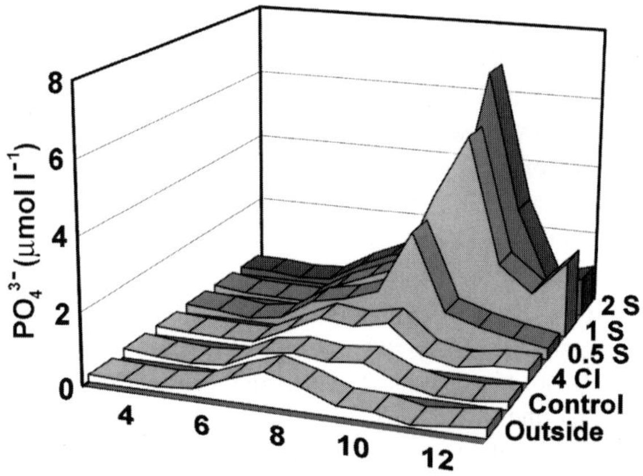

Figure 6. Enclosure experiment in a fen ditch, in which either 4 mmol l^{-1} chloride (4 Cl), 0.5 mmol l^{-1} sulphate (0.5 S), 1 mmol l^{-1} sulphate (1 S), or 2 mmol l^{-1} sulphate (2S) has been added to the surface water. All sulphate treatments cause P eutrophication of the surface water (vertical axis), while the control enclosures (Control) and chloride treatment show the normal development of the P concentration throughout the year as measured outside the enclosures (Outside).

mobilisation of PO_4^{3-} Fe∼P compounds when oxygen becomes depleted and Fe is reduced. Temporal lowering of the water table, which is natural for summer, leads to the oxidation of a greater part of Fe and concomitant stronger PO_4^{3-} binding. In carrs where the water table is raised and artificially maintained high during summer, PO_4^{3-} concentration increase considerably. This leads to a complete domination of the vegetation by *Lemna* spp., at the expense of the original vegetation (A. Boxman & E. Lucassen, pers. com.). In contrast, high water tables during winter will not lead to internal eutrophication, because of the lower microbial activity in winter. Although the hydrology of many fens in the Netherlands has always been a consequence of human activity, we plead for applying a more natural water table management in fens: somewhat lower water tables during summer and higher tables during winter. An additional advantage is the improved conditions (oxygen and light; pers. obs.; Coops & van der Velde, 1995) for germination of submerged plants and helophytes. Whether or not this more natural regime is feasible, will depend on the plant communities involved and hydrological considerations other than nature management. A potential risk of water table reduction seems to be the effect on mineralisation rates. Although constant long-term desiccation (several years) may indeed lead to increased availability of N but not of P (Grootjans et al., 1986; Berendse et al., 1994; Oomes et al.,

1997), short-term lowering of the water table (months) in fen peat showed hardly any effect on N availability (Lamers, 2001). Moreover, Oomes et al. (1997) found increased N mineralisation only after 2 years of constantly lowered water table (30 cm below surface level) in fen soils. The application of a more natural hydrological regime in the management of fens requires, however, further research.

In addition, the increase of phytotoxic compounds have apparently become increasingly important in the malfunctioning of fens. Free sulphide (H_2S, HS^-, S^{2-}) is one of the most important biogenic phytotoxins (Koch et al., 1990) and its concentrations have invariably increased due to sulphur pollution in freshwater fens. Even low levels of sulphide harm the roots of *Statiotes aloides* (Smolders & Roelofs, 1996) and decrease the biomass of e.g. *Carex* spp (Lamers et al., 1998b). The competitive advantage of sulphide-resistant, fast growing species will, therefore, increase considerably. Field observations suggest that this may stimulate the formation of monotonous stands of species like *Glyceria maxima*. Free Fe, capable of sequestering toxic sulphide by the formation of iron sulphides, is an important modifier of the sulphide toxicity (Smolders et al., 1995). Loss of groundwater discharge rich in Fe due to desiccation may lead to sulphide toxicity. On the other hand, free Fe, abundant in fens that receive anaerobic groundwater, may itself be toxic to non-characteristic fen plants (Lucassen et al., 2000). Some fen communities, therefore, seem to be governed by toxicity rather than by nutrient availability.

The decline of fen-dwelling fauna has to be attributed to biotope destruction, eutrophication, habitat fragmentation and even toxification (e.g. pesticides and heavy metals). This not only applies to mammals like *Lutra lutra* (Otter) and *Microtus oeconomus* (Northern vole), but also to characteristic birds like *Acrocephalus arundinaceus* (Great reed warbler) and *Botaurus stellaris* (Bittern) (Jefferies, 1988; Kruuk, 1995; van Turnhout & Hagemeijer, 2001). Both bird species, for instance, need extensive *Phragmites australis* zones along fen waters as a habitat. There has, however, been a strong decline of *Phragmites* in recent decades (van der Putten, 1997). Butterflies, such as *Lycaena dispar* (Large copper) and *Clossiana selene* (small pearl-bordered fritillary) depend on very specific vegetation structures within their habitat range, where host plants not only offer opportunities for feeding, but also for oviposition and shelter (Bink, 1992). Extended floating vegetation of *Stratiotes aloides* is needed by the dragonfly *Aeshna viridis* (Green hawker) for oviposition, and the strong decline of this species has been attributed to the decline of healthy populations of its host plant (Higler, 1977). In addition to the above changes, eutrophication of surface waters has also had a negative impact on food availability (e.g. macrofauna) for several bird species and other carnivorous species (van Turnhout & Hagemeijer, 2001). In conclusion, the decline of fen-dwelling fauna has been caused by the strong decline of fen area in the Netherlands and its fragmentation. Plant communities and vegetation structure have become unfit for critical species, and the deterioration of water and sediment quality has had both direct (toxicity) and indirect negative effects (food quantity and quality).

Restoration objectives

In all cases, restoration aims at the recovery of original plant communities and their fauna, often semi-natural, as known from the times before undesirable anthropogenic disturbances such as desiccation or eutrophication. Restoration should, as Grootjans & van Diggelen (1995) stated, 'be aimed at restoring the fen system, not at "restoring" fen species'. Note that restorers do not use a 'pre-human' concept for most fen types, but rather a cultural-historical concept, using the landscape of the 19th century or the beginning of the 20th century as a benchmark. The idea that not only (so-called) natural, but also semi-natural systems in the Netherlands are well worth conserving, was initiated by Westhoff in 1945. The concept implies the re-establishment of traditional management. This type of restoration, in which vegetation redevelopment forms the main objective, can be termed renaturation (Wheeler & Shaw, 1995). The restoration of peat accumulation, i.e. a 'true' system restoration approach (regeneration), is generally not the main goal, but will often be achieved through vegetation development. If, however, restoration implies resetting the (natural) succession, e.g. by sod cutting, restoration will even counteract regeneration.

As desiccation is often the principal cause for the decline of fens, it is not surprising that efforts to recover these systems generally begin with the restoration of the 'original' hydrology. 'Original' generally refers to the conditions thought to be optimal for the restoration of the *desired* plant communities. Which communities are considered *desired* depends not only on historical frames, but often also on the chosen restoration target. This may even lead to conflicts, e.g. between botanists and bird conservationists, as both groups do not always strive for the same restoration goals. Monotonous reed stands or *Salix* shrubs can be very valuable for the restoration of bird populations, but generally not from a vegetation perspective (Wheeler & Shaw, 1995).

Water quality is just as important as the appropriate water quantity and level. Most fens require a certain mineral enrichment, and the groundwater or surface water supply of these areas should provide the systems with adequate buffering against acidification. The level of buffering has to be very precise, as changes in ANC will lead to changes in the vegetation. In addition, the nutrient flux (determined by nutrient concentration and water flux) should be within the ranges required by the vegetation concerned. For submerged plant communities, it needs to be ensured that the water quality enables appropriate light conditions. The complete coverage of the water surface with floating-leaved species or the massive development of algae in eutrophic fens will obstruct all light. It follows that while hydrological restoration is possible, high nutrient runoff from arable land makes it difficult to restore a fen. As stated above, the recovery of flow of anoxic water that carries Fe might be an additional prerequisite for the recovery of carrs and other fen types. In general, it is easier to restore the local hydrological systems than the regional systems, because in the latter case large areas of land and therefore many different parties are involved. In the Netherlands with its extremely intensive forms of land use, it is almost always an arduous task to raise groundwater tables in the surrounding agricultural areas.

In eutrophied fens, it is essential that the nutrient concentrations of the inflowing surface water or groundwater are minimised so as to combat eutrophication. Internal eutrophication can be alleviated by employing hydrological alternative for the allochthonous, alkaline and sulphate-enriched river water. In (semi-)terrestrial fens that have been heavily fertilised in the past, it may even be necessary to remove the strongly eutrophied top layer, in order to expose the deeper and cleaner peat layers as a substratum for fen vegetation. High concentrations of toxic metals in the soils and sediments will strongly increase the costs of these measures.

In summary, fen restoration aims at the recovery and conservation of characteristic, often semi-natural fen systems (flora and fauna), by restoring the hydrology, hydrochemistry and the sediment considered optimal for the fens under restoration.

Successful restoration

For the development of submerged plant communities, dominated by Characeae, clear water is an essential requisite. This condition may be achieved via a bottom-up approach, leading to a significant reduction of the nutrient availability, or alternatively by a more top-down oriented approach like biomanipulation (Shapiro et al., 1975; Jeppesen et al., 1990; Scheffer et al., 1993; Hosper, 1997; Gulati & van Donk, 2002). The bottom-up approach will be discussed below. Biomanipulation is defined as the deliberate exploitation of the interactions between components of the aquatic ecosystem, in order to reduce the algal biomass (Shapiro et al., 1975). The main aim is to increase zooplankton (mainly larger-bodied *Daphnia* spp.) grazing on phytoplankton and concomitantly change the turbid- water state to an alternative clear-water state, even though the nutrient concentrations are similar (Scheffer et al., 1993). To achieve this, the system has to be 'pushed' through the hysteresis effect that prevents an easy transition from one state to the other. Once a submerged vegetation has established, it helps to maintain the clear-water equilibrium by providing a habitat for zooplankton, preventing resuspension of sediment particles, competing with algae for nutrients and depressing algal growth by the excretion of allelopathic substances (Scheffer et al., 1993; van Donk & van de Bund, 2002). In many shallow lakes, the desired change to clear water has been achieved by a drastic reduction of the planktivorous and benthivorous fish stocks (Shapiro et al., 1975; see several papers in Gulati et al., 1990; Jeppesen et al., 1990; Moss, 1998; Meijer et al., 1999; Gulati & van Donk, 2002). This amounts to a reduction in the standing crops of planktivorous and benthivorous fish to 10–15 kg ha^{-1} and 15–25 kg ha^{-1}, respectively (generally corresponding to a minimal reduction of 75%). In addition, piscivorous fish as *Esox lucius* (Northern pike) and *Stizostedion lucioperca* (Pikeperch) may be (re-)introduced. In smaller turbaries and fen lakes, however, biomanipulation in surface waters of fens and marshes appeared to be far less successful compared with other waters in the Netherlands (Meijer et al., 1999; Meijer, 2000). Only in a few cases, as in lake Duinigermeer (Overijssel), a clear-water equilibrium could be established. In this lake, extensive

Chara communities developed after removing about 75% of the fish stock. In a few other fen lakes, the decrease in turbidity proved to be insufficient for colonisation by submerged plants. The causes for failures of biomanipulation as a restoration technique, some of these specific to peatlands, will be discussed in the next section ('Failures').

Unlike aquatic fen communities, (semi-)terrestrial fens are generally threatened by acidification. When the groundwater tables and fluctuations characteristic for the plant communities concerned can be restored, the restoration of acidified fen meadows is often very successful (de Graaf et al., 1994; Jansen et al., 1996; Grootjans et al., 2001). This implies the 'recharge' of the CEC in the rhizosphere with base cations like Ca during the winter, restoring the base saturation and thus the acid neutralising capacity. Litter fens, for instance, recovered when groundwater could reach the rhizosphere again during winter and early spring. When the locations were strongly dominated by graminoids due to eutrophication, sod-cutting (5–10 cm depth) was necessary in order to remove excess nutrients and create chances for the diaspore bank (de Graaf et al., 1994; Jansen & Roelofs, 1996; Jansen et al., 1996; Grootjans et al., 2001). Fast growing, highly competitive graminoids, as *Holcus lanatus, Elymus repens* and *Calamagrostis canescens*, are removed by this measure, and already after one year, characteristic and endangered species like *Cirsium dissectum*, *Parnassia palustris* and *Pinguicula vulgaris* may return. A prerequisite, however, is the occurrence of moderate desiccation of the top layer during summer. The accumulation of NH_4^+ under anaerobic conditions seems to cause problems for species such as *Cirsium dissectum* that prefer NO_3^- as N source (Jansen & Roelofs, 1996; de Graaf et al., 1998).

In fen meadows on mineral soils that only suffered from eutrophication, sod cutting proved to be an adequate restoration measure. If acidification is also a concern, restoration will only be successful if measures are taken to restore the ANC of the top layer. The same processes hold for floating fens, in which sod removal leads to a significant reduction of the availability of both N and P (Beltman et al., 1996ab; Fig. 7). Mineralisation rates for both these nutrients showed the same pattern. However, to prevent rapid re-growth of an undesired *Sphagnum-Polytrichum* carpet after a few years, it appeared necessary to construct a small drainage ditch to remove excessive rainwater. This method corresponds with the restoration of ANC by hydrological management in litter fens and other semi-terrestrial fen types. Loss of ANC due to the isolation from telluric water has been successfully combated by this renewed access to the floating peat of base-rich water, for instance in Ilperveld (N-Holland; brackish fen) and in Westbroek (Utrecht; freshwater fen). However, the construction of small ditches through floating peat is only effective for renewed 'mineral reload' if there is a relatively dense network, accompanied by the construction of shallow trenches to remove excess rainwater (Bootsma, 2000). Contrary to the general assumption, the rainwater is barely or not acidic in the Netherlands (due to the high ammonia emission rates), and the acidifying effect of extra nitrification is minimal in the buffered fens. The greatest problem is that the rainwater body on top of the telluric water mass prevents the efficient infiltration of base-rich water to the top layer of the peat. The re-establishment of base-rich water infiltration can only succeed if the intruding water does not contain or generate high nutrient concentrations in the rhizosphere.

The direct application of lime to offset the acidification appeared adequate in restoring alkalinity and pH, but only if the topsoil were removed (Bootsma, 2000). However, base saturation did not increase. Liming without sod-cutting showed no effect on alkalinity or pH, because of rapid reacidification due to the *Sphagnum* mat. Given the expensive and complicated nature of sod cutting, trench digging, or liming, or both, it is unlikely that these measures will be applied to large deteriorated fens. The only alternatives seem to either allow flooding (which is impossible for floating fens) or allow a succession towards more ombrotrophic plant communities or carr woods. Flooding of fen systems may be an appropriate measure to keep the ANC high enough, although the intrusion of surface water into the peat often appeared to be too shallow for sufficient recharge of the base saturation (Vermeer & Joosten, 1992; De Mars et al., 1996; Beltman et al., 2000). It, however, is accompanied by the high risk of (internal) eutrophication. This risk is expected to be low if the intruding water is mesotrophic and does not lead to internal eutrophication, or if a broad buffer zone ('natural helophyte filter') is present, like for instance in Polish river fens. Fens that have developed in the past in flooded areas can only have become mesotrophic under one or both of these conditions. At present, however, most Dutch fens are small fragments fed by alkaline, eutrophic surface water, which generally contains much higher concentrations of SO_4^{2-}, leading to internal eutrophication. The experience with a successful restoration of mesotrophic,

Figure 7. Restoration of quagfen vegetation by the mechanical removal of the top layer. In this way, succession is reset. Photo: B. Beltman.

semi-terrestrial fen ecosystems on thick peat layers is, however, still limited, and therefore more work is needed.

Eutrophication caused by external nutrient loading can only be prevented by reducing the concentrations of the influx water, or by using a less nutrient-rich water source. The former can be achieved by chemical removal of nutrients (e.g. PO_4^{3-}-stripping using Fe compounds) or by biological removal with helophyte filters. In many cases, however, this is not enough to prevent further eutrophication. High bicarbonate and sulphate levels in the influx water promote internal eutrophication in fens, fen lakes and ditches. Strong preference is therefore given to the restoration of the former hydrology, i.e. with reduced use of allochthonous water and wherever feasible by amelioration of groundwater seepage. This may be achieved by establishing a more natural water-table regime (see below) and if possible by increasing water tables in the environs. For lakes with high PO_4^{3-} loading rates from the peaty sediment to the water layer, it may even be necessary to partly or wholly remove this nutrient source by dredging (compare restoration measures for heathland pools: Roelofs et al., 2002). This has been carried out in fen lakes, e.g. Molenpolder near Utrecht.

The addition of Fe to sediments in order to bind PO_4^{3-} shows a strong de-eutrophication effect in both lakes with mineral sediment (Boers, 1991) and in fens (De Bruuk in Gelderland; Smolders et al., 1995). If, however, there is a constant supply of SO_4^{2-} enriched water, the response is only transient because Fe consumption is extremely high. In such a case, Fe addition would both be very laborious and costly and therefore not recommended as a general, large-scale restoration measure (Smolders et al., 1995). The addition of lime in order to bind phosphate will not be effective, because of the strong stimulation of P mineralisation in fen peat (see above), counteracting the possible effect of chemical binding of P (Roelofs, 1991; Smolders, 1995).

In fens that have become terrestrialised to a eutrophic Alder carr or reed vegetation, efforts have been made to restore species-rich, earlier successional stages by renewed peat removal. The Dutch State Forestry Service has even developed a boat specially designed to create new peat holes (Fig. 8). The conversion of carr to open water is, however, the most expensive measure employed in fens, amounting to euro 11 000 per ha. Given the period needed for succession to semi-terrestrial vegetation (20–30 years;

Figure 8. The '*Kraggenvreter*' ('Scragh Wolf'), a floating device developed by the State Forestry Service to create new peat holes in fen peatlands. Peat including its vegetation is removed by the machine's harvester arm, mixed with water and piped to a depot. Inlay: the wolf's head (harvester arm) busy under water. Photo: L. Brouwer, State Forestry Department De Weerribben.

Bakker et al., 1994), however, this measure is fairly cheap compared with the costs of annual mowing and harvesting.

On the fertilised meadows acquired by nature conservation agencies or by the state, by temporarily keeping the water level low, basins were (re)created using draglines and bulldozers. After refilling with water, succession began again within a few years with aquaphyte- and helophyte communities (Vermeer & Joosten, 1992; van den Broek & Beltman, 1995; Beltman et al., 1996). It is, however, not clear whether the conditions are appropriate for the formation of mesotrophic floating fens. In many case, eutrophication due to nutrient influx from the present or former agricultural land, or by rapid nutrient mobilisation from the sediment, impedes succession towards this type of mesotrophic fens.

For plant communities that require human interference (mostly mowing), nature management needs to mimic the cultural activities that were once common in these types of peatlands. The other, relatively less expensive, possibility is to leave it to nature and allow succession towards plant communities that are often, but not always, considered less interesting by botanists. The rather monotonous plant communities as *Phragmites australis* stands or *Salix* shrubs can be very important habitat for wetland fauna like waterfowl. In Westbroek (Utrecht) *Salix* fens have been conserved to protect nesting facilities for *Luscinia svecica* (Bluethroat), whereas *Alnus* trees have been removed.

Although the restoration of the original (seminatural) hydrology, both at the landscape level and at the habitat level is preferred, a more artificial hydrological system may also provide suitable conditions in some situations for which this option is not possible (Boeye et al., 1995; Wassen et al., 1996). An example of this is the seepage of canal water through a bank to the adjacent peatland. In brackish fens, restoration of the original hydrology refers to the recovery of brackish water influx, which is generally hampered by the agricultural use of adjacent land.

A rather new viewpoint is to consider the development of bog vegetation in fens by natural succession, like in some floating fens of De Stobbenribben (Overijssel; G. van Wirdum, pers. com.; Schouwenaars et al., 1997). The traditional viewpoint was to see this vegetation development as undesirable, and to counter it by mowing and *Sphagnum* removal and by digging new ditches to allow the infiltration of telluric water. The development might erroneously be interpreted as anthropogenic acidification, caused by the high input of acidifying deposition. Unlike acid-sensitive ecosystems (e.g. heaths, moorland pools), fens have a high

ANC that is constantly or regularly 'recharged', and the development of acidophilic plant communities often reflects a natural isolation from telluric water. It has become clear that characteristic hummock-hollow communities in fens develop much faster than in many (expensive) restoration projects on cutover bogs. A new hypothesis concerning this rapid development is the supply of carbon dioxide from deeper peat layers, which is much higher in peat influenced by calcareous water (Lamers et al., 1999b). This may induce the rapid formation of *Sphagnum* hummocks because the growth appears to be C (co-)limited.

Until now, hardly any research has focused on the restoration of fauna populations. A rule of thumb for the optimal development of fen fauna seems to rests on a differentiated and diverse type of landscape, providing different habitats for all different animal groups (see above; H. Esselink, pers. com.). Fen restoration has, however, in most cases only focused on the restoration of plant communities, rather than on the restoration of animal groups and their food webs. In order to ensure an ecosystem approach, future research should pay more attention to both plant and animal communities and foodweb aspects.

Failures

Unfortunately, the cases of literature reporting restoration failures are (for obvious reasons) scarce. In contrast to other lake types, the reduction of turbidity in fen lakes by external or internal measures appeared to be often unsuccessful. Water remained turbid or became turbid again quickly after taking the measures, and submerged plant communities did not develop. In hypertrophic lakes, turbidity is the only possible stable situation. Biomanipulation will only work after PO_4^{3-} concentrations are reduced to a concentration range in which two alternative stable states are possible (Scheffer et al., 1993). The threshold value seems to be around 7 μmol total P l^{-1}, although shallow lakes may become clear even at much higher concentrations (Jeppesen et al., 1990; Klinge et al., 1995). This can be attributed to the strong competition of submerged macrophytes with phytoplankton for P, as explained earlier. Poor success with biomanipulation in fens well within the range of two alternative stable states is often caused by insufficient reduction of the fish stock, or by massive fish remigration from small interconnected ditches or through fish exclosures (Meijer, 2000). Especially in fens comprised of a large network of interconnected peat excavation pools and ditches (which is the general peat excavation pattern in the Netherlands), this is expected to become a problem. So far, however, no information is available on these fen types. In large lakes, resuspension of sediment by wind action may induce turbidity; this particularly holds for lakes with peaty sediments. Finally, mass development of cyanobacteria like *Plankthiotrix* spp. in hypertrophic waters might hamper restoration, because they can form inedible and toxic colonies that complicate grazing by *Daphnia* (Gliwicz, 1990; Laurén-Määttä et al., 1997). Also lumping of the filaments will interefere with their grazing by *Daphnia* spp., in addition to their poor food value for the growth of the daphnids (Gulati & van Donk, 2002). Only individual cyanobacteria and small colonies can be, however, consumed by *Daphnia*, in contrast to larger colonies (E. Van Donk, pers. com.). In the evaluation of biomanipulation projects in the Netherlands (Meijer, 2000), however, evidence collected for the edibility, or to the contrary, of these filaments by the daphnids is inadequate (R.D. Gulati, pers. com.).

A major problem in Dutch fens is the fact that terrestrialisation of open water rarely occurs (Verhoeven & Bobbink, 2001). In hypertrophic waters, the low density of macrophytes most likely hampers peat formation. However, in mesotrophic and eutrophic waters the absence of succession towards terrestrial fens remains unclear probably because the rate of peat formation in open water is low. Peat formation rates in the aquatic phase and in floating mats in fens near Utrecht (Molenpolder and Westbroek) were one-quarter to one-half, respectively, of those in semi-terrestrial fens (1 kg m^{-2} yr^{-1}; Bakker et al., 1997). The high peat accumulation rate in semi-terrestrial fens explains its short 'turnover time' of 12 years, as determined by GIS interpretation of aerial photos (Bakker et al., 1994; Verhoeven & Bobbink, 2001). The rapid formation of carr woods can only be counteracted by mowing and hay removal.

An additional explanation may be the absence of keystone species like *Stratiotes aloides* and *Calla palustris,* initiating terrestrialisation by the formation of floating mats. This is caused by the high degree of fragmentation of fens in the Netherlands, which complicates the dispersal of diaspores of keystone or threatened species. The old-fashioned statement that 'everything is everywhere and nature selects' is rather oversimplified. Characteristic and threatened species are often absent in the diaspore bank of the restoration area, and the closest remnant populations are many kilometres away (T. van den Broek, pers. com.). The interconnection of scattered grazing areas, like pastures in fens, by flocks of sheep (Poschlod, 1995) has disappeared nowadays, although modern agricultural machines may inadvertently have taken over a small part of the cultural diaspore dispersal. Tackling this problem requires either new corridors connecting nearby populations (risking eutrophication for wet corridors) or active re-introduction of plants, seeds or spores. Experiments with the re-introduction of seeds and plants of the terrestrial species *Succissa pratensis* have shown positive results, provided the environmental conditions are restored first (P. Vergeer, pers. com.). If the donor population was small, however, germination levels proved to be lower, indicating possible inbreeding effects. In contrast, efforts to re-introduce aquatic macrophytes showed little or no success, often due to algal blooms (several managers of fen reserves, pers. com.). Therefore, re-introduction of diaspores should only be considered after the appropriate water quality has been restored. Fragmentation is also one of the main problems for fen-dwelling fauna. For this reason, corridors have been constructed between fen reserves and some species, like *Lutra lutra*, *Clossiana selene* and *Lycaena dispar*, have been or will be actively re-introduced.

Unlike aquatic fens, terrestrial and semi-terrestrial fens often cope with acidification. In projects where restoration of the ANC was implemented in eutrophic water, base saturation was restored but nutrient availability became too high simultaneously. The vegetation developed from acidophilic to minerotrophic but also to eutrophic – which, of course, was not the restoration goal. This is also often the case with the restoration of fens by renewed peat extraction. If initial conditions are eutrophic due to nutrient loading from the environs or high nutrient concentrations in the remaining sediment, restoration of mesotrophic floating fens is extremely difficult. It may be well worth to consider the conservation of carr woodlands rather than the employment of such radical restoration measures in these situations.

If surface water is used to compensate for the lowering of the groundwater table, problems may arise due to Fe depletion. As surface water is generally oxygenated, it contains only low concentrations of Fe. The high Fe input via anoxic groundwater in many fens plays a very important role in the binding of PO_4^{3-}, maintaining a low PO_4^{3-} availability to the vegetation. The use of surface water as a substitute will therefore inevitably result in (internal) eutrophication, even if the PO_4^{3-} concentrations in the water are low e.g. by P stripping methods. As the infiltration of surface water into fen soils appears to be very shallow (Koerselman et al., 1990b), large areas of fen peat can become influenced by the adverse water quality. In peatlands that have developed in the past under the influence of surface water, rather than groundwater, the situation is different. If the surface water is not eutrophicated and will not lead to internal eutrophication, or if the buffer zone is broad enough, mesotrophic fens may well develop and thrive after restoration.

Efforts to restore the base saturation in fens by applying lime have so far been not always successful (van Diggelen et al., 1996; Beltman et al., 2000). The major cause for this seems to be a strong acceleration in the rates of decomposition, and thereby mineralisation and eutrophication, of the peat (see Fig. 5). Liming has already for a long time been used for the purpose of raising the nutrient availability. In medieval times, monks applied lime to fish ponds to increase the production. Arable land and compost piles are limed for the same reason. Liming of fens may easily result in the development of a vegetation dominated by common, fast growing species such as *Juncus effusus* and *Glyceria maxima*.

A restoration measure that has unfortunately been employed in several fen types is the establishment of a high water table throughout the year. This artificial situation is generally far from the original state. Already within one year, the water in Alder carrs was completely covered by *Lemna* spp., while characteristic species like *Caltha palustris* and *Calla palustris* were drowned. PO_4^{3-} concentrations increased several-fold because of prolonged anoxic conditions (A. Boxman & E. Lucassen, pers. com.). On these locations there was even a die-off of *Alnus glutinosa*. It is clear that this artificial hydrological regime, which may be termed 'over-rewetting', is very detrimental and undesirable. The same holds for fen meadows, which need aerobic conditions in the topsoil during summer, presumably also because the vegetation strongly prefers NO_3^- to NH_4^+ as N source (de Graaf et al., 1998). The re-establishment of a more natural water table regime, with higher water tables during winter and lower during summer, is expected to provide better perspectives than the compensation of lower summer tables using riverine water.

In all restoration efforts, the diaspore bank or the dispersion of diaspores may form a significant constraint. After the re-establishment of optimal abiotic conditions, this can be an important pitfall for fen res-

De Graaf, M. C. C., R. Bobbink, J. G. M. Roelofs & P. J. M. Verbeek, 1998. Differential effects of ammonium and nitrate on three heathland species. Plant Ecol. 135: 185–196.

Den Held, A. J., M. Schmitz & G. van Wirdum, 1992. Types of terrestrializing fen vegetation in the Netherlands. In Verhoeven, J. T. A. (ed.), Fens and Bogs in the Netherlands: Vegetation, History, Nutrient Dynamics and Conservation. Kluwer Academic Publishers, Dordrecht: 237–321.

De Lyon, M. J. H. & J. G. M. Roelofs, 1986. Waterplanten in relatie tot waterkwaliteit en bodemgesteldheid. Deel 1 & 2. Laboratorium voor Aquatische Oecologie, Katholieke Universiteit Nijmegen, Nijmegen: 106 + 126 pp.

De Mars, H., A. C. Garritsen & C. van Elswijk, 1996. Van veenrivier tot boezemwater; de Greft en het behoud van de Kamerikse Nessen. Landschap 13: 207–221 (Abstract in English).

Frazier, S., 1999. Ramsar sites overview. Wetlands International + 1999 Edition of the directory of wetlands of international importance: 42 pp.

Gliwicz, Z. M., 1990. Why do cladocerans fail to control algal blooms? Hydrobiologia 200/201: 83–97.

Gore, A. J. P. (ed.), 1983. Mires – Swamp, Bog, Fen and Moor. Ecosystems of the World, 4A. Elsevier, Amsterdam: 440 pp.

Grootjans, A. P., J. P. Bakker, A. J. M. Jansen & R. H. Kemmers, 2002. Restoration of brook valleys in the Netherlands. Hydrobiologia 478/Dev. Hydrobiol. 166: 149–170.

Grootjans, A. P. & R. van Diggelen, 1995. Assessing the restoration prospects of degraded fens. In Wheeler, B. D., S. C. Shaw, W. J. Fojt & R. A. Robertson (eds), Restoration of Temperate Wetlands. John Wiley & Sons Ltd, Chichester: 73–90.

Grootjans, A. P., P. C. Schipper & H. J. van der Windt, 1986. Influence of drainage on N-mineralisation and vegetation response in wet meadows II. Cirsio-Molinietum stands. Acta oecol. Oecol. plant. 7(21): 3–14.

Gulati, R. D., E. H. R. R. Lammens, M.-L. Meijer & E. van Donk, 1990. Biomanipulation Tool for Water Management. Developments in Hydrobiology 61. Kluwer Academic Publishers, Dordrecht: 628 pp. Reprinted from Hydrobiologia 200/201.

Gulati, R. D. & E. van Donk, 2002. Lakes in the Netherlands, their origin, eutrophication and restoration: state-of-the art review. Hydrobiologia 478/Dev. Hydrobiol. 166: 73–106.

Higler, L. W. G., 1977. Macrofauna-cenoses on *Stratiotes* plants in Dutch broads. Ph.D. thesis, Research Institute for Nature Management, Leersum: 86 pp.

Hosper, S. H., 1997. Clearing lakes – An ecosystem approach to the restoration and management of shallow lakes in the Netherlands. Ph.D. thesis Wageningen Agricultural University, Wageningen: 168 pp.

Houdijk, A. L. F. M. & J. G. M. Roelofs, 1991. Deposition of acidifying and eutrophicating substances in Dutch forests. Act. bot. neerl. 40: 245–255.

Jansen, A. J. M. & J. G. M. Roelofs, 1996. Restoration of Cirsio-Molinietum wet meadows by sod cutting. Ecol. Eng. 7: 279–298.

Jansen, A. J. M., M. C. C. de Graaf & J. G. M. Roelofs, 1996. The restoration of species-rich heathland communities in the Netherlands. Vegetatio 126: 73–88.

Jefferies, D. J., 1988, The changing otter population of Britain 1700–1989. Biol. J. linn. Soc. 38: 61–69.

Jeppesen, E., J. P. Jensen, P. Kristensen, M. Søndergaard, E. Mortensen, O. Sortkjaer & O. K. Olrik, 1990. Fish manipulation as a lake restoration tool in shallow, eutrophic, temperate lakes 2: threshold levels, long-term stability and conclusions. Hydrobiologia 200/201: 219–227.

Kemmers, R. H., 1996. Humusprofielen en bodemprocessen: beoordeling van mogelijkheden voor wateraanvoer. Landschap 13/3: 157–168 (Abstract in English).

Klinge, M., M. P. Grimm & S. H. Hosper, 1995. Eutrophication and ecological rehabilitation of Dutch lakes: presentation of a new conceptual framework. Wat. Sci. Technol. 31: 207–218.

Koch, M. S., I. A. Mendelssohn & K. L. McKee, 1990. Mechanism for the sulfide-induced growth limitation in wetland macrophytes. Limnol. Oceanogr. 35: 399–408.

Koerselman, W & J. T. A. Verhoeven, 1992. Nutrient dynamics in mires of various trophic status: nutrient inputs and outputs and the internal nutrient cycle. In Verhoeven, J. T. A. (ed.), Fens and Bogs in the Netherlands: Vegetation, History, Nutrient Dynamics and Conservation. Kluwer Academic Publishers, Dordrecht: 397–432.

Koerselman, W. & J. T. A. Verhoeven, 1995. Eutrophication of fen ecosystems: external and internal nutrient sources and restoration strategies. In Wheeler, B. D., S. C. Shaw, W. J. Fojt & R. A. Robertson (eds), Restoration of Temperate Wetlands. John Wiley & Sons Ltd, Chichester: 91–112.

Koerselman, W., H. de Caluwe & W. M. Kieskamp, 1989. Denitrification and dinitrogen fixation in two quaking fens in the Vechtplassen area, the Netherlands. Biogeochemistry 8: 153–165.

Koerselman, W., S. A. Bakker & M. Blom, 1990a. Nitrogen, phosphorus and potassium budgets for two small fens surrounded by heavily fertilized pastures. J. Ecol. 78: 428–442.

Koerselman, W., D. Claessens, P. ten Den & E. van Winden, 1990b. Dynamic hydrochemical and vegetation gradients in fens. Wetl. Ecol. Managem. 1: 73–84.

Kok, C. J. & B. J. van de Laar, 1991. Influence of pH and buffering capacity on the decomposition of *Nymphaea alba* L. detritus in laboratory experiments: a possible explanation for the inhibition of decomposition at low alkalinity. Verh. int. Ver. theor. angew. Limnol. 24: 2689–2692.

Kooijman, A. M. & C. Bakker, 1995. Species replacement in the bryophyte layer in mires: the role of water type, nutrient supply and interspecific interactions. J. Ecol. 83: 1–8.

Koops, J. G., O. Oenema & M. L. van Beusichem, 1996. Denitrification in the top and sub soil of grassland on peat soils. Plant Soil 184: 1–10.

Kruuk, H., 1995. Wild Otters – Predation and Populations. Oxford University Press, Oxford: 300 pp.

Lamers, L. P. M., A. J. P. Smolders, E. Brouwer & J. G. M. Roelofs, 1996. Sulfaatverrijkt water als inlaatwater? – De rol van waterkwaliteit bij maatregelen tegen verdroging. Landschap 13, 169–180 (Abstract in English).

Lamers, L. P. M., S. M. E. van Roozendaal & J. G. M. Roelofs, 1998. Acidification of freshwater wetlands: combined effects of non- airborne sulfur pollution and desiccation. Water Air Soil Pollut. 105: 95–106.

Lamers, L. P. M., H. B. M. Tomassen & J. G. M. Roelofs, 1998b. Sulfate-induced eutrophication and phytotoxicity in freshwater wetlands. Environ. Sci. Technol. 32: 199–205.

Lamers, L. P. M., A. J. P. Smolders & J. G. M. Roelofs, 1999a. Hoe gevoelig is natte natuur voor grondwaterverontreiniging? – Op zoek naar sturende processen en factoren. Landschap 16/3: 15–25 (Abstract in English).

Lamers, L. P. M., C. Farhoush, J.M. van Groenendael & J. G. M. Roelofs, 1999b. Calcareous groundwater raises bogs; the concept of ombrotrophy revisited. J. Ecol. 87: 639–648.

Lamers, L. P. M., 2001. Tackling some biogeochemical questions in peatlands. Ph.D. thesis, University of Nijmegen, Nijmegen.

Laurén-Määttä, C., J. Hietala & M. Walls, 1997. Responses of *Daphnia pulex* populations to toxic cyanobacteria. Freshwat. Biol. 37: 635–647.

Lucassen, E. C. H. E. T., A. J. P. Smolders & J. G. M. Roelofs, 2000. Increased groundwater levels cause iron toxicity in Glyceria fluitans (L.). Aquat. Bot. 66: 321–328.

McKinley, V. L. & J. R. Vestal, 1982. Effects of acid on plant litter decomposition in an arctic lake. Appl. environ. Microbiol. 43: 1188–1195.

Meijer, M.-L., I. De Boois, M. Scheffer, R. Portielje & H. Hosper, 1999. Biomanipulation in shallow lakes in The Netherlands: an evaluation of 18 case studies. Hydrobiologia 408/409: 13–30.

Meijer, M.-L., 2000. Biomanipulation in the Netherlands – 15 years of experience. Ph.D. thesis, Wageningen Agricultural University, Wageningen: 208 pp.

Moss, B., 1998. Ecology of Fresh Waters – Man and Medium (3rd edn). Blackwell Scientific, Oxford: 557 pp.

Oomes, M. J. M., P. J. Kuikman & F. H. H. Jacobs, 1997. Nitrogen availability and uptake by grassland in mesocosms at two water levels and two water qualities.Plant Soil 192: 249–259.

Pegtel, D., 1983. Ecological aspects of a nutrient-deficient wet grassland (Cirsio-Molinietum). Verh. ges. Ökologie X: 217–228.

Pons, L. J., 1992. Holocene peat formation in the lower parts of the Netherlands. In Verhoeven, J. T. A. (ed.), Fens and Bogs in the Netherlands: Vegetation, History, Nutrient Dynamics and Conservation. Kluwer Academic Publishers, Dordrecht: 7–79.

Poschlod, P., 1995. Diaspore rain and diaspore bank in raised bogs and implications for the restoration of peat-mined sites. In Wheeler, B. D., S. C. Shaw, W. J. Fojt & R. A. Robertson (eds), Restoration of Temperate Wetlands. John Wiley & Sons Ltd, Chichester: 471–494.

Roelofs, J. G. M., 1991. Inlet of alkaline river water into peaty lowlands: effects on water quality and *Stratiotes aloides* L. stands. Aquat. Bot. 39: 267–293.

Roelofs, J. G. M., E. Brouwer & R. Bobbink, 2002. Restoration of aquatic macrophyte vegetation in acidified and eutrophicated shallow softwater wetlands in the Netherlands. Hydrobiologia 478/Dev. Hydrobiol. 166: 171–180.

Schaminée, J. H. J., E. J. Weeda & V. Westhoff, 1995. De vegetatie van Nederland. Deel 2. Plantengemeenschappen van wateren, moerassen en natte heiden. Opulus Press, Uppsala, Leiden: 360 pp.

Schaminée, J. H. J., A. H. F. Stortelder & E. J. Weeda, 1996. De vegetatie van Nederland. Deel 3. Plantengemeenschappen van graslanden, zomen en droge heiden. Opulus Press, Uppsala, Leiden: 356 pp.

Schaminée, J. H. J., E. J. Weeda & V. Westhoff, 1998. De vegetatie van Nederland. Deel 4. Plantengemeenschappen van de kust en van binnenlandse pioniermilieus. Opulus Press, Uppsala, Leiden: 346 pp.

Scheffer, M., S. H. Hosper, M.-L. Meijer, B. Moss & E. Jeppesen, 1993. Alternative equilibria in shallow lakes. TREE 8: 275–278.

Schouwenaars J. M., H. Esselink, L. P. M. Lamers & P. C. van der Molen, 1997. Hoogveenherstel in Nederland, bestaande kennis en benodigd onderzoek. University of Groningen, University of Nijmegen, Stichting Bargerveen, Bureau GroenHolland: 122 pp.

Shapiro, J., V. Lamarra & M. Lynch, 1975. Biomanipulation: an ecosystem approach to lake restoration. In Brezonik P. L. & J. L. Fox (eds), Water Quality Management Through Biological Control. University of Florida, Gainesville: 85–96.

Smolders, A. J. P., 1995. Mechanisms involved in the decline of aquatic macrophytes; in particular of *Stratiotes aloides* L. Ph.D. thesis University of Nijmegen, Nijmegen: 159 pp.

Smolders, A. J. P. & J. G. M. Roelofs, 1993. Sulphate mediated iron limitation and eutrophication in aquatic systems. Aquat. Bot. 46: 247–253.

Smolders, A. J. P. & J. G. M. Roelofs, 1995. Internal eutrophication, iron limitation and sulphide accumulation due to the inlet of river Rhine water in peaty shallow waters in the Netherlands. Arch. Hydrobiol. 133: 349–365.

Smolders, A. J. P. & J. G. M. Roelofs, 1996. The roles of internal iron hydroxide precipitation, sulphide toxicity and oxidizing ability in the survival of *Stratiotes aloides* roots at different iron concentrations in sediment pore water. New Phytol. 133: 253–260.

Smolders, A. J. P., R.C. Nijboer & J. G. M. Roelofs, 1995. Prevention of sulphide accumulation and phosphate mobilization by the addition of iron(II) chloride to a reduced sediment: an enclosure experiment. Freshwat. Biol. 34: 559–568.

Stortelder, A. H. F., J. H. J. Schaminée & P. W. F. M. Hommel, 1999. De vegetatie van Nederland. Deel 5. Plantengemeenschappen van ruigten, struwelen en bossen. Opulus Press, Uppsala, Leiden: 376 pp.

Stumm, W. & J. J. Morgan, 1981. Aquatic Chemistry, an Introduction Emphasizing Chemical Equilibria in Natural Waters. 2nd edn. John Wiley & Sons, New York: 780 pp.

Van den Broek, T. & B. Beltman, 1995, Mesotrofe verlandingsvegetaties, nieuwe kansen in het Noorderpark. Landschap 12(5): 29–38 (Abstract in English).

Van der Putten, W. H., 1997. Die-back of Phragmites australis in European wetlands: an overview of the European Research Programme on Reed Die-back and Progression. Aquat. Bot. 59: 263–275.

Van de Ven, G. P. (ed.) 1994. Man-Made Lowlands – History of Water Management and Land Reclamation in the Netherlands. Matrijs, Utrecht: 293 pp.

Van Diggelen, R., W. J. Molenaar & A. M. Kooijman, 1996. Vegetation succession in a floating mire in relation to management and hydrology. J. Veg. Sci. 7: 809–820.

Van Donk, E. & W. J. van de Bund, 2002. Impact of submerged macrophytes including charophytes on phyto- and zooplankton communities: allelopathy *versus* other mechanisms. Aquat. Bot. 72: 261–274.

Van Duren, I. C., D. Boeye & A. P. Grootjans, 1997. Nutrient limitations in an extant and drained poor fen: implications for restoration. Plant Ecol. 133: 91–100.

Van Turnhout, C. & W. Hagemeijer, 2001. Long-term population developments in typical marshland bird species in the Netherlands. Ardea, in press.

Van 't Veer, R., B. van Geel, J. P. Pals & D. van Smeerdijk, 2000. Fossiele plantengemeenschappen als referentiekader voor moderne moerasontwikkeling. In Schaminée, J. & R. van 't Veer (eds), 100 Jaar op de knieën – De geschiedenis van de plantensociologie in Nederland. Opulus Press, Noordwolde: 174–189.

Van Wirdum, G.,1991. Vegetation and hydrology of floating richfens. Ph.D. thesis, University of Amsterdam, Amsterdam: 310 pp.

Van Wirdum, G., A. J. den Held & M. Schmitz, 1992. Terrestrializing fen vegetation in former turbaries in the Netherlands. In Verhoeven, J. T. A. (ed.), Fens and Bogs in the Netherlands: Vegetation, History, Nutrient Dynamics and Conservation. Kluwer Academic Publishers, Dordrecht: 323–360.

Verhoeven, J. T. A. & W. M. Liefveld, 1997. The ecological significance of organochemical compounds in Sphagnum. Acta Bot. Neerl. 46: 117–130.

Verhoeven, J. T. A. & R. Bobbink, 2001, Plant diversity of fen landscapes in The Netherlands. The role of water chemistry, succession and management. Aquat. Bot., in press.

Verhoeven, J. T. A., A. M. Kooijman & G. van Wirdum, 1988. Mineralization of N and P along a trophic gradient in a freshwater mire. Biogeochemistry 6: 31–43.

Verhoeven, J. T. A., W. Koerselman & A. F. M. Meuleman, 1996. Nitrogen- or phosphorus-limited growth in herbaceous, wet vegetation: relations with atmospheric inputs and management regimes. TREE 11: 494–497.

Vermeer, J. G., 1986. The effect of nutrient addition and lowering of the water table on shoot biomass and species composition of a wet grassland community (Cirsio-Molinietum Siss. et de Vries, 1942). Acta oecol. 7: 145–155.

Vermeer, J. G. & J. H. J. Joosten, 1992. Conservation and management of bog and fen reserves in the Netherlands. In Verhoeven, J. T. A. (ed.), Fens and Bogs in the Netherlands: Vegetation, History, Nutrient Dynamics and Conservation. Kluwer Academic Publishers, Dordrecht: 433–478.

Wassen, M. J. & A. Barendregt, 1992. Topographic position and water chemistry of fens in a Dutch river plain. J. Veg. Sci. 3: 447–456.

Wassen, M. J., H. G. M. Olde Venterink & E. O. A. M. de Swart, 1996. Nutrient concentrations in mire vegetation as a measure of nutrient limitations in mire ecosystems. J. Veg. Sci. 6: 5–16.

Waughman, 1980. Chemical aspects of the ecology of some south German peatlands. J. Ecol. 68: 1025–1046.

Westhoff, V., P. A. Bakker, C. G. van Leeuwen & E. E. van der Voo, 1970. Wilde planten: flora en vegetatie in onze natuurgebieden. Deel 1: Algemene inleiding, duinen, zilte gronden. Natuurmonumenten, 's-Gravenland: 320 pp.

Wheeler, B. D. & S. C. Shaw, 1995. Restoration of damaged peatlands, with particular reference to lowland raised bogs affected by peat extraction. HMSO, London: 211 pp.

Wheeler, B. D. & M. C. F. Proctor, 2000. Ecological gradients, subdivisions and terminology of north-west European mires. J. Ecol. 88: 187–203.

Wiegers, J., 1992. Carr vegetation: plant communities and succession of the dominant tree species. In Verhoeven, J. T. A. (ed.), Fens and Bogs in the Netherlands: Vegetation, History, Nutrient Dynamics and Conservation. Kluwer Academic Publishers, Dordrecht: 361–396.

Hydrobiologia **478:** 131–148, 2002.
P.H. Nienhuis & R.D. Gulati (eds), Ecological Restoration of Aquatic and Semi-Aquatic Ecosystems in the Netherlands (NW Europe).

Towards a decision support system for stream restoration in the Netherlands: an overview of restoration projects and future needs

P.F.M. Verdonschot & R.C. Nijboer
Alterra , Department of Freshwater Ecology, P.O. Box 47, 6700 AA Wageningen

Key words: decision support system, stream restoration, key factor, inquiry, water management, lowland stream

Abstract

Stream restoration is one of the answers to the lowland stream deterioration. For making proper choices in stream restoration; one firstly needs to understand the complex spatial and temporal interactions between physical, chemical and biological components in the stream ecosystem. Several ecological concepts on the four dimensions, scale and hierarchy in a stream ecosystem are integrated into the 5-S-model. This model provides the theoretical backbone of the first outline of a decision support system for stream restoration. Stream restoration is developing fast in the Netherlands. In 1991, 70 projects were counted, in 1993 there were 170, and this number increased in 1998 to 206. Positive signs in this increase in the number of stream restoration projects are the increase in the amount of money, in background studies, in improvement of the selection process of stretches to be tackled, and the broadening of the objectives and measures. Negative signs are amongst others that measures often deal only with stream hydrology and structures in-stream. The catchment takes no part. Furthermore, bottlenecks often relate to finances and agreement between people and/or organisations. Finally, the first steps towards a decision support system for stream restoration are made. The system presented provides only information based on which measures should be taken. 'Where and how' these measures need to be taken remains a challenge for the future.

Introduction

Lowland streams are characterised by a gentle slope of the terrain (zero to five per mil) and a sandy soil. They occur in the flat lowland areas of the Western European plain. They are fed by rainwater, and often lack a well-defined source. Thus, their discharge shows a smoothed relation with the amount and frequency of precipitation in the various seasons. Such streams occur in both the eastern and southern part of the Netherlands. Their current velocity varies from 5 to 30 cm s^{-1} in summer and early autumn and from 30 to 60 cm s^{-1} in late autumn to spring. Often the rainwater fed upper courses dry up in summer, though sometimes they are fed by a helocrene spring and then show a more constant discharge pattern (Verdonschot, 1990).

After a long time-period of adapting lowland streams and their catchment to agricultural, domestic, drinking water and industrial needs, one became aware of the damages of these alterations. In the Netherlands, only about 4% of the streams still have a natural hydro-morphology. During the last 10 years, the ecological importance of streams became more and more apparent.

Nowadays, stream restoration is one of the answers to the lowland stream deterioration. In order to make the proper choices for stream restoration; one firstly has to understand the complex spatial and temporal interactions between physical, chemical and biological components. The success of restoration depends on steering the appropriate key factor(s), which differ from stream to stream and site to site. The most important stream ecology concepts are those concerning dimensions and hierarchy. Ward (1989) introduced the concept of the four dimensional nature of stream ecosystems with a longitudinal, lateral, vertical and temporal component. The major idea was the attempt to identify the controlling processes in a stream as integral part of a catchment. Frissell et al. (1986) organized the controlling factors from catchment to stream habitat in a hierarchical space and time framework. Knowledge of this hierarchy allows us to infer the direction and magnitude of potential changes (alteration as well as restoration) due to human activities. But theories on stream ecology are complex and not easy to use in the practice of stream management, as is evident from an evaluation of a large number of surveys about stream restoration. The main problem in stream restoration is the conflict between the theory of integrated approach *versus* the limited practical possibilities on the square meter.

Stream managers need a simple decision support system to handle the ecological complexity for an effective restoration plan at a site. This paper provides such a system, giving an opportunity to go through the most important steps in stream restoration and to extract the factors in the catchment that need to be tackled. Although the site and streams differ, the planning for successful restoration should be the same. The decision support system is based on the theories of dimensions, scale and hierarchy and water manager is obliged to include the catchment in the restoration plan. This section one in this paper discusses the stream ecology concepts, the section two deals with the experiences of Dutch water managers with stream restoration and, finally, in section three the first steps towards a decision support system for stream restoration are made. This last section is based on the knowledge available in the concepts on stream ecology, comprised in the 5-S-model (Verdonschot et al., 1998) and combines these with the daily practice and experiences of stream management.

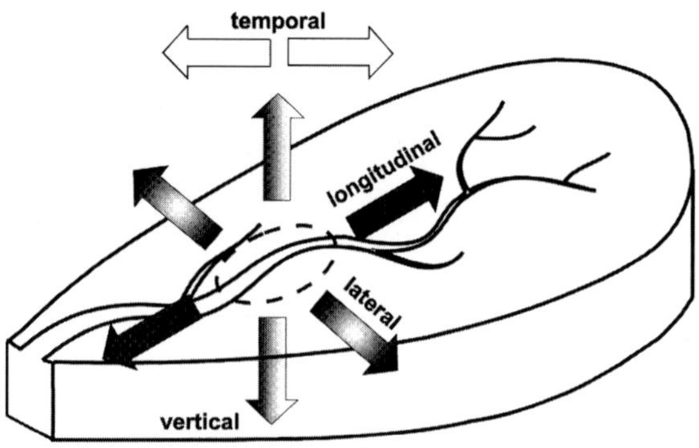

Figure 1. The concept of the four-dimensional nature of stream ecosystems with a longitudinal, lateral, vertical and temporal component (Ward, 1989).

Lowland stream ecology

Theoretical concepts in stream ecology

As most ecosystems, lowland stream ecosystems are composed of groups of interacting and interdependent parts (e.g., species, resources) linked to each other by the exchange of energy and matter. Linkage not only occurs between different parts in the transversal profile of a stream but also between upstream and downstream parts of a stream. For a long time, the longitudinal component of a stream was seen as a sequence of inter-linked zones (Illies & Botosaneanu, 1963; Hawkes, 1975) or as a longitudinal continuum (Wallace et al., 1977; Vannote et al., 1980). But exchange of energy and matter is not limited to the stream itself. Hynes (1975) was the first one who included the catchment. Stream ecosystems are considered to be complex because their functioning is not limited by the stream itself and the banks but it stretches out all over the catchment. Within the catchment as a whole, streams are characterised by strong interactions between components, feedback loops, significant interdependencies in time and space, discontinuities, thresholds, and limits (Costanza et al., 1993). To extricate this complexity, Ward (1989) introduced the concept of the four dimensional nature of stream ecosystems with a longitudinal, lateral, vertical and temporal component (Fig. 1). Except from this theory about dimensions, a second theory concerning hierarchy is important. Processes in streams are important at different scales. The organisms in a stream are dependent on habitat characteristics. These characteristics in turn depend on morphology and hydrology of a stream, which in turn depend on geomorphologic structure and climate in the catchment. In addition to the dimension and scale theories, some other concepts on streams should be taken into account when dealing with lowland streams. The most important concepts on stream ecology are summarised in Table 1.

Stream catchment in four dimensions

The overall concept "the four dimensional nature of lotic systems" identifies in general terms all interactions and functioning of the stream as an integral part of the catchment (Fig. 1).

(1) The *longitudinal* component covers the whole catchment. There are important upstream–downstream linkages in the stream as well as in the catchment as a whole. The functioning of a lowland stream strongly depends on the inputs, transport and finally outputs of water. Water moves downstream and transports energy and matter. There are differ-

Table 1. Some characteristics of major ecological concepts in stream ecology

Ecological concept	Key theme	Direction	Scale	Reference
	Dimensions			
Four dimensional nature of lotic systems	Longitudinal, lateral, vertical and temporal exchange	Longitudinal, lateral, vertical, temporal	Catchment	Ward (1989)
River Continuum Concept	Longitudinal gradient	Longitudinal, lateral	Stream, floodplain	Vannote et al. (1980)
Nutrient Spiralling Concept	Longitudinal nutrient cycling	Longitudinal, lateral	Stream, floodplain	Wallace et al. (1977)
Flood Pulse Concept	Lateral exchange of substances	Lateral, temporal	Lower reach, floodplain	Junk et al. (1989)
	Hierarchy			
Hierarchy	Hierarchy and feed-back	Top–down and bottom–up	Fine to coarse	Frissell et al. (1986)
	Response of species			
Dynamic Equilibrium Model	Dynamic equilibrium in systems	Conceptual, temporal	Independent	Huston (1979)
Intermediate Disturbance Hypothesis	Non-equilibrium maximises diversity	Conceptual, temporal	Independent	Ward & Stanford (1983a)
Habitat Template Concept	*r*, *k*, a selection in space and time	Spotwise	Stream section	Southwood (1977)
Patch Dynamics Concept	Competition versus disturbance	Spotwise	Stream section	Townsend (1989)
	Human influence			
Serial Discontinuity Concept	Discontinuity through human interference	Longitudinal	Stream	Ward & Stanford (1983b)

ent longitudinal zones (Hawkes, 1975). There are the river continuum concept of Vannote et al. (1980) and the Serial Discontinuity Concept of Ward & Stanford (1983b) which stresses the importance of the stream as a continuum with or without discontinuities. All these concepts imply a more or less gradual shift in abiotic stream characteristics and species composition along the stream gradient. Furthermore, also the effect of downstream activities on up-stream stretches must be included in the longitudinal component. Channelisation down-stream affects the discharge and erosion patterns up-stream (Schumm, 1977). A weir down-stream acts as a migration barrier and thus changes biotic components and interactions up-stream. Another important feature of the longitudinal transport of water is nutrient spiralling. This concept explains that a nutrient cycle is completed while the nutrient is transported downstream (Wallace et al., 1977).

(2) The *lateral* component includes the whole of interactions transversal through the catchment and covers the floodplain. The lateral component concerns the interaction within and between stream, riparian zone, floodplain and catchment. The interactions in-stream concern turbulence in the water column and erosion-deposition processes in the a-symmetric transversal profile. The interactions between stream and riparian zone got more and more attention in the last two decades (Petersen et al., 1987; Naiman & Décamps,

1990). The stream–floodplain interaction is amongst others described in the Flood Pulse Concept (Junk et al., 1989). The interactions concern exchange of silt and substances through direct surface and sub-surface runoff but also exchange of biota through inundation and migration.

(3) The *vertical* component concerns the soil beneath and the air above the stream. The vertical component stresses the groundwater–surface water interaction (Brunke & Gonser, 1997), expressed in the hyporheic zone (Stanford & Ward, 1988). The latter zone is of little or no importance in lowland streams. The vertical interaction also includes the exchange processes between stream and air, such as evaporation, deposition of substances (e.g., Kristensen & Hansen, 1994) and biotic interactions. Some animals use the atmosphere for movement in the adult phase of their life cycle (reproduction, dispersion).

(4) The *temporal* component includes changes in time. The temporal component includes processes going on, such as the length of the organisms life histories, morphological changes in meander patterns over long periods of time or abrupt changes through channelisation (Boon, 1992). Also within a stream, the processes in time and space are conceptualised, such as in the patch dynamics concept (Townsend, 1989). But the temporal component should also include historic developments (Kondolf & Larson, 1995).

The catchment approach includes all four dimensions

This concept could be used as a frame for an integrated catchment management. The transport property of a stream is the most important process and directly depends on the catchment (spatial component). Because of the open character of the stream, it reflects the past and present structure and functioning of the whole catchment and thus includes the temporal component. Water that infiltrates in the catchment can have a long retention time before it enters the stream. In a catchment approach the longitudinal and transversal components also include the 'dry' floodplain and the (infiltration) areas at a higher altitude in the catchment. In fact, infiltration areas affect the stream water quality and land use in these areas influence, amongst others, transport of substances towards the stream. The deep groundwater flow, which connects infiltration areas to the streams, is important in lowland streams and differs in the different reaches. Upper courses often only receive subsurface and less deep flow, middle reaches can receive subsurface flow but are also often infiltrating, lower reaches almost always receive deep, old groundwater. The water enters the stream in a more vertical direction as seepage. A stream is, therefore, part of its catchment and can not be studied without looking along these four dimensions.

Scale and hierarchy

Large watersheds are comprised of tributaries and their catchment, tributaries contain multiple stream reaches, each of which potentially includes different habitats, and all these habitat contain multiple microhabitats (Frissell et al., 1986; Sedell et al., 1990). The multitude of processes that operate in a stream systems exist within a hierarchical framework (Allan & Starr, 1982; Frissell et al., 1986). The hierarchy theory provides a framework for the description of the components of an ecosystem and their scaled relations (O'Niell et al., 1986, Jensen et al., 1996). The four dimensions in streams concern aspects of spatial and temporal scale and hierarchy (Fig. 2). The temporal component is not always independent from the spatial ones and can be added to each of them.

(1) The longitudinal component stretches out over the whole spatial area of the catchment. It more often concerns processes acting over a long-term period, such as deep groundwater flow and processes of longitudinal meandering. But there are also examples of shorter term like nutrient spiralling and fish migration. The longitudinal component can be related to a coarse spatial and a different temporal scales.

(2) The lateral component interacts at the spatial scale of the flood plain and concerns processes like inundation and (sub-) surface runoff. These interactions more often act over a shorter time period. The lateral components also include the creation and evolution of oxbows or marshes; they act over a long term. Thus, the lateral component can be related to an intermediate spatial and again different temporal scales.

(3) The vertical component includes the riparian zone or the wooded bank as well as the thin more or less oxygenated substrate layer on the stream bottom. Its interactions more often are for short time duration, such as the exchange of gasses between atmosphere and water column, the emerging and reproduction of adult insects in the overhanging trees or the (bio-) turbation of the stream bottom substrate. On the other hand, the vertical component is highly influenced by processes that operate at a long temporal scale, such as erosion and deposition resulting from stream incision.

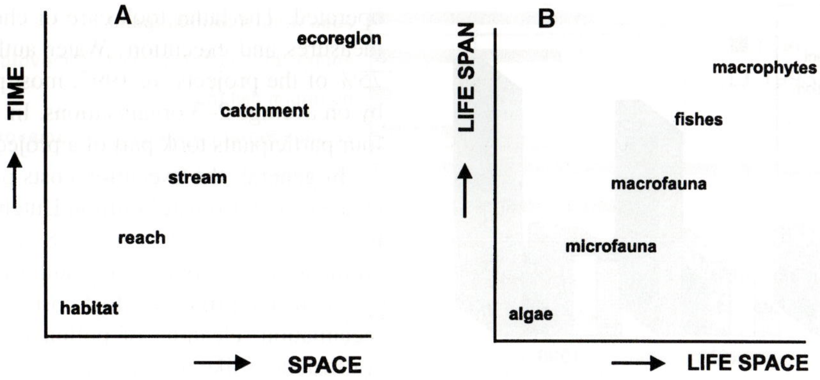

Figure 2. Spatial and temporal scale and hierarchy (A), and the response scale of different taxonomical groups (B) (after Verdonschot, 1999).

The vertical component can be related to fine spatial and different temporal scales.

A catchment approach acts at different scales and includes hierarchy

There is a hierarchy between the three components in space and time whereby the longitudinal component (coarser scale) bounds the range of ecological features of the lateral and vertical ones (finer scales), but also the vertical one (finer scales) affects the lateral and longitudinal components (coarser ones). Stream functioning acts at multiple spatial and temporal scales with 'top down' and 'bottom up' controls often termed dominance and feedback. Integrated ecological approach in stream restoration should include more than one spatial (include at least one lower scale) and temporal scale (to include system dynamics) dependent on the objective which is addressed. Looking at stream functioning always should cover a fine, intermediate and coarse scale in space and time.

Including the whole catchment in stream ecology and restoration of streams implies working in hierarchical order. It is of not much use to start at a small scale (certain habitat in a stream) if there are problems on the large scale (in the infiltration area of the catchment). In a catchment approach processes at different scales in the catchment varying from microhabitat to catchment are included.

Stream restoration in the Netherlands

Present condition

Lowland streams in the Netherlands belong to the most threatened ecosystems. Ninety six percent of all Dutch streams are directly impacted by human activities (Verdonschot et al., 1995; Nienhuis et al., 2001). The changes relate to the length and transversal profile through channelisation, affecting the structures in the stream. It also concerns the discharge pattern and water level by construction of weirs and other artificial constructions, water extraction and drainage. All these factors affect the stream hydrology. Changes in the shape and hydrology of the lowland stream leads to an increase in discharge fluctuations and an erosive character of the stream. The latter is responsible for a deep bed-incision of most lowland streams. Bed-incision, in turn, is the cause of drought in the floodplain. Thereby, direct discharge of wastes, surface runoff of agricultural land and water inlet affect the chemical composition of the water. It is estimated that between 60 and 80% of the in stream nutrients originate from agricultural sources, and only between 10 and 20% from other discharges. Agricultural activities also affect the quality of the groundwater, a future threat of the stream environment. Only 4% of Dutch lowland streams are near natural, indirectly affected by atmospheric pollution, groundwater extraction and changes in the catchment. All near natural stream parts are only located in the upper courses of streams.

A comparison between large data banks including macro-invertebrates and environmental variables of the 1980s and 1990s led to the following conclusions (Verdonschot et al., 1995):

- The increase in drainage intensity and the management directed towards quickened discharge of water have caused a decrease in current velocities during base flows and an increase during peak flows. Furthermore, they resulted in an increase in drought especially in the floodplain (not shown), and an increase in acidification.

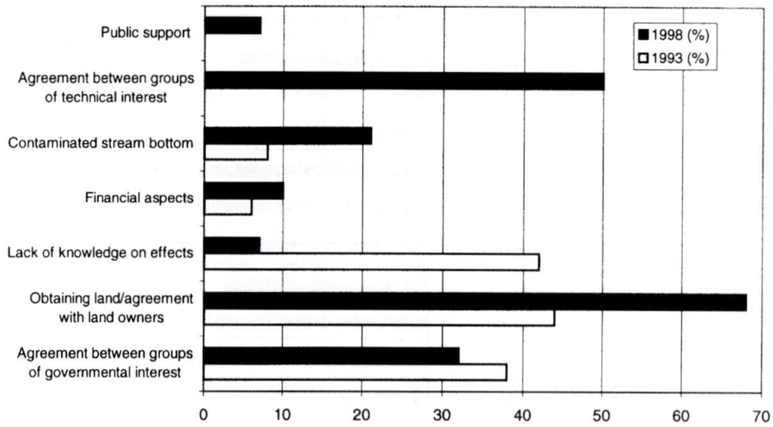

Figure 6. Problems experienced during stream restoration projects.

Table 2. The most often executed measures per category (%)

Hydrology	**1993 (%)**	**1998 (%)**
Removal of drainage	21	32
Improvement of infiltration	0	32
Forest development	14	36
Development inundation zone	17	36
Increase water retention	14	54
Removal weir	14	29
Structures	**1993 (%)**	**1998 (%)**
Digging meanders	38	57
Decrease stream profile	7	46
Construction a-symmetric profile	0	36
Construction wooded banks	45	43
Construction fish ladder	31	50
Construction pools in stream valley	0	46
Water quality	**1993 (%)**	**1998 (%)**
Decrease nutrient supply	10	21
Removal sewer overflow	14	18
Lowering floodplain level	3	32
Construction buffer zone	7	21
Reducing toxic pollution	7	19
Species	**1993 (%)**	**1998 (%)**
Re-introduction species	10	7

From the biological measures, only 'introduction of species' is sometimes applied. Among the measures applied 'digging of meanders' and 'increase in water retention' are most frequent, followed by 'construction of fish ladders'.

Positive and negative tendencies

Positive aspects

The increase in number of stream restoration projects (Fig. 3) and in the available amount of money are positive signs. Also positive is the increase in background studies that support an improvement in restoration quality. Streams were more often selected because of present and potential values, which added to the successes of restoration. The objectives mainly focussed on stream hydrology, structures and ecological values (Fig. 4). All stretches of lowland streams were taken under consideration (Fig. 5), although the length of the stretches was relatively short within a catchment perspective. The applied measures were based on combinations of ecological and non-ecological demands. The measures often dealt with stream hydrology and structures (Table 2). All these features indicate a positive tendency in lowland stream restoration in the Netherlands. But some caution is needed in generalisation all such information. The aspects that need improving emerge from the bottlenecks as illustrated in Figure 6 and points that need future attention are summarised in Table 3.

Problems in stream restoration projects

All indicated bottlenecks deal with finances and agreement between people and/or organisations. In 1998, in only 7% of the projects the respondents stated lack of knowledge as the problem. From the most restoration projects it became clear that they were restricted to a limited part of the catchment, mostly only the stream and its first 5–10 m of the riparian zone. Measures almost never dealt with, nor were executed in the catchment as a whole. Thus, coarse and intermediate scales were not involved in the restoration projects. Furthermore, the impact of diffuse pollution from agricultural land use as well as the increasing pollution of the groundwater was neglected. Looking at the major points of attention, only once this problem was indicated with a high priority (Table 3).

Ecological effectiveness

Beside the positive and negative tendencies in restoration projects also interesting are the ecological effects of the measures taken. Therefore, some projects were selected, which were executed five or more years ago. This time period was chosen because a restored stream ecosystem needs time to recover. Only a limited number of recorded evaluation results are available. This indicates the first important but negative sign; projects have no or only limited monitoring programmes. An example, of this is the number of restored wooded banks in the 1960s. All evaluations dealt with the landscape improvement, none monitored the in-stream ecosystem (Werkgroep Beekbegeleidende Beplantingen, 1989).

About all projects show an initial increase in diversity the first years after the project took place. This feature is a consequence of colonisation processes and not of recovery. Two projects dealt with the restoration of riparian wetlands. In both projects, the first results were hopeful, but within about 5 years silt and nutrients deposits changed the vegetation into a disturbed and degraded state (Baker, 1993; Zonderwijk, 1994).

In general, little is yet known about the ecological effectiveness of restoration measures. The general impression is that a number of restoration projects yield limited amount of ecological improvement because of problems encountered (Table 3), spatial limitations and a fragmented systems approach. Spatial limitations imply a restricted approach not including the whole catchment. The fragmented systems approach implies the restriction of many projects to restore only one or a few number of disturbed key factors. The ecological profit expresses itself mainly in an increase of biodiversity within the stream and the riparian zone.

Future restoration projects

In future restoration projects, more attention must be paid to the processes in the whole catchment especially those acting at wider spatial scales, which are often neglected. The main problem is obtaining land in the catchment area because the intensive agriculture practice in the Netherlands needs much land area. Secondly, a co-ordinated involvement of the people is needed. It takes a lot of effort to reach a shared agreement on a plan. Public support, not only from government but also from the users of the area is very important. Legal procedures can take a long time and only retards a faster application of measures in the area.

Table 3. Points of attention in future projects (in order of importance with a priority score of 1 (low) to 5 (high)

Points of attention	Number of times noted	Priority (average score)
Obtaining of land	11	3
Agreement within plans	10	3
Legal procedures	9	3
Contaminated soils	8	3
Public support	11	2
Human power	1	1
Diffuse sources of pollution	1	4

Theory versus *practice*

The inquiries on stream restoration gave interesting insights into the state of the running projects. But from the answers it did not become clear how much ecologically profitable they were. The objectives and ambitions on the ecological aspects showed high expectations, but little was done to evaluate the measures taken. To get a detailed idea of the status of the ecological effects of the projects, three examples were chosen, which together represent an average overview of stream restoration in The Netherlands. They indicate clearly what is expected from an ecological point of view of looking in detail at the positive and negative aspects.

Example 1: Re-meandering of the stream "Vloedgraaf"

Study area

The stream "Vloedgraaf" is a channelised middle to lower course of a lowland stream. The water quality is moderately affected by the input of sewage and effluent of a purification plant. The water does not meet the standard for phosphate, nitrogen, ammonia, copper, zinc, lindane, and sulphate. PAC's and PCB's

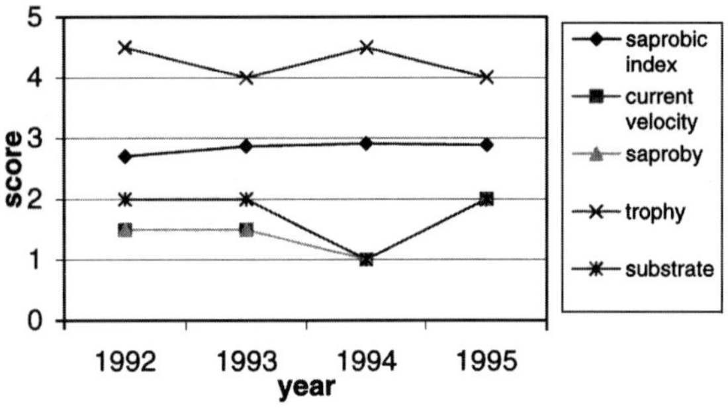

Figure 7. Quality scores based on macro-invertebrates over the years 1992–1995 (after Maris et al., 1998, 1992 indicates the score before restoration).

pollute the bottom. Before restoration the channel was straight, and fixed by stones and concrete. The streams hydrology showed extreme discharge peaks due to runoff from paved areas.

Measures

Originally, the stream "Vloedgraaf" was reconstructed to optimise discharge capacity but in such a way that it meets ecological demands for both the landscape and the stream. The channelised stream was reconstructed in 1992–1993 by:

- digging a meandering longitudinal profile over a length of 1140 m;
- constructing a transversal profile with a varying bottom width between 3 and 7 m and a double profile in order to create swampy, inundated areas, about 75 m wide, whereby locally the surrounding grounds were lowered;
- digging of pools and oxbow lakes; and
- planting of trees at some stretches.

Present condition

The first monitoring results became available 5 years after reconstruction (Maris et al., 1998). The physical-chemical water quality was not improved. This was be expected because this problem was not tackled during restoration. The macro-invertebrate assemblages did not improve despite the increase in variation in current velocities and substrates, e.g. the construction of gravel beds (Fig. 7). Organic wastes polluted part of the new or deposited substrates. This was especially true for the new pools and oxbows. Both were heavily polluted and did not support the intended communities. Only the riparian zone, which was extensively grazed by horses, showed an increase in diversity.

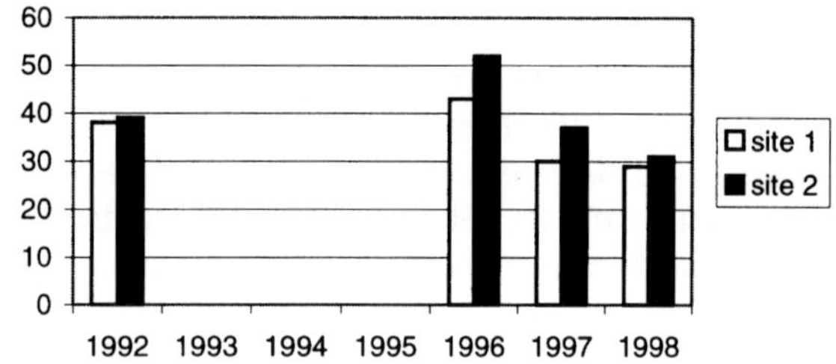

Figure 8. Number of target species since 1992 (after Duursema, 1999).

Conclusions

The stream "Vloedgraaf" restoration project showed that ecological development can be combined with other functions of streams. However, when this restoration is not done from an integrated ecological approach of the catchment and the relevant key factors, the results are disappointing. Not tackling the problem of water quality and not taking measures to reach a more natural hydrology resulted in a may be visually changed stream bed but with ecologically marginal improvements.

Example 2: Water retention in the stream "Gasterense diep"

Study area

The stream "Gasterense diep" is a middle course of a half-natural, unshaded lowland stream. The water quality is good. The stream drains a natural area and an extensively used agricultural area. Before restoration the channel was slightly meandering, the water was shallow and the substrates were diverse.

Measures

Firstly, in 1993 a sewer discharge was removed. Secondly, in 1997 the stream "Gasterense diep" was reconstructed to heighten the groundwater levels in the near surroundings of the stream channel. The objective was to restore the former wet riparian areas parallel to the stream and to improve the conditions for groundwater dependent vegetation. The measures consisted of constructing six submersed weirs into the stream channel. The expectation was that the weirs would with hold sand that is transported by the stream. After a few years, sand deposition in the stretch upstream from a weir has filled up the stream until the level of the weir. Furthermore, no submersed vegetation will be removed from the stream anymore.

Present condition

The monitoring results became available 5 years after sewer discharge removal and 2 years after construction of the (Duursema, 1999). This evaluation focussed on the in-stream effects. The number of target species of macro-invertebrates increased after removal of the sewer discharge (Fig. 8). But the in-stream effect of the weirs tended to be negative. The number of target species dropped.

Conclusions

Removal of organic pollution on the stream "Gasterense diep" showed a positive effect on the number of rare and important macro-invertebrate species. The construction of weirs, on the other hand, showed a negative effect. By constructing weirs the current velocity dropped and the substrate diversity decreased. The bottom became covered by silt. Again, stream restoration by only looking at certain components of the whole stream ecosystem restricted the success of the measures.

Example 3: Wetland construction along the stream "Midden Regge"

Study area

The stream "Midden Regge" is a channelised lowland stream. The stream receives discharges from several purification plants and sewer systems. The water quality is moderate. The sewer-system discharges also cause fluctuations in stream discharge during rainy periods. The stream bottom is polluted, comparable to the stream "Vloedgraaf".

Measures

To improve natural values in the stream valley of the "Midden Regge" the following measures were taken in 1988 and 1989:

- construction of a gradient between land and water to give freedom to erosion and deposition processes in a riparian zone of 2–6 m;
- removal of the top soil layer of 20 cm in this riparian zone which is contaminated and enriched;
- increase of stream bottom width and depth; and
- decrease of maintenance frequency.

Present condition

The condition of the stream "Midden Regge" was reported in 1994, 6 years after measures were executed (Zonderwijk, 1994). As most measures were directed to the development of the riparian zone, the evaluation is mainly focussed to this area. Figure 9 shows the number of plant species within the major riparian habitats. After implementation of the measures, the number of species increased up to 1992. But in 1993, the number of species decreased again and this decrease pointed towards an increase in eutrophication of the soil. The latter is due to sand and silt deposition in the riparian zone. The author concludes that only hydrological isolated habitats can be sustainable. The same conclusions accounted for fishes and amphibians. The increase in profile width and depth lead to a decrease in current and an increase in alga development and oxygen depletion.

Conclusions

Re-construction of the riparian zone of the "Midden Regge" initially led to improvement of the natural values in the terrestrial, amphibian and aquatic habitats. But this was only true for the first few years after construction. The last year showed again deterioration of the vegetation patterns due to deposition of contaminated and enriched silt. Profile changes also had a negative effect on the stream community.

General tendencies

Some general conclusions follow from the examples above, which more in general account for a high number of restoration projects included in the inquiries:

1. Often a combination of measures is necessary to improve the stream ecosystem, i.e. measures that focus on both different components of the whole system and tackle most stressing conditions.
2. Sometimes not the hierarchically most important measure was taken. For the stream "Vloedgraaf", it becomes clear that even if the morphology of a stream has improved, the ecological stream quality will not improve if hydrology and or water quality, or both, are insufficient. A similar conclusion can be drawn for the stream "Midden Regge".
3. The relation between catchment and stream is often not included. For the streams "Gasterense diep" and "Midden Regge" one focussed on the stream floodplain rather than the stream itself.
4. Only a small stream reach was restored. In case of "Vloedgraaf", restoration was only focussed on about 1 km of the stream without accounting for the upstream region and its characteristics.

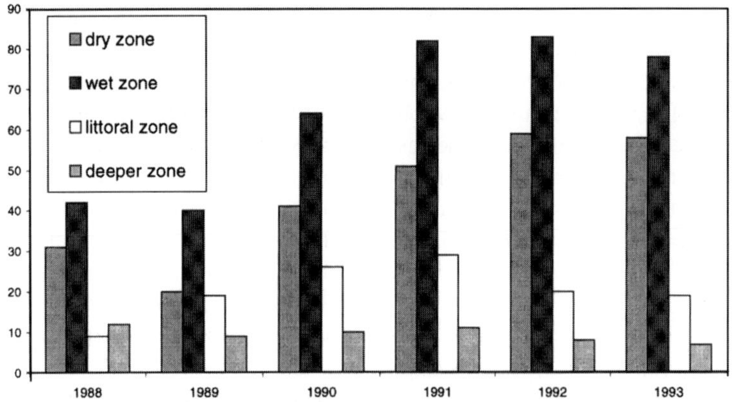

Figure 9. Average number of plant species in different habitats in the riparian zone of the stream 'Midden Regge' (Zonderwijk, 1994).

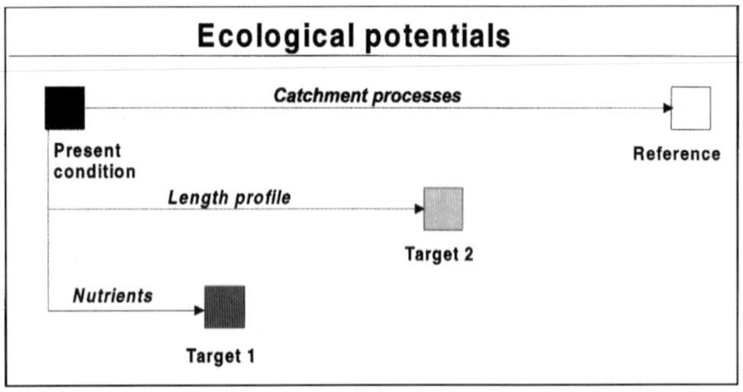

Figure 10. Ecological potentials in stream restoration. Management goals set targets and references.

Towards a decision support system for stream restoration

Potentials

The success of stream restoration depends on the societal and ecological potentials of the area to be restored. The societal potentials depend on political and management choices. The water managers involved in case of Vloedgraaf project most probably did know beforehand that the bad water quality could be a problem for ecological rehabilitation. This is a clear example of lack of power (political) and financial means to also simultaneously tackle this quality problem. It shows that political and management choices really are a major bottleneck for an integrated approach of stream restoration. These aspects are not furthered in this paper. The ecological potentials depend on the conditions in the catchment and the stream and on the reference condition. For the last named, restoration needs clear goals that are defined in terms of reference and target conditions. These reference conditions refer to natural or pristine stream ecosystems, which can be accomplished in the long term (25–50 years). The target conditions represent conditions in between the present and the reference condition and can be accomplished in a middle to long term (5–25 years). From a management point of view, these targets have an important practical use in monitoring and evaluation. Species composition and ranges of abiotic variables describe targets and references. To set management goals, it is important to know what the stream type is and how it should look like in the most optimal ecological situation. For stream management, it is necessary to identify the key factors that are disturbed by human influences. The higher the level of the key factor that will be managed the closer it is to approach the reference condition (Fig. 10).

Key factors

It is clear from the above that, first, stream restoration at presents lacks a catchment perspective. Secondly, measures are often taken because of practical motives such as non-ecological demands, land availability, maintenance of the stream channel, etc. Therefore, an integrated, ecological approach is needed in which ecological concepts; threats and practical experiences are combined. This synthesis leads to a decision support system for stream restoration in The Netherlands. The key themes in theoretical stream ecology deal with: the four dimensions, hierarchy, response of species and human influence (Table 1). A catchment approach includes all these themes and can be applied at different scales. When these theoretical considerations are confronted with the practical execution of stream restoration a number of problems occur. To improve this practical approach and include all dimensions relevant to a stream restoration project, from a catchment and landscape ecological point of view, the key factors in stream restoration were ordered.

For making proper choices in stream and catchment management; one has to understand the functioning and interactions (feed back) of the controlling key factors. To simplify the ecological complexity of the controlling key factors, and the concepts on hierarchy and scale as described in the first section of this paper, the *5-S-Model* was formulated. This conceptual model integrates the four dimensions in streams, scale and hierarchy and provides guidelines for management (Verdonschot et al., 1998). Aspects of scale and hierarchy relating to key factors are included in the model. The five main components (from high to low level) are:

(1) *System conditions* comprise the factors related to climate (temperature, rainfall), geology and geomorphology (like slope, soil composition). System conditions are composed of ultimate controlling factors and are boundary conditions for a stream. The system conditions set the possibilities and limits for stream ecosystem functioning. Ultimate controlling factors continuously interact with a stream at a high hierarchical scale in space (the catchment), as well as in time (±100 years). Generally, management can not change the system conditions. Human activities influence this level through, for example, atmospheric deposition and climate change. Stream rehabilitation does not focus on these factors but one has to consider the effects of these boundary conditions as well as the long-term effects of change. All system conditions together determine the type of stream. A stream is seen as a whole and as part of its catchment. The catchment is composed of gradients. For restoration purposes, it is necessary, one way or another, to identify and arrange stream types as discrete entities which can be dealt with. For the water manager, who can not deal with a gradient in environmental circumstances, this more or less arbitrary identification of entities is of high practical value.

(2) *Stream hydrology* characteristics are set by the system conditions. Stream hydrology comprises, at the scale level of catchment, the processes, like infiltration, ground water flow, seepage, run off and discharge. At the level of stream and habitat, stream hydrology comprises hydraulic processes, like current velocity and turbulence. Stream hydrology refers to the water quantity parameters. The direction of the water flow strongly influences the direction of all other parameters in the system. The two main directions of flow are: (1) lateral, which one runs from the boundary of the catchment towards the stream; and (2) longitudinal, which runs from the source to the mouth of the stream.

(3) *Structures* of the stream valley and the stream itself are strongly determined by the hydrological and hydraulic processes of stream hydrology. Structures imply the morphological features of the longitudinal and transversal shape of the stream bottom, banks and bed, as well as the substrate patterns within. Structures also refer to old meanders, terrestrialisation, sand deposits and others in the stream valley. The dynamics of these structures directly relate to the dynamics of hydrology and hydraulics. Structures in streams provide habitats for organisms.

(4) *Substances* include the dissolved materials like nutrients, organic matter, oxygen, major ions and contaminants, and they directly follow the water flow. They are called water quality parameters. There are both lateral and longitudinal increases in the streams of the dissolves substances i.e. from the catchment boundary towards the stream and from the source to mouth, respectively. Stream hydrology, structures and substances together compose the group of controlling factors that directly determine how the stream community functions. These factors are included in the decision support system (see next paragraph).

(5) *Species* are the response to the functioning of all above-mentioned groups of controlling factors. Species and their communities are the actual goal of ecological stream management and rehabilitation. Controlling and response characteristics are not solely related to one of the mentioned groups of factors. There are mutual interactions. Structure, for example, can respond to the action of stream hydrology but can, conversely, also reduce discharge fluctuations. Species can be adapted to stream hydrology but, for example, trees can steer stream flow and bed morphology. Despite a dominant hierarchical effect, a feed back is

always present. Thus, factors interact on different hierarchical scales and with different intensity.

Human influences

Knowledge of the hierarchy in factors and processes acting in space and time in streams, allows also inferring the direction and magnitude of potential changes due to human activities. The changes refer to alteration as well as to restoration, and the time involved/needed. Human alterations can be seen as a sixth 'S'; the 'S' of *Steering.* The alteration and restoration of streams steer the ecosystem in a negative (alteration) or positive (restoration) direction.

Decision support system for stream restoration

In Figure 11, the present macrofauna assemblage is taken as a starting point for the decision support system. The macrofauna assemblage is composed of all macro-invertebrate species and their numbers of individuals present at the site or stream reach under consideration. From the macrofauna assemblage, the required information on the state of the stream environment is extracted by using the ecological information of each individual species, such as saprobic valence, current velocity, trophic state, substrate composition, etc. Verdonschot & Nijboer (2000) give examples of a quantitative approach to this process. When there are one stressor at the same time, more often the assemblage will indicate this possibility, except when the disturbance is too intensive. Severely impacted stream sites, where stressors are mostly very obvious, need an additional approach. Thus, the macrofauna assemblages tell about the state of the stream environment. For the decision support system the following major steps are taken:

(1) The macrofauna firstly refers to the four major groups of key factors, as described in the 5-S-model, which comprises four decision sub-keys in the decision support system. These major groups of key factors can be specifically managed and refer to; water quantity (stream hydrology), water quality (substances), habitat variability (structures), and direct human interference. System conditions mainly act at a coarse spatial and temporal scale. Stream hydrology, structures and substances mainly act at an intermediate scale, and species function at a fine scale though indicates the state of the first three named. Of course, exceptions to this rule show patterns of dominance and reaction whereby even species can dominate system conditions.

(2) Within each decision sub-key the macrofauna is related to three items: (a) the controlling processes responsible for the state of the stream under study, (b) the specified groups of macrofauna indicative of these processes, and (c) the state of some relevant indicative environmental parameters.

(3) The macrofauna composition, environmental parameters and controlling processes lead to the following questions concerning major human influences. The potential human causes of alteration detected instream within each decision sub-key, are arranged a hierarchically. The key starts asking questions at the scale of the catchment as a whole or more specifically at the infiltration area. Then it focuses towards finer scales: surroundings, riparian zone, towards in-stream.

(4) In the following step the human causes detected are related to major groups of actions to be taken (Table 4).

(5) Finally, in Table 4, the actions decided in the decision key (Fig. 11) are translated into measures to be taken, dependent on which specific key parameters are disturbed and on the specific causes of alteration.

This decision support system can be used to identify the effect of human influences upon key factors to be steered. Furthermore, the controlling processes are hierarchically ordered. The figure and table translate the 5-S-model, scale and hierarchy and the four dimensions together into a frame that can be used to improve restoration planning. The decision support system (combined Fig. 11 and Table 4) is based on knowledge on the ecology of macrofauna as well as on measures or knowledge of a number of key factors and key parameters, their reference or target value and their present state. The difference between the latter two indicates the processes that are important and may be disturbed by the indicated human activities. With this relationship it is possible to choose the relevant actions to be undertaken.

Discussion

The decision support system presented provides information only on measures that should be taken and not on 'where' the measure should be taken and 'how' intensive or strict the measure should be. For example, information is need as to how much agriculture, drainage and emissions are acceptable in a catchment and where, in order to come to an integrated approach

Table 4. Actions and stream restoration measures

Action	Restoration measures
A1 Water quality improvement (point sources)	Discharge reduction of sewage
	Discharge reduction of effluent
	Improvement purification plants
	Construction of natural purification filters (helophytes)
	Construction of horse-shoe wetlands
	Construction of buffer zones
A2 Water quality improvement (non-point sources)	Discharge/use reduction of toxic substances
	Discharge reduction of manure and nutrients
	Diversion of polluted flows
	Construction of natural purification filters (helophytes)
	Construction of horse-shoe wetlands
	Construction of buffer zones
B Restoration groundwater supply and flow	Removal of surface and subsurface drainage
	Improvement of infiltration
	Change of (ground-)water extraction
	Afforestation of the catchment
	Construction of hydrological buffers
	Infiltration of purified effluent
	Creation of inundation areas
	Improvement of water retention
	Reconstruction of natural catchment
C Length profile adjustment	Natural re-meandering
	Digging new meanders
	Construction of in-channel meanders
	Removal of weirs
D Transverse profile restoration	Construction of a-symmetric profile
	Creation of overhanging banks
	Profile narrowing and stream bottom raising
	Removal of profile consolidation
	Bottom silt removal
	Construction of by-passes and secondary channels
	Create berms to take high flows
E Riparian zone restoration	Creation of wooded banks
	Digging of pools
	Reconstruction/opening of old meanders
	Lowering of adjacent land
	Establishing agricultural free zone
F Habitat improvement	All measures listed under B, C, D and E
	Reconstruct habitats for specific species
	Creation of riffles and pools
	Construction of fish ladders
	Introduction of objects into the channel (trees, stones)
G Maintenance adaptation	Reduction of maintenance frequency
	Reduction of maintenance intensity
	Spotwise maintenance
H Re-introduction/removal of species	Removal by fishing
	Re-introduction programmes
	Maintenance adaptation (G)

Furse (eds), Assessing the Biological Quality of Fresh Waters: RIVPACS and Other Techniques. Freshwater Biological Association, Ambleside, Cumbria, U.K. The RIVPACS International Workshop, 16–18 September 1997, Oxford, U.K. Chapter 17: 241–262.

Wallace, J. B., J. R. Webster & W. R. Woodall, 1977. The role of filter feeders in flowing waters. Arch. Hydrobiol. 79: 506–532.

Ward, J. V., 1989. The four dimensional nature of lotic ecosystems. J. n. am. Benthol. Soc. 8(1): 2–8.

Ward, J. V. & J. A. Stanford, 1983a. The intermediate-disturbance hypothesis: an explanation for biotic diversity patterns in lotic ecosystems. In Fontaine, T. D. & S. M. Bartell (eds), Dynamics of Lotic Ecosystems. Ann Arbor Science, Collingwood: 347–356.

Ward, J. V. & J. A. Stanford, 1983b. The serial discontinuity concept of lotic ecosystems. In Fontaine, T. D. & S. M. Bartell (eds), Dynamics of Lotic Ecosystems. Ann Arbor Science, Collingwood: 29–42.

Werkgroep Beekbegeleidende Beplantingen, 1989. Hoofdrapport van de werkgroep beekbegeleidende beplantingen. Beplantingen langs waterlopen als beheersmaatregel. Mededelingen Landinrichtingsdienst, Utrecht. 179: 1–60 (in Dutch).

Zonderwijk, M., 1994. Monitoring van flora en fauna, morfodynamiek, waterbodemkwaliteit en beheerin beekproject "Midden-Regge fase I". Waterschap Regge & Dinkel, Almelo: 95 pp (in Dutch).

Hydrobiologia **478:** 149–170, 2002.
P.H. Nienhuis & R.D. Gulati (eds), Ecological Restoration of Aquatic and Semi-Aquatic Ecosystems in the Netherlands (NW Europe).

Restoration of brook valley meadows in the Netherlands

A.P. Grootjans[1], J.P. Bakker[1], A.J.M. Jansen[2] & R.H. Kemmers[3]
[1]*Laboratory of Plant Ecology, University of Groningen, P.O. Box 14, 9750 AA Haren, The Netherlands*
E-mail: A.P.Grootjans@biol.rug.nl
[2]*WMO, Water Company Overijsel, P.O. Box 10.005, 8000 GA Zwolle, The Netherlands*
[3]*Alterra, P.O. Box 25, 6700 AC, Wageningen, The Netherlands*

Key words: fen meadows, hydrology, Red List species, restoration management, rewetting, sod cutting

Abstract

Until recently, restoration measures in Dutch brook valley meadows consisted of re-introducing traditional management techniques, such as mowing without fertilisation and low-intensity grazing. In the Netherlands, additional measures, such as rewetting and sod cutting, are now carried out on a large scale to combat negative influences of drainage and acidifying influences by atmospheric deposition. An analysis of successful and unsuccessful projects shows that restoration of brook valley meadows is most successful if traditional management techniques are applied in recently abandoned fields that had not been drained or fertilised. Large-scale topsoil removal in former agricultural fields that had been used intensively for several decades is often unsuccessful since seed banks are depleted, while hydrological conditions and seed dispersal mechanisms are sub-optimal. In areas with an organic topsoil, long-term drainage had often led to irreversible changes in chemical and physical properties of the soil. Successful sites were all characterised by a regular discharge of calcareous groundwater provided by local or regional hydrological systems, and, where not very long ago, populations of target species existed. On mineral soils, in particular, sod removal in established nature reserves was a successful measure to increase the number of endangered fen meadow species. It is argued that attempts to restore species-rich meadows should be avoided on former agricultural fields, where pedological processes have led to almost irreversible changes in the soil profile and where soil seed banks have been completely depleted. From a soil conservation point of view, such areas should be exploited as eutrophic wetlands that are regularly flooded.

Introduction

Fen meadows or litter meadows are nutrient-poor grasslands which can be derived from groundwater fed mires after modest drainage (Ellenberg, 1978), but can also occur on groundwater fed mineral soils. Such meadows may produce 1–4 tons dry weight ha^{-1} yr^{-1} (Klapp, 1965). Hay meadows, usually derived from eutrophic marshes in flood plains, may produce 4–8 tons ha^{-1} yr^{-1}. The most species-rich fen meadows are classified as Cirsio-Molinietum and harbour a large number of species that have been listed on the 'Red List' of highly endangered species. Examples are *Cirsium dissectum, Carex dioica, C. pulicaris, C. hostiana, Viola persicifolia, Parnassia palustris* and orchid species, such as *Gymnadenia conopsea, Dactylorhiza incarnata*. Hay meadows are usually less species-rich and have less Red List species. Target species here are: *Caltha palustris, Crepis paludosa, Dactylorhiza majalis, Juncus acutiflorus, Lychnis flos cuculi, Carex aquatilis, Pedicularis palustris* (see Appendix 1). In Western Europe, such species-rich meadows have become increasingly rare as a consequence of changes in agricultural practices. Estimates from the United Kingdom are that 95–98% of species-rich hay meadows that were present before 1940, have been lost due to intensification of agricultural exploitation or due to abandonment and subsequent development of forest (Garcia, 1992). In the Netherlands, a similar process has occurred. Most species-rich meadows that exist today are managed by nature conservation organisations or have been restored after intensive agricultural use (Bakker & Olff, 1995). The establishment of nature reserves, however,

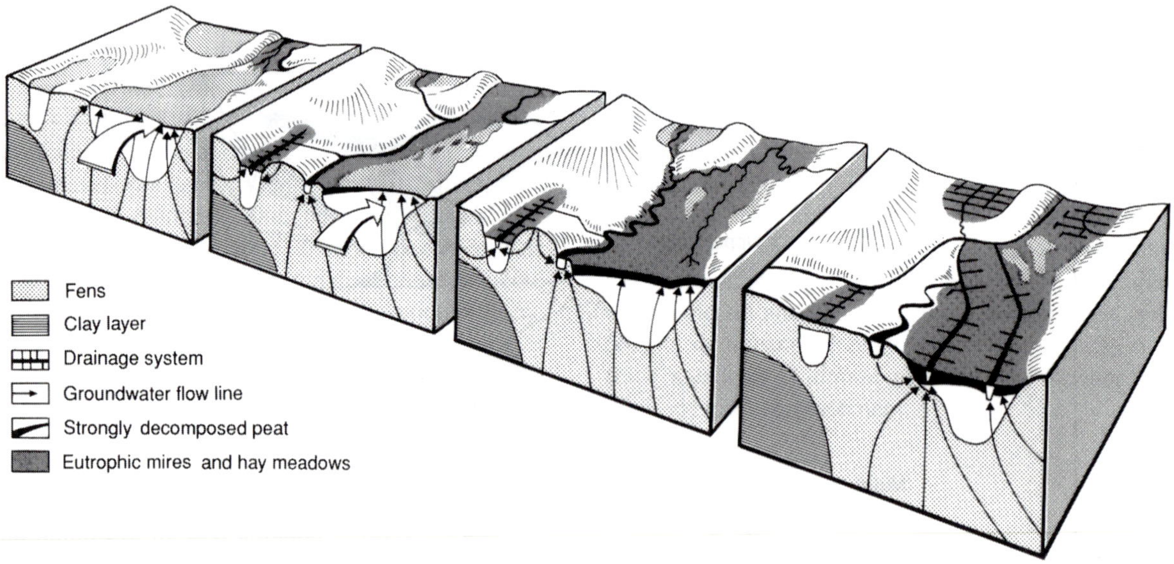

Figure 2. Changes in time in a brook valley landscape due to changed agricultural practices. First, natural mires, such as bogs and fens were changed into semi-natural fen meadows. This was followed by ecological degradation due to intensive drainage and fertilisation (adapted from Grootjans & van Diggelen, 1995).

The transition from the semi-natural to the present cultural period in NW-Europe was caused by the introduction of artificial fertiliser. The large-scale reclamation and subsequent fertilisation of common grassland and heathland occurred after 1920 when mechanisation in agriculture started. It resulted in the development of the cultural landscape, in which not only the vegetation but also the flora and fauna became heavily influenced by man. Indigenous species were eradicated by herbicides and non-indigenous species were introduced. These landscapes represent the cultural period.

These changes in land use had strong repercussions for brook valley systems. The large-scale reclamation of heathlands also led to many short cuts in the hydrological cycle, which increased flooding of meadows in the middle- and lower courses of the brook valleys after heavy rainfall. This in turn led to large-scale interference with the hydrological conditions in the early 1960s. It often resulted in the complete disappearance of all natural watercourses and in deep drainage of all peat soils. This resulted in subsidence of the peat, increased mineralisation, eutrophication of surface water, replacement of calcareous groundwater by rainwater, and subsequent acidification of the topsoil (van Wirdum, 1991; Wassen et al., 1996). All aforementioned processes resulted in a dramatic decrease of species of the former semi-natural landscape. Many plant species became endangered and were restricted to marginal environments in a fragmented landscape.

Nutrient cycling

Pedogenetic processes in brook valleys are mainly controlled by hydrological conditions. Both quantitative and qualitative aspects of the hydrology are conditioning site factors, for instance by controlling decomposition and humification processes in the organic topsoil. Consequently, the balance between input of litter and these processes determines whether peat or humus is formed. This balance between litter input and consequent decomposition or humification is displayed in the humus form of a site (Klinka et al., 1981), consisting of a sequence of soil horizons. The morphology of such horizons is often very informative for the type of nutrient cycling that is going on in the soil (Kemmers et al., 1995). Fibric humus forms consist of poorly decomposed fibric plant remains indicating accumulation of nutrients, slow nutrient cycling and poor N-availability. Potassium, however, is easily available from rapid cytolysis. Very wet sites with fibric humus forms are usually N-limited and respond to drainage with increased above-ground production (Grootjans et al., 1985). Humic humus forms consist of organic matter that is already decomposed and humified. Such humus forms have a rapid nutrient cycling and the availability of nutrients is high. However, the productivity of sites with humic soil horizons may gradually decrease due to continued removal of biomass by haymaking. Very old fen meadows, for instance, have developed P-limitation due to long-term removal of phosphorus by mowing (Egloff, 1986; Koerselman et al., 1990; Pegtel et al., 1996). Even potassium may become in short supply (Kapfer, 1988), particularly after deep drainage (de Mars et al., 1996; van Duren et al., 1997a,b; van Duren & Pegtel, 1999). Such meadows respond to drainage with a decreased above-ground biomass (Grootjans et al., 1986). Potassium deficiency favours grasses while inhibiting the development of leguminosae seedlings and other herbs (Mengel & Kirkby, 1982; Pegtel, 1983). If the potassium level drops below 0.5% K dry weight, fertilisation with K is necessary to enable establishment of target species, while (co-) limitation of P and N stimulates the establishment of many fen meadow species (Biewer, 1997).

Mesic humus forms have an intermediate character and usually occur in wet sites influenced by base-rich surface or groundwater.

During the 'natural' period, typical species of fen meadows were probably only present along a small fringe of brook valleys where humic-soil types could develop (Fig. 3). During the 'semi-natural' period, most of the natural fens were slightly drained and sites with humic topsoils developed, causing *Junco-Molinion* species to spread in the new meadows. Simultaneously more productive *Calthion meadows* developed from *Magnocaricion* mires with fibric topsoils, which changed into mesic sites.

At present *Junco-Molinion*, *Parvocaricion* and *Calthion* meadows are associated with humics, mesics and fibrics, respectively (Kemmers et al., 1995). Fen meadows (*Junco-Molinion*) developed where humics persisted from the natural period to present. These low productive meadows are the most species-rich and harbour a large number of endangered (Red List-) species. The difference in productivity between mesotrophic fen meadows (*Junco-Molinion*) and eutrophic hay meadows (*Calthion* and *Magnocaricion*) is clearly illustrated in Table 1. It shows above-ground biomass production and nutrient content of three meadow communities (Olde Venterink, 2000): fen meadows (*Junco-Molinion*), *Caltha palustris* hay meadows, and *Glyceria maxima* flood meadows. The fen meadows have the lowest and the *Glyceria maxima* dominated meadows have the highest above-ground biomass. An interesting feature is that the *Junco-Molinion* fen

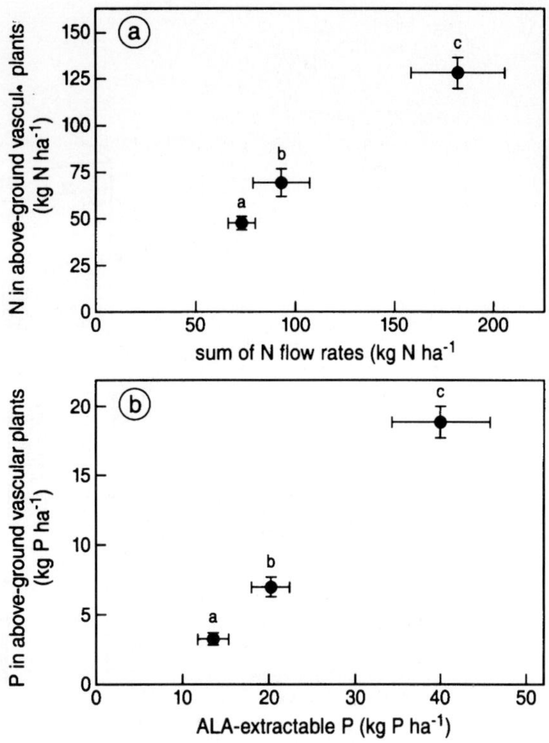

Figure 4. Amounts of N and P in the above-ground biomass of vascular plants at the peak of the growing season (July 1995) vs. assessed N and P availabilities for plants. The three meadow communities investigated are *Junco-Molinion* (a), *Calthion palustris* (b), and a *Glyceria maxima* dominated meadow type (c). Figure 4A shows the N in vegetation vs. the sum of annual flow rates of atmospheric deposition, net annual N-mineralisation, net losses via denitrification and input and output of N via ground and surface water. Figure 4B shows the amount of extractable P-pool measured in acetic-acid/lactic-acid extraction (ALA-extractable P). Changed after Olde Venterink (2000).

meadows differ from the *Calthion palustris* hay meadows only in phanerogam biomass, not in total biomass including bryophytes. Furthermore, the fen meadows have lower phosphate contents than the hay meadows. Nitrogen levels were also lower, but no significant differences were found in potassium levels. These findings are consistent with experiments indicating that well developed *Junco-Molinion* fen meadows are P-limited (Pegtel, 1983; Egloff, 1986), while most hay meadows (*Calthion palustris* and *Magnocaricion*) are N-limited (van Duren & Pegtel, 2000). Figure 4 shows the relation between N and P content in the above-ground biomass and the N and P availabilities in the soil (Olde Venterink, 2000). N-availability was calculated as the sum of annual flow rates of atmospheric deposition, net annual N-mineralisation, net losses via denitrification and input and output of N via ground

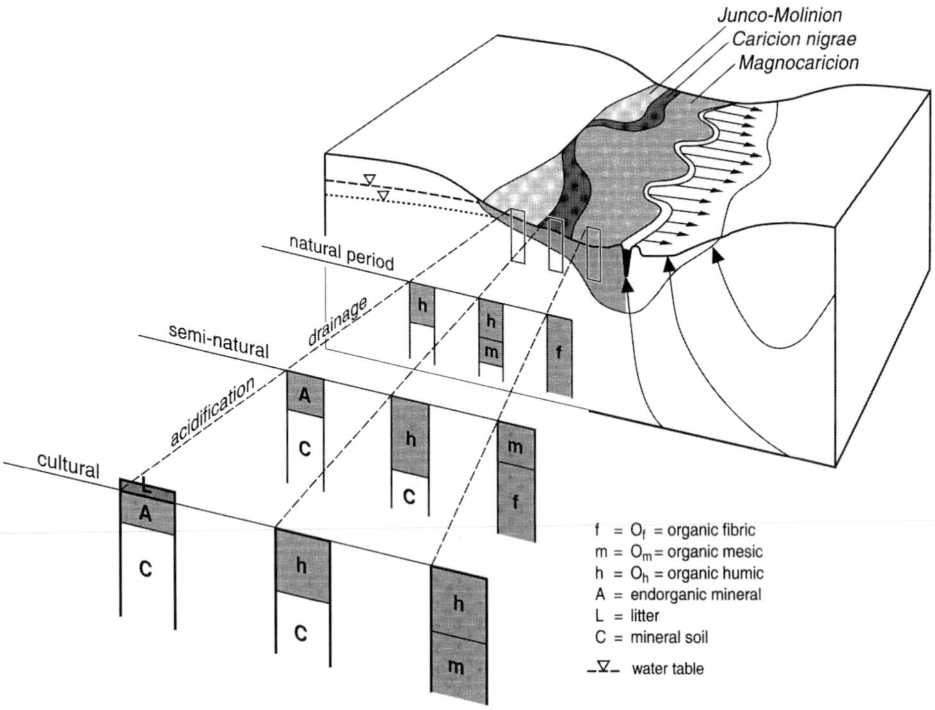

Figure 3. Typical brook valley system with mesotrophic fen meadows on the valley margins influenced by discharge of calcareous groundwater, bordered by sedge communities which are flooded by surface water. Pedological changes in the organic topsoil are shown in three transects, indicating the soil transformation after drainage during the semi-natural period and the cultural period.

and surface water (Fig. 4A). Figure 4B shows the extractable P-pool measured in acetic-acid/lactic-acid extraction (ALA-extractable P). A close relationship exists between N and P availability in the topsoil (10 cm) and N and P content at peak standing crop. Olde Venterink (2000) found that most of the available N and K was stored in above ground biomass (75% and 57%, resp.). Storage of P in the above-ground biomass was less. Mowing, therefore, removes relatively more N and K then P when fertilisation is stopped. However, since nutrient pools in the soil are generally much higher than plant available N and K (Güsewell, 1997) the soil is not easily depleted. As mentioned earlier, P-depletion can occur in fen meadows that have been mown for a long time without fertilisation (Koerselman et al., 1990). Olde Venterink (2000) also found that the availability of both P and K was primarily determined by internal nutrient cycling (labile pools, mineralisation), although in flood meadows almost 20% of the K originated from floodwater. Atmospheric N-deposition was the most important N source in low productive meadows, while N from mineralisation was more important in high productive meadows.

The above mentioned studies indicate that productivity and consequently also species composition in hay meadows is primarily determined by internal nutrient cycling within the ecosystem, which is governed by the groundwater fluctuation pattern and ionic composition of the groundwater. In low productive fen meadows, atmospheric deposition of N and S may result in eutrophication and increased alkalinity (Lamers et al., 1998). The latter might trigger an increased P-availability and a shift from P to N-limitation. Many present-day fen meadows are also influenced by drainage from surrounding agricultural areas, and acidify rapidly. Their topsoil is developing fibric characteristics indicating nutrient accumulation and decreased productivity. Pedologically these soils irreversibly change towards podzolic systems, in which iron is leached by a process of ferrolyses. Such soils can not easily be restored by hydrological measures intended to restore the original high base status (Faulkner & Richardson, 1989; Kemmers et al., 2000).

Table 1. Above-ground biomass and nutrient contents of phanerogams and mosses of three meadow communities, measured in July 1995): *Junco-Molinion*, *Calthion palustris* and *Glyceria maxima* dominated flood meadow (after Olde Venterink, 2000). Significant differences ($p<0.05$) between plant communities are indicated with different characters

Plant community	*n*	Above ground biomass (g m^{-2})	N (kg ha^{-1})	P (kg ha^{-1})	K (kg ha^{-1})
Molinia coerulea meadow					
Phanerogams	8	352^c	47^c	3^d	30^b
Mosses	8	131^p	19^p	2^p	12^p
Caltha palustris meadow					
Phanerogams	7	479^b	69^b	7^c	32^b
Mosses	7	24^q	5^q	1^p	2^q
Glyceria maxima meadow					
Phanerogams	9	915^a	132^a	18^a	75^a
Mosses	9	3^r	1^r	0^r	0^r

Seed set, seed banks and seed dispersal

Haymaking is a dominant feature of meadows and the period of cutting depends on the productivity of the ecosystem. In general, the least fertile meadow types (fen- or litter meadows) are mown late, while the more productive hay meadows are cut early in the summer, and they may be cut twice. Nutrient-poor fen meadows are usually mown in September and many characteristic species set seed in late summer. Most hay meadow species set seed between May and July. Mowing occurs in June/July. Some productive *Magnocaricion* meadows are cut twice (early June and late August). Ter Borg (1972) showed that some species have adapted the period of seed set to the mowing period. She found genetic differentiation within *Rhinanthus angustifolius* with respect to the period of seed set. Early flowering populations were found in relatively productive grasslands mown in May. Intermediate flowering populations produced seed before the first cut in July, while a late flowering population, found in a large sedge meadow that was mown twice a year, flowered after the first cut and produced ripe seeds before the second cut.

The period of mowing is important for seed dispersal in yet another way. Nowadays species-rich meadows are generally mown by machines that go from one field to another on the same day. As an experiment, Strijkstra et al. (1997) cleaned different parts of the mower between two fields in order to estimate seed transport. The mower has a horizontal safety skirt covering the mowing disk, which traps a lot of seeds. When the machine leaves the field, the part covered by the safety skirt is put vertically, and all the seeds drop onto the soil. The skirt is, therefore, only important for seed transport within fields. The skid disk of the mower, at soil level, remains in the same position and is, therefore, important for seed transport between fields. Strijkstra et al. (1997) estimated that transport by haymaking machinery could amount to over 1 000 000 seeds during haymaking.

Seeds that are not removed from the field by mowers drop onto the soil. The seeds that are not predated or do not decay become incorporated into the soil seeds bank. Seeds in the seed bank can be classified according to their estimated longevity. Species only present in the established vegetation and not in the seed bank are referred to as having a transient or short-lived seed bank. The same holds for species that only have viable seeds in the topsoil. Species with many viable seeds in deeper soil layers are classified as having a long-term persistent seed bank, since it is assumed that it takes time to become buried in the deeper soil layers (Bekker et al., 1998). Van Diggelen (1998) used the soil seed data bank of Thompson et al. (1997) and found that in comparison to natural communities, meadow communities seem to consist mainly of species with a longer-lived seed bank. Fifty seven percent of the characteristic meadow species show indications of still having viable seeds after 5 years and 30% even after 20 years. These results suggest that meadows can be easily restored from the soil seed bank after they have been lost. This was, however, true for common hay meadow species, not for the rare ones. Field studies showed that most Red List species of fen meadows (*Junco-Molinion*; Maas & Schopp-Guth,

1995) hay meadows (*Calthion palustris*; Biewer & Poschlod, 1995) and flood plain meadows (*Magnocaricion*; McDonald et al., 1996) have short-lived seed banks.

Restoration of species-rich meadows: why and how?

From the beginning of the 20th century onwards, areas have been acquired by private organisations for landscape and nature conservation purposes, more recently also by the State.

The reason for the maintenance or restoration of brook valley systems is their typical semi-natural character; a complex historical matrix superimposed on a very complex geological matrix. The occurrence of a large variety of environmental gradients ranks brook valley systems in the top of species-rich environments in the Netherlands.

Mowing without fertilisation

The most common technique to restore species-rich meadows is to re-introduce a mowing regime in former agricultural fields or abandoned meadows without additional input of fertiliser (Bakker & Olff, 1995; Muller et al., 1998). Grassland communities that have high species richness achieve a maximum standing crop, including litter, of between 4 and 6 tons ha^{-1} (Grime, 1979). Both fertilised grasslands (because of nutrient input) and abandoned grasslands (because of litter accumulation) have a much higher above-ground standing crop. Therefore, it is anticipated that a reduction of the maximum above-ground biomass will trigger an increase of species richness, provided that the target species are indeed present in the soil seed bank, or can be dispersed from the actual species pool. Grasslands under a restoration management, indeed, showed a decrease of above-ground standing crop, after the cessation of fertiliser application (Olff & Bakker, 1991).

In order to investigate how the result of the restoration management depends on the fertilisation history, fields have been selected with similar soil and hydrological conditions, but with different fertilisation histories. All fields were cut once a year in late July/early August. Fertiliser application ceased in different years in the selected fields (Fig. 5): 1985 (field B), 1967 (field F) and 1945 (field D), respectively. Differences between fields are assumed to have been established since then. The species composition of the fields was assessed in 1994. *Poa trivialis* and *Lolium perenne* dominated in field B, *Rhinanthus angustifolius* in field F, and *Juncus acutiflorus* in field D. Only field D had many target species. Taking 1945 as a starting point, when the input of artificial fertiliser was low, an estimation can be made of the cumulative input of nitrogen (including atmospheric deposition) and output via cuttings. The difference between input and output estimates enables a cumulative nitrogen balance (CNB) to be calculated for different fields (Bakker & Olff, 1995). It is obvious that field D shows a negative CNB from the beginning of restoration management.

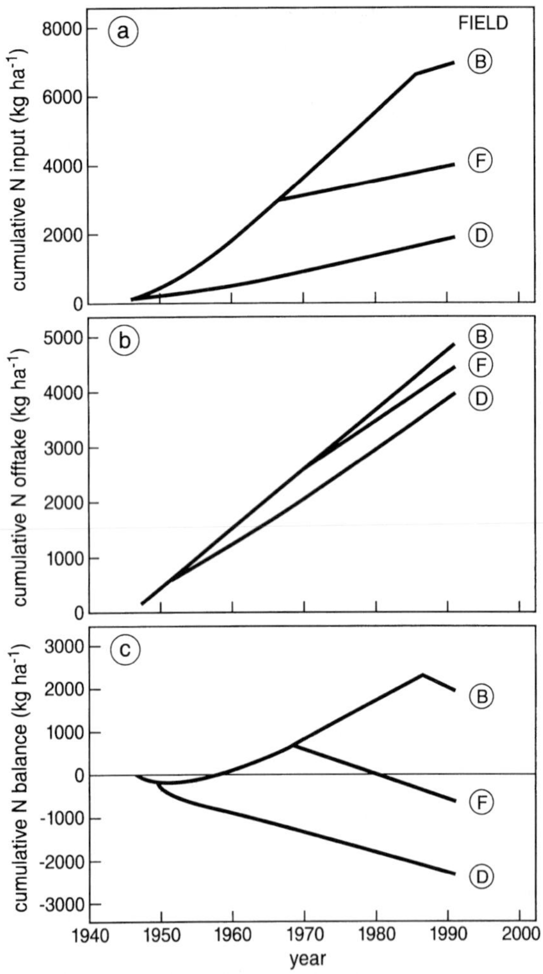

Figure 5. Estimated cumulative N input (a), (fertiliser application and atmospheric deposition since 1947), cumulative N off-take by mowing (b) and cumulative N balance (c) (difference between input and off-take) in three grasslands differing in time under restoration management (not fertilised for 45 (D), 20 (F) and 2 (B) years, respectively, changed after Bakker & Olff, 1995).

Field F reached the turning point from positive to negative CNB in 1980, whereas field B still has many years to go to achieve a negative CNB. Differences in CNB coincided well with differences in above-ground standing crop in July and accounted better for the number of species per plot than did the number of years without fertiliser application (Bakker & Olff, 1995). Concluding, it is important to consider the fertilisation history in restoration management.

The low yields of hay obtained from unfertilised species-rich meadows are uneconomical for farmers to harvest without receiving financial compensation from public sources. These meadows need specialised nature management using equipment especially designed to mow extensive wet areas with little manpower.

Grazing in wet meadows

The introduction of extensive grazing systems, which is promoted in many European countries to sustain or restore species-rich grasslands (Pfadenhauer & Grootjans, 1999), is not applied on a regular basis in Dutch wet meadow ecosystems. Most often it is combined with a mowing regime; in particular young cattle may graze wet meadows in late summer after the meadows have been mown. Light cattle breeds are sometimes used in large eutrophic wetlands on mineral soils. Small brook valleys with wet organic soils are considered unsuitable for cattle grazing by the Dutch nature conservation organisations. Cattle prefer dry places within a wetland leading to a rapid growth of tall herbs in the wet areas (Bakker & Grootjans, 1991; Bakker, 1998). Similar effects of grazing were observed by Kaiser (1995) in eutrophic wet meadows in the former German Democratic Republic where a restoration management by extensive grazing (stocking rate of 1 animal ha^{-1}) was applied in an intensively used agricultural field of 37 ha. After four years of grazing (1991–1995), ruderal species, such as *Cirsium arvense*, were spreading in dry and moist areas, while tall grasses, such as *Phalaris arundinacea*, *Elymus repens* and *Agrostis stolonifera* became dominant in the wet parts of the meadows.

Grazing in very low productive *Cirsio-Molinietum* fen meadows may cause health problems for cattle, particularly in autumn, when *Molinia caerulea* reallocates nutrients from the leafs and stores them in the basal nodes of the tussock to be used again for the next season. Fen meadows dominated by *Molinia caerulea* become very unsuitable for grazing in autumn, and the health of live stock may suffer severely (Tallowin & Smith, 1996).

Rewetting

Rewetting often requires more than just raising local water levels in nature reserves. If, for example, water levels in the surroundings have been lowered such measures may lead to acidification of the top layer since only precipitation water will be stored within the reserve (Grootjans et al., 1996; van der Hoek & Kemmers, 1998). Sometimes the closing of deep ditches to promote seepage of calcareous groundwater must be followed by re-installing shallow ditches to prevent inundation with acid precipitation water. Restoration of the whole hydrological system is sometimes necessary (van Diggelen et al., 1994), for instance, by eliminating drainage systems or closing down groundwater abstraction facilities. But often this is not an option. Technical measures, such as isolating the effects of drainage by placing artificial barriers against groundwater flow, is possible but is very costly. New techniques, e.g. deep well infiltration to produce drinking water without lowering water tables is available but application is mostly restricted to dune areas (Janssen & Salman, 1995). Rewetting by initiating flooding of surface water is also a method to restore hay meadows. Although originally artificial flooding with surface water was applied to increase the productivity of the hay meadows (Burny, 1999), it is now applied in nature reserves to prevent desiccation and acidification of the meadows (Spieksma et al., 1994). In nutrient-poor ecosystems, however, flooding with nutrient-rich or sulphate-rich surface water may lead to severe eutrophication (Lamers et al., 1998). This is why helophyte filters or large ditch systems are often used to purify the water (Meuleman, 1990; van Duren et al., 1998). On one occasion, groundwater from deep aquifers was pumped to the surface to mimic former flooding practises (van der Hoek & Kemmers, 1998).

Sod cutting

Sod cutting used to be a traditional measure to increase the soil fertility in small arable fields by taking sods from the common heathlands, which were mixed with dung. Later, nature conservation applied sod cutting on a large scale in Dutch nature reserves in order to combat grass encroachment in heathland vegetation. Sod cutting is now also applied in fen meadow reserves to counteract the effects of acidification and atmospheric nitrogen deposition (Jansen & Roelofs, 1996). The most drastic impoverishment can be achieved by complete removal of the nutrient-rich topsoil. This measure is often applied in former agricultural fields now destined to become nature reserves. In some cases the topsoil is removed up to a depth of 10–40 cm (van Diggelen et al., 1997; Jansen, 2000). As a disadvantage it must be mentioned that sod cut sites often have obtained a low pH buffer capacity due to removal of the (organic) cation exchange complex. Such soils are inclined to rapid acidification if they are not buffered by base-rich groundwater (Jansen, 2000). In addition to this radical depletion of nutrients, undesirable competitive elements are removed, although they may redevelop from the soil seed bank when sod cutting was too shallow (van Diggelen et al., 1997). In extreme cases, topsoil removal eliminates practically the whole soil seed bank (Klooker et al., 1999).

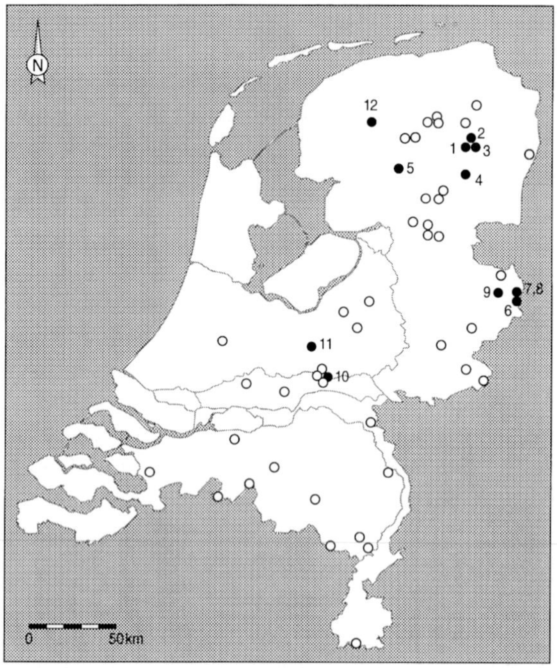

Figure 6. Areas where restoration measures have been applied recently in existing hay- and fen meadow reserves in the Netherlands (changed after van der Linden et al., 1996). Study areas discussed in the present paper are indicated in black. 1=Taarlo, 2=Anloo, 3=Eexterveld, 4=de Reitma, 5=de Barten, 6=Punthuizen, 7=Stroothuizen, 8=Groener, 9=Lemselermaten, 10=Veenkampen, 11=Grootzandbrink, 12=Wyldlannen.

Successful projects

In the Netherlands, a restoration project is considered a success if many Red List species or target species, which were common in meadows that existed half a century ago, re-establish. In a sense, successful restoration is like constructing a field museum for preserving a living part of a lost cultural heritage. This idea of success can very different in other countries (Pfadenhauer & Grootjans, 1999) and even in other areas of the Netherlands, where natural processes can form new ecosystems little disturbed by man.

In the following, we will use the number of target species that has established since the restoration measures to judge the success of restoration projects. The list of target species (see Appendix 1) shows officially recognised Red List species and characteristic species of wet meadow ecosystems (Schaminée et al., 1996).

Fen and hay meadow sites where recent restoration projects have been carried out are shown in Figure 6. Projects discussed in the present paper are indicated in black.

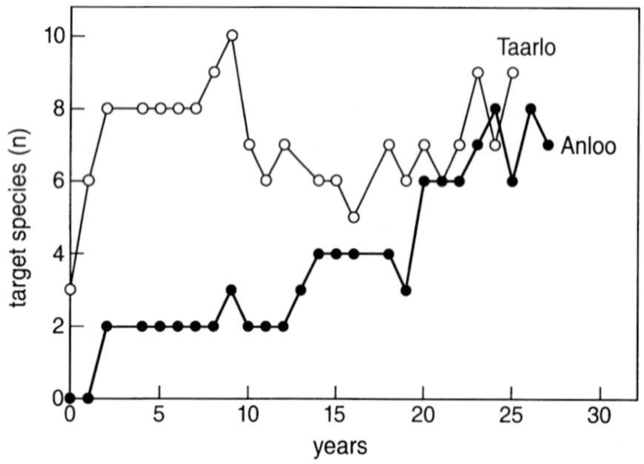

Figure 7. Increase of target species monitored in three permanent plots during 25 years in two brook valley meadows in the Drentse Aa area. The meadow near the village of Taarlo was fertilised only little before abandonment followed by restoration management, while the meadow near Anloo had been fertilised for several decades before a mowing regime without fertilisation was installed.

Mowing without fertilisation in wet meadow reserves ('Drentse Aa' valley)

The Drentse Aa brook valley reserve is the best preserved valley system in the northern part of the Netherlands. This landscape reserve was established in 1965 when it was decided that 3500 ha (of the 30 000 ha catchment area) would eventually be acquired by the State and managed by the State Forestry Service in order to preserve and restore the semi-natural landscape. The hydrology of the reserve is, however, heavily influenced by agricultural drainage systems along the

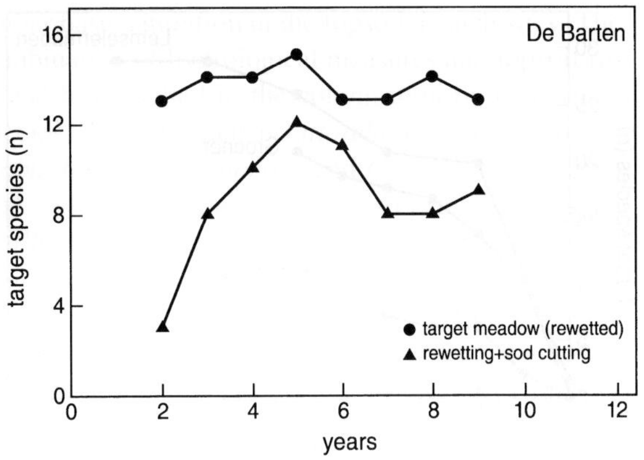

Figure 8. Changes in the number of target species in the hay meadow reserve 'de Barten' after sod cutting and rewetting.

valley flanks and by abstraction of groundwater on behalf of the public water supply (Grootjans et al., 1993). After 30 years, practically all remaining heathlands and almost all wet grasslands in the valleys are now owned by the State. The process of restoring species-rich meadows is being carried out on a large scale since 1971. Until recently, the approach was to rewet the less intensively used grasslands and re-introduce a regular mowing regime without fertiliser application. Some results of such a restoration management over 25 years are shown in Figure 7. The wet grassland areas close to the brook near the village of Taarlo (Fig. 6) had not been used very intensively, because they were too wet for intensive agricultural use. The restoration of such grasslands by mowing without fertilisation was relatively easy. In another area near the village of Anloo (Fig. 6), the meadows had been fertilised for several decades. The increase in target species was much slower, because of their unfavourable cumulative nitrogen balance (CNB: Bakker & Olff, 1995). It took almost 20 years before an *Alopecurus geniculatus/Holcus lanatus* dominated community, changed into a *Juncus acutiflorus* stage with many orchids (*Dactylorhiza majalis*). Most target species that were absent in the field at the beginning of restoration management were also lacking in the soil seed bank and must have been imported from elsewhere, most likely by hay-making machinery (Strijkstra et al., 1997).

Rewetting and sod cutting in desiccated hay meadows 'De Barten' (Linde valley)

'De Barten' is a complex of hay meadows, which escaped amelioration due to the intensive discharge of groundwater in the area. The hay meadow reserve is situated in the Linde valley (Fig. 6) and is bordered by drained agricultural fields on either side of the reserve. A straight water channel has replaced the original brook. As a consequence the hay meadows showed severe signs of desiccation in most of the reserve. Nature conservancy tried to rewet the meadows in 1990 by closing ditches in the reserve and by placing plastic shields along one of the main drainage channels to prevent water losses to agricultural areas. These measures led to a pronounced rise in groundwater levels in most of the reserve. Between 1994 and 1996, the mean highest and mean lowest water levels varied between 30 cm above and 25 cm below the soil surface. As an experiment, topsoil removal was applied in 1991 in two fields. One had been a hay meadow subjected to restoration management for 20 years already, but showed clear signs of desiccation. The other field had developed into an alder carr. Vegetation changes and abiotic conditions were monitored from 1991 to 1999. Figure 8 shows the average number of target species in the two sod cut areas compared to those in the adjacent uncut meadow. The restoration measures were quite successful, particularly in the sod cut areas. Within 5 years, the target area had the same number of target species as the rewetted hay meadows. Species that increased in numbers were characteristic hay meadow species, such as *Senecio aquaticus, Caltha palustris, Lychnis flos cuculi*, but also some Red List species, such as *Carex diandra, C. aquatilis* and *C. oederi* appeared in the sod cut areas. The increase in water levels caused an increased discharge of calcareous groundwater that reached the topsoil in wet periods. This led to an increase in pH and calcium saturation on the exchange complex, both in the former desiccated meadows and in the sod cut areas. The number of target species decreased again during the last few years, probably due to much precipitation causing acidification again in the topsoil.

Rewetting in fen meadow reserve 'de Reitma' (Elper stroom)

The fen meadow reserve 'de Reitma' was severely affected by desiccation between 1971 and 1985 due to large-scale drainage activities in the surrounding agricultural areas. Grootjans et al. (1986) have reported on the decline of several characteristic fen meadow species, such as *Carex hostiana, C. pulicaris,* and *Parnassia palustris*. They measured very high mineralisation rates in the desiccated peat soil and observed a dramatic decline in productivity, which they attrib-

Table 2. Changes in nutrient conditions in the topsoil of the hay meadow reserve Veenkampen after two different hydrological restoration measures (after van der Hoek & Kemmers, 1998). Values are given in percentages (relative to the control where no hydrological restoration measures were taken)

	Compartment	1986	1989	1991	1993	1997
Exch. Ca	Conservation	100	114	103	95	89
	Irrigation	100	98	86	96	91
Org. Matter	Conservation	100	91	89	98	97
	Irrigation	100	102	107	105	106
N-tot	Conservation	100	105	104	95	95
	Irrigation	100	112	105	101	107
P-tot	Conservation	100	105	120	108	108
	Irrigation	100	87	86	86	95
P-exch.	Conservation	100	169	132	135	141
	Irrigation	100	69	56	46	46

Rewetting and sod cutting in eutrophicated fen meadows ('Veenkampen')

The 'Veenkampen' is situated in the Gelderse Vallei in the lower Rhine area (Fig. 6), which once was famous for its wealth of species-rich fen meadows. Only small remnants have remained. The Veenkampen is a complex of former agricultural fields where the regeneration of species-rich meadows has been studied in experiments since 1986. Mowing without fertilisation has been applied in part of the complex. The other parts have been additionally rewetted and sod cut. Vegetation response on changed geochemical and hydrological conditions has been studied for more than a decade (Berendse et al., 1994; Oomes et al., 1996, 1997; van der Hoek & Kemmers, 1998). Rewetting was carried out by conserving precipitation water within some fields and by subsurface irrigation of calcium-poor groundwater (Ca-content=10 mg l^{-1}) from an artesian well, supplying groundwater from a confined aquifer at 50 m below the surface. The soils of the experimental fields consisted of peat with a shallow clayey top layer. They have been seriously affected by 40 years of intensive agricultural use. Inorganic phosphate accumulation had occurred and could not be reduced to its original levels by the restoration measures applied (Table 2). The study clearly showed that rewetting by conserving precipitation water in the field actually increased the P-availability for the vegetation, even after 10 years. Subsurface irrigation with calcium-poor groundwater from an artesian well decreased P-availability considerably after 5 years, but could not prevent further acidification of the toplayer. The decrease in P-availability was caused by the dissolution of iron-phosphates under anaerobic conditions and subsequent leaching of P to the subsoil. The conservation of precipitation water appeared to be an inadequate measure to restore species-rich grasslands. Almost no target species re-appeared in the experimental fields. Fields that were sod cut to drastically remove the high nutrient stocks were more successful in establishing populations of target species (Fig. 12), but most of the species were pioneer species, which probably emerged from long-persistent seed banks, even after 40 years.

Rewetting and sod cutting in eutrophicated fen meadows ('Stroothuizen')

After the nature reserve 'Stroothuizen' (c. 40 ha.) was founded in the 1960s, the manager decided to construct a dike along its western border in order to maintain high water levels. As a result, many depressions were inundated with base-poor precipitation water for at least 4 months. Calcareous groundwater could no longer reach the surface due to the presence of drainage ditches and drainage pipes in adjacent arable fields. This led to infiltration of precipitation water and subsequent acidification and eutrophication.

To stop the decline of mesotrophic fen meadow species, sod cutting was applied in large parts of the reserve. This measure was successful in restoring species-rich dwarf-rush communities, wet heathlands and small-sedge marshes, but not in restoring fen meadows (Fig. 9). Whereas the restored wet heathlands are now situated at the upper part of the height gradient, the small-sedge communities occur in the lower part. The failed restoration of the fen meadow vegetation might be due to increased P-levels in the topsoil because of prolonged inundation leading to dissolution of Fe-P and Al-P minerals under anaerobic conditions (Boeye et al., 1997; Jansen, 2000).

Rewetting without sod cutting in the degenerated fen meadow led to very incomplete recovery compared to the original fen meadow that existed some decades ago (Jansen, 2000). This might be due to irreversible changes in the structure of the organic topsoil layer. Long-term drainage probably had led to iron depletion (de Mars, 1996).

Acidified fen meadows ('Wyldlannen')

The fen meadow complex 'Wyldlannen' consists of c. 60 ha of degenerated and species-poor *Cirsio-Molinietum* vegetation in which *Agrostis canina* and *Phalaris arundinacea* are dominating on clay–peat soil. Characteristic species of the community, e.g. *Cirsium dissectum, Carex panicea* and *Molinia caerulea*, are still present, although in low numbers. The meadows were inundated by stagnating precipitation water for several months during winter and early spring. Calcareous groundwater could no longer reach the nature reserve, since the surrounding polder areas had been deeply drained for several decades. These deep drainage practices had caused a severe lowering of the peat surface in all polder areas, leaving the fen meadow reserve as the highest area in the landscape.

Attempts to restore the fen meadows started in 1986 when the water levels in the ditches were raised. This led to some rewetting in the summer, but not to an increase of characteristic fen meadow species. An old practice to flood the meadow with surface water was resumed in 1991 to stop the acidification. A small wetland (helophyte filter) of c. 1 ha was used to reduce nutrient availability in the surface water. Sod cutting was applied to a small area of 0.5 ha to investigate if topsoil removal of c. 20 cm would produce a less acidified environment for fen meadow species. Sod cutting, however, exposed a former *Sphagnum*-peat, with quite different chemo-physical characteristics.

At some distance from the degraded fen meadow complex, a small area ('Ulekrite') exists where a well-developed *Cirsio-Molinietum* still harbours species, e.g. *Pedicularis palustris, Carex hostiana, C. flacca, C. echinata* and *C. lasiocarpa*.

Van Duren et al. (1998) attempted to restore the fen meadow by means of rewetting, sod cutting, liming and introduction of target species. An evaluation was done in the framework of a monitoring project

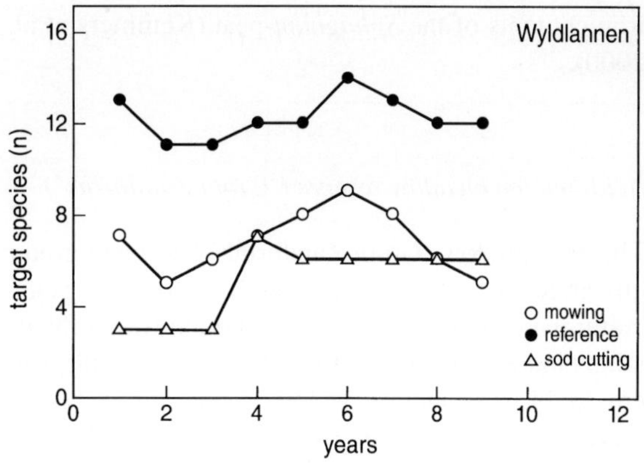

Figure 13. Changes in the number of target species in the fen meadow reserve 'Wyldlannen' after carrying out restoration measures, such as sod cutting and re-flooding with purified surface water.

carried out at three sites: a degenerated, a sod cut and a reference site. Groundwater levels in the degenerated and the sod cut sites differed slightly from that of the reference site but were within the range of *Cirsio-Molinietum* requirements as known from the literature. Topsoil removal considerably had reduced macro-nutrient contents and caused a pronounced P-deficiency. No increase in soil pH was observed immediately after the topsoil removal. A slight increase in soil pH was found, however, after 5 years in the sod cut site, indicating a slight recovery from acidification. Liming after topsoil removal had little effect on the soil pH after 6 months but stimulated the growth of established species as *Agrostis canina*. Few viable seeds of characteristic fen meadow species were present in the seed banks of both degenerated and well-developed fen meadow sites and the species composition after topsoil removal closely reflected the seed bank of the degenerated fen meadow. Establishment of characteristic fen meadow species was poor, even after 9 years (Fig. 13). The introduced species survived during the first year in both the degenerated and the sod cut sites. Liming stimulated the growth of two of the introduced species. In the second year, however, all introduced species showed considerable growth reduction and the positive effect of liming disappeared. After 2 years, none of the introduced species survived in the sod cut site, indicating that the environmental conditions were still unfavourable. It was concluded that the restoration measures failed so far, apparently due the continuation of deep drainage in the surrounding agricultural areas. Restoration of base-rich conditions is further hampered by the very low iron contents of the *Sphagnum*-peat (Kemmers et al., 2000).

Acidified fen meadow reserve ('Groot Zandbrink')

The fen meadow 'Groot Zandbrink' had been gradually affected by deep drainage in the surrounding agricultural area in the past 25 years. Groundwater fluctuations increased from 70 to 110 cm. As a result, several Red List species on the lowest sites, e.g. *Schoenus nigricans, Gymnadenia conopsea, Dactylorhiza incarnata* and *Plathantera bifolia* disappeared; others, as *Carex pulicaris, C, hostiana* and *Parnassia palustris* declined. A sharp increase of *Cirsium dissectum*, both in frequency and abundance had been recorded. Most vulnerable species are still present, but common meadow species are becoming more frequent. Originally the fen meadow was influenced by upward seepage of calcareous groundwater causing a high base saturation of the exchange complex of the soil (humic Ah-horizons). Rain water could not penetrate into the soil and left the meadows through small depressions. Deep drainage has now turned the whole meadow complex into an infiltration area and rain water found its way to deeper soil layers or formed small pools. Depending on their acid buffering capacity, the sites acidified by cation leaching to a base saturation level of less than 25%. At these levels, litter decomposition and humification becomes hampered, resulting in the formation of fibric horizons. This pedogenetic process appeared to be accompanied by leaching of iron from the fibric horizons (Kemmers et al., 2000). This fundamental pedogenic process turns these gleyic soils into podzolic soils.

To prevent further infiltration of rain water into the soil during wet periods, shallow trenches were dug in 1991 as a restoration measure. These measures were intended to increase the discharge of base-rich groundwater again and boost the degree of base saturation. The increased seepage flux of calcium, however, neutralised the acidic inputs derived from the atmosphere but did not lead to higher pH values in the topsoil.

The number of target species, appeared to increase in the best preserved parts of the reserve immediately after the restoration measures had been carried out in 1991, but remained more or less at the same level after 8 years (Fig. 14). None of the declining basiphilous species responded positively. No increase in target species was observed in the most degenerated parts of the reserve.

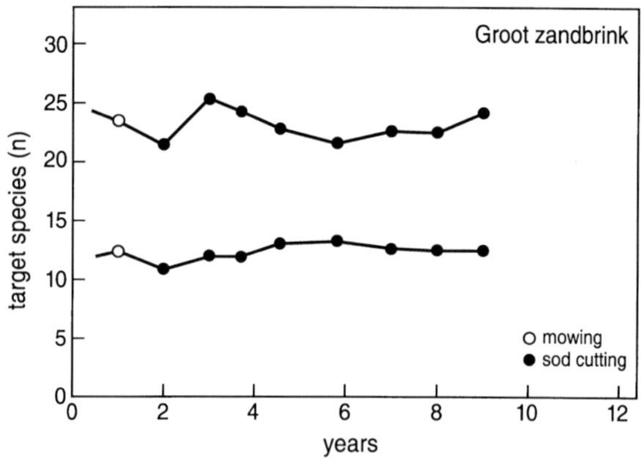

Figure 14. Changes in the number of target species in two different parts of the fen meadow reserve 'Groot Zandbrink' after carrying out restoration measures, such as rewetting and digging shallow ditches to prevent stagnation of rain water.

Most probably the newly formed fibric horizons fail to restore a high base status under the anaerobic conditions due to lack of iron, indicating an irreversible change in the organic soil compartment.

Sheep grazing and topsoil removal in a former agricultural field 'Eexterveld'

The 'Eexterveld' is a wet heathland reserve in the source area of the Drentse A brook valley system. The area is very wet during the winter due to the presence of semi-pervious clay layers in the shallow subsoil, but it can be very dry during summer. The area was once famous for its high biodiversity (van Andel et al., 1945). Most of the species-rich heathland, however, has now been reclaimed for intensive agriculture. Some *Junco-Molinion* relics are still present in small patches, but from 27 target species present in 1945, only 18 have remained. The species that have become extinct from the area include *Arnica montana, Parnassia palustris, Sagina nodosa* and *Scutellaria minor.*

Sheep grazing started in some reclaimed grasslands from 1970 onwards, when the fields were acquired and fertilisation was stopped. After 15 years, the number of target species, however, was very small; only a few individuals of *Calluna vulgaris, Erica tetralix, Juncus conglomeratus* and *J. acutiflorus* were found (Bakker, 1989).

Sod cutting in 1994 in an adjacent agricultural field resulted in a steady increase in the number of target species from four to eight species within 5 years (Fig. 10). Most of the characteristic fen meadow species, such as *Cirsium dissectum, Carex hostiana*, however, did not re-appear. Many species that emerged after topsoil removal (*Juncus bufonius, Rorippa palustris, Alopecurus geniculatus* and *Holcus lanatus*, were also dominant in the soil seed bank (Klooker et al., 1999). Other species (*Juncus effusus, Ranunculus repens*) probably recovered from root fragments that were not removed. *Erica tetralix* was the only wet heathland species still present is the soil seed bank. This species was not only found in the topsoil, but also at depths between 20 and 50 cm. Species that could reach the target area by wind dispersal were practically the same species as the ones present in the seed bank (*Alopecurus geniculatus, Poa annua, Juncus bufonius* and *Sagina procumbens*. The only target species that was found in the local seed rain was *Viola palustris*. Environmental conditions, immediately after topsoil removal were not favourable for acidophilous species. The pH remained relatively high (c. 5.5) and also the phosphorus levels remained high. These high values still reflect the effect of intensive fertilisation before 1995.

Evaluation

The case studies discussed above show that the time that has elapsed since the degradation of a fen meadow is critical for the restoration prospects of a site. Sometimes restoration is achieved with relatively little effort, while in other cases, restoration fails completely despite large costs on restoration measures.

Conceptual model of fen meadow restoration

A conceptual model of fen meadow restoration summarising successes and failures of restoration measures is presented in Figure 15. In most species-rich meadow reserves in the Netherlands, a traditional mowing regime without fertilisation was continued or re-instated by state or private nature conservation organisations. A success story is, for instance, the creation of the largest meadow reserve in the Netherlands in the Drentse Aa catchment. Most of such meadows responded very well to a restoration management (Fig. 15: 1) after a short period of abandonment at time T_1. Parts affected by drainage, however, responded well in the beginning, but lost many target species after some 10 years of restoration management (2; example: 'Taarlo drained'). Rewetting and/or sod cutting was quite successful in cases, where the process of de-

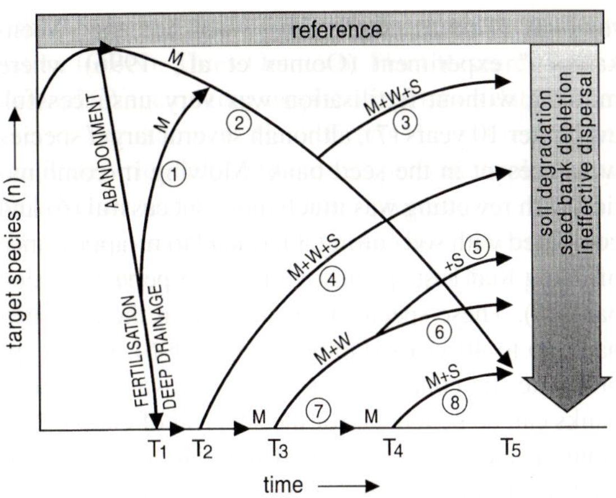

Figure 15. Conceptual model of occurrence of target species in fen and wet hay meadows under restoration management. Continuation of the traditional management (mowing without fertilisation) in meadow reserves can not not always prevent the extinction of many endangered species negative influences from surrounding agricultural areas. Restoration measures, e.g. rewetting and sod cutting are less effective after long-term exposure to these influences. Resuming a traditional management after a short period of abandonment is often very successful with respect to establishing of target species. If restoration management is resumed shortly after cessation of agricultural use, the restoration success is usually high compared to situations where intensive fertilisation has taken place for a long time. M=mowing without fertilisation, W=(re)wetting, S=sod-cutting. The numbers correspond to the case studies mentioned in the text.

siccation had proceeded only for a few decades (3; examples; 'Barten', 'Reitma'). When restoration management starts in meadows, which had been subjected to intensive fertilisation for 10–30 years (4), the increase of target species (starting at T_2), depends on the amount of nutrients accumulated during former fertiliser application (Bakker & Olff, 1995; example 'Anloo'). It may take 5–15 years (T_3-T_4), before the nutrient output has reduced the nutrients stocks to a level in which the productivity of dominant species no longer prevents the establishment of target species. Even more target species established in former agricultural areas on mineral soils after top soil removal, but only when former hydrological conditions were restored, seed banks had not been depleted and where dispersal mechanisms (flooding) from adjacent nature reserves were effective (Jansen et al., 2000; examples 'Stroothuizen', 'Punthuizen', 'Lemselermaten' and 'Groener').

Resuming the traditional management (grazing, mowing without fertilisation) in grasslands that have been fertilised intensively in former days, will not lead to rapid success without rewetting and topsoil

De Mars, H., M. J. Wassen & W. H. M. Peeters, 1996. The effect of drainage and management on peat chemistry and nutrient deficiency in the former Jegrzenia floodplain (N.E. Poland). Vegetatio 126: 59–72.

Dijk, E. & N. D. Eck, 1995. Effects of mycorrhizal fungi on *in vitro* nitrogen response of some Dutch indigenous orchid species. Can. J. Bot. 73: 1203–1211.

Egloff, T. B., 1983. Der Phosphor als primair limitierender Nährstoff in Streuwiesen (*Molinion*); Düngungsexperiment im unteren Reustal. Ber. Geobot. Inst. ETH Stiftung Rübel, Zürich 59: 119–148.

Ellenberg, H., 1978. Vegetation Mitteleuropas mit den Alpen. Ulmer. Everts, F. H. & N. P. J. De Vries, 1991. The vegetation development in stream valley systems; a landscape ecological study of Drenthian stream valleys. Ph.D. thesis, Univ. of Groningen, Historische Uitgeverij Groningen (in Dutch with English summary).

Faulkner, S. P. & C. J. Richardson, 1989. Physical and chemical characteristics of freshwater wetland soils. In Hammer, D. A. (ed.), Constructed Wetlands for Wastewater Treatment. Lewis, Publ., Chelsea, Michigan: 41–72.

Garcia, A., 1992. Conserving the species-rich meadows of Europe. Agric. Ecosyst. Envir. 40: 219–232.

Gough, R. H. & R. H. Marss, 1990. A comparison of soil fertility between semi-natural and agricultural plant communities: implications for the creation of species-rich grassland on abandoned agricultural land. Biol. Cons. 51: 83–96.

Grant, V., 1971. Plant Speciation, Columbia Univ. Press.

Grime, J. P., 1979. Plant Strategies and Vegetation Processes. Wiley, Chichester.

Grootjans, A. P., 1980. Distribution of plant communities along rivulets in relation to hydrology and management. In Wilmanns, O. & R. Tüxen (eds), Epharmonie, Berichte der Int. Symp. der I V.F.V. Cramer Verlag, Vaduz: 143–165.

Grootjans, A. P. & R. van Diggelen, 1995. Assessing the restoration prospects of degraded fens. In Wheeler, B. D., S. C. Shaw, W. J. Foyt & R. A. Robertson (eds), Restoration of Temperate Wetlands. Wiley, Chichester: 73–90.

Grootjans, A. P., P. C. Schipper & H. J. van der Windt, 1985. Influence of drainage on N-mineralisation and vegetation response in wet meadows. I: *Calthion palustris* stands. Oecol. Plant. 6: 403–417.

Grootjans, A. P., P. C. Schipper & H. J. van der Windt, 1986. Influence of drainage on N-mineralisation and vegetation response in wet meadows. II: *Cirsio-Molinietum* stands. Oecol. Plant. 7: 3–14.

Grootjans, A. P., G. J. R. Allersma & C. Kik, 1987. Hybridisation of the habitat in disturbed hay meadows. In van Andel, J., J. P. Bakker & R. W. Snaydon (eds), Disturbance in Grasslands. Geobotany 10: 67–77.

Grootjans, A. P., L. F. M. Fresco, C. C. de Leeuw & P. C. Schipper, 1996. Degeneration of species-rich *Calthion palustris* hay meadows; some considerations on the community concept. J. Veg. Sci. 7: 185–194.

Grootjans, A. P., W. Bijkerk, F. H. Everts, M. Jongman, M. Salomons & M. E. Tolman, 1997. Monitoring van effectgerichte maatregelen tegen verzuring 1994–1996. Lab. Plant Ecology, Haren/Bureau Everts & De Vries, Groningen.

Grootjans, A. P., R. van Diggelen, F. H. Everts, P. C. Schipper, J. Streefkerk, N. P. J. De Vries & A. Wierda, 1993. Linking ecological patterns to hydrological conditions on various spatial scales, a case study of small valleys. In Vos, C. C. & P. Opdam (eds), Landscape Ecology of a Stressed Environment. Chapman and Hall, London: 60–76.

Güsewell, S., 1997. Evaluation and management of fen meadows invaded by common reed (*Phragmites australis*). Ph.D. thesis, ETH Zürich.

Hobbs, R. J. & D. A. Norton, 1996. Toward a conceptual framework for restoration ecology. Rest. Ecol. 4: 93–110.

Jansen, A. J. M., 2000. Hydrology and restoration of wet heathland and fen meadow communities. Ph.D. thesis, Univ. of Groningen.

Jansen, A. J. M. & J. G. M. Roelofs, 1996. Restoration of *Cirsio-Molinietum* wet meadows by sod cutting. Ecol. Engen. 7: 279–298.

Jansen, A. J. M., M. C. C. de Graaf & J. G. M. Roelofs, 1996. The restoration of species-rich heath land communities in the Netherlands. Vegetatio 126: 73–88.

Jansen, A. J. M., A. P. Grootjans & M. H. Jailing, 2000. Hydrology of Dutch *Cirsio-Molinietum* meadows: Prospects of restoration. Appl. Veg. Sci. 3: 51–64.

Jansen, A. J. M., A. Th. W. Eysink & C. Maas, 2001. Hydrological processes in a *Cirsio-Molinietum* wet meadow: implications for restoration. Ecol. Engin. 17: 33–47.

Janssen, M. P. J. M. & A. H. P. M. Salman, 1995. A national strategy for dune conservation in the Netherlands. In Salman A. H. P. M., H. Berends & M. Bonazountas (eds), Coastal Management and Habitat Conservation. EUCC, Leiden: 215–230.

Kaiser, T., 1995. Vegetationskundliche Untersuchungen auf reliefiertem Niedermoor bei Extensivweide. Zeitschrift für Kulturtechnik und Landentwicklung 36: 175–177.

Kapfer, A., 1988. Versuche zur Renaturierung gedüngten Feuchtgrünlands – Aushagerung und Vegetationsentwicklung. Diss. Bot. 120: 1–144.

Kemmers, R. H., P. C. Jansen & P. Mekkink, 1995. Humus form profiles and ecohydrological system analysis. In Schouten et al. (eds), Scenario Studies for the Rural Environment. Kluwer Academic Publishers, Dordrecht: 489–494.

Kemmers, R. H., P. C. Jansen & S. P. J. Van Delft, 2000. De regulatie van de basentoestand in kwelafhankelijke schraalgraslanden en laagvenen. Report Ministry of Agriculture, Nature and Fisheries/Alterra, Wageningen.

Klinka, K., R. N. Green, R. N. Trowbridge & L. E. Lowe, 1981. Taxonomic classification of humus forms in ecosystems of British Columbia, Ministry of Forests, U.S.A.

Klötzli, F., 1991. Möglichkeiten und erste Ergebnisse mitteleuropäischer Renaturierungen. Verh. Ges. Ökol. 20: 229–242.

Koerselman, W., S. A. Bakker & M. Blom, 1990. Nitrogen, phosphorus and potassium mass balances for two small fens surrounded by heavily fertilised pastures. J. Ecol. 70: 428–442.

Koerselman, W. & J. T. A Verhoeven, 1995. Eutrophication of fen ecosystems: external and internal nutrient sources and restoration strategies. In Wheeler, B. D., S. C. Shaw, W. J. Foyt & R. A. Robertson (eds), Restoration of Temperate Wetlands. Wiley, Chichester: 91–112.

Lamers, L. P. M., H. B. M. Tomassen & J. G. M. Roelofs, 1998. Sulphate-induced eutrophication and phytotoxicity in freshwater wetlands. Envir. Sci. Technol. 32: 199–205.

Landolt, E. & F. Grossman, 1968. Zur vermutlich hybridogenen Entstehung von Wiesenpflanzen des schweizerischen Mittellandes. Verh. Schweiz. Naturf. Ges. 1968: 114–117.

Louwe Kooijmans, L. P., 1974. The Rhine/Meuse delta. Ph.D. Thesis, Univ. of Leiden.

Mengel, K. & E. A. Kirkby, 1982. Principles of Plant Nutrition, 3 ed. International Potas Institute, Bern, Switserland.

Meuleman, A. F. M., 1999. Performance of treatment wetlands. Ph.D. Thesis, Univ. of Utrecht.

Mulder, J., N. van Breemen & H. C. Eijck, 1989. Depletion of soil aluminium by acid deposition and implications for acid neutralisation. Nature 337: 247–249.

Muller, S., T. Dutoit, D. Alard & F. Grevilliot, 1998. Restoration and rehabilitation of species-rich grassland ecosystems in France; a review. Rest. Ecol. 6: 94–101.

Olde Venterink, H. 2000. Nitrogen, phosphorus and potassium flows controlling plant productivity and species richness; eutrophication and nature management in fens and meadows. PhD Thesis, Utrecht University.

Olff, H. & J. P. Bakker, 1991. Long term dynamics of standing crop, vegetation composition and species richness after cessation of fertiliser application to hay-fields. J. Appl. Ecol. 28: 1040–1052.

Olff, H., D. M. Pegtel, J. Van Groenendael & J. P. Bakker, 1994. Germination strategies during grassland succession. J. Ecol. 82: 69–77.

Oomes, M. J. M., H. Olff & H. J. Altena, 1996. Effects of vegetation management and raising the water table on nutrient dynamics and vegetation change in a wet grassland. J. Appl. Ecol. 33: 576–588.

Oomes, M. J. M., P. J. Kuikman & F. H. H. Jakobs, 1997. Nitrogen availability and uptake in mesocosms at two water levels and two water qualities. Plant Soil 192: 249–259.

Parker, V. T. & S. T. A. Pickett, 1997. Restoration as an ecosystem process: implications of the modern ecological paradigm. In Urbanska, K. M., N. R. Webb & P. J. Edwards (eds), Restoration Ecology and Sustainable Development. Cambridge Univ. Press: 17–32.

Patrick, W. H. Jr. & R. A. Khalid, 1974. Phosphate release and sorption by soils and sediments: effect of aerobic and anaerobic conditions. Science 186: 53–55.

Pegtel, D. M., 1983. Ecological aspects of nutrient deficient wet grassland (*Cirsio-Molinietum*). Verhandl. Ges. Ökol 10: 217–228.

Pegtel, D. M., J. P. Bakker, G. L. Verwey & L. F. M. Fresco, 1996. N, K and P deficiency in chronosequential cut summer-dry grasslands on gley podsol after cessation of fertiliser application. Plant Soil 178: 121–131.

Pfadenhauer, J. & A. P. Grootjans, 1999. Wetland restoration in Central Europe: aims and methods. Appl. Veg. Sci. 2: 95–106.

Poschlod, P., J. P. Bakker, S. Bonn & S. Fischer, 1996. Dispersal of plants in fragmented landscapes; changes of dispersal problems in the actual and historical man-made landscape. In Settle, J., C. R. Margules, P. Poschlod & K. Henley (eds), Species Survival in Fragmented Landscapes. Kluwer Academic Publishers, Dordrecht: 123–127.

Poschlod, P. & S. Bonn, 1998. Changing dispersal processes in the central European landscape since the last ice age: an explanation for the actual decrease of plant species richness in different habitats? Acta bot. Neerl. 47: 27–44.

Runhaar, J., 1999. Impact of hydrological changes on nature conservation areas in the Netherlands. Ph.D. thesis, Univ. of Leiden.

Runhaar, J., C. R. Van Gool & C. L. G. Groen, 1996. Impact of hydrological changes on nature conservation areas in the Netherlands. Biol. Cons. 76: 269–276.

Putman, R. J., A. D. Fowler & S. Tout, 1991. Patterns of use of ancient grassland by cattle and horse and effects on vegetational composition and structure. Biol. Cons. 56: 329–347.

Schaminée, J. H. J., A. H. F Stortelder & E. J. Weeda, 1996. De vegetatie van Nederland part 3. Opulus Press, Uppsala/Leiden.

Strijkstra, R. J., G. L. Verweij & J. P. Bakker, 1997. Seed dispersal by mowing machinery in a Dutch brook valley system. Acta bot. Neerl. 46: 307–401.

Tallis, J. H., 1991. Plant Community History. Chapman & Hall.

Tallowin, J. R. B. & R. E. N. Smith, 1996. Management options to conserve a *Cirsio-Molinietum* and integrate its use into productive livestock systems. Asp. appl. Biol. 44: 203–210.

Ter Borg, S. J., 1972. Variability of *Rhinanthus serotinus* in relation to the environment. Ph.D. thesis, Univ. of Groningen.

Thompson, K., J. P. Bakker & R. M. Bekker, 1997. The Soil Seed Banks of North West Europe: Methodology, Density and Longevity. Cambridge Univ. Press.

Turner, S. D. & C. F. Friese, 1998. Plant-Mycorrhizal community dynamics associated with a moisture gradient within a rehabilitated prairie fen. Rest. Ecol. 6: 44–51.

Van Andel, M., T. van Andel & T. Waterbolk, 1945. Bodem & plantengroei in het dal van het Anderense Diep. Kruipnieuws 7: 1–48.

Van Breemen, N., 1987. Effects of redox processes on soil acidity. Neth. J. agric. Sci. 35, 275–279.

Van der Hoek, D. & R. H. Kemmers, 1998. The effectiveness of rewetting. The impact of rewetting measures on key processes in the topsoil of the nature restoration project 'De Veenkampen'. Landschap 15: 211–224 (in Dutch with English summary).

Van der Linden, M. et al. 1996. Herstel van natte en vochtige ecosystemen. Report 9-1, National Research Programme on Desiccation (in Dutch, with English summary).

Van Diggelen, R., 1998. Moving gradients: assessing restoration prospects of degraded brook valleys. Ph.D. thesis, Univ. of Groningen.

Van Diggelen, R., A. P. Grootjans & R. Burkunk, 1994. Assessing restoration perspectives of disturbed brook valleys: the Gorecht area, the Netherlands. Rest. Ecol. 2: 87–96.

Van Diggelen, R., J. P. Bakker & J. Klooker, 1997. Topsoil removal: new hope for threatened plant species? In Cooper, A. & J. Power (eds), Species Dispersal and Land Use Processes. Proceedings Sixth Annual IALE (U.K.) Conference, Coleraine, 9–11th September 1997: 257–263.

Van Duren, I. C. & D. M. Pegtel, 2000. Nutrient limitation in wet, drained and rewetted fen meadows: Evaluation of methods and results. Plant Soil 220: 35–47.

Van Duren, I., D. Boeye & A. P. Grootjans, 1997a. Nutrient limitation in an extant and drained poor fen: implications for restoration. Plant Ecol. 133: 91–100.

Van Duren, I. C., D. M. Pegtel, B. A. Aerts & J. A. Inberg, 1997b. Nutrient supply in undrained and drained *Calthion* meadows. J. Veg. Sci 8: 829–838.

Van Duren, I. C., R. J. Strijkstra, A. P. Grootjans, G. N. J. Ter Heerdt & D. M. Pegtel, 1998. A multidisciplinary evaluation of restoration measures in a degraded fen meadow (*Cirsio-Molinietum*). Appl. Veg. Sci. 1: 115–130.

Van Wirdum, G., 1991. Vegetation and hydrology of floating rich-fens. Ph.D. thesis, University of Amsterdam. Datawyse, Maastricht.

Wassen, M. J., R. van Diggelen, J. T. Verhoeven & L. Wolejko, 1996. A comparison of fens in natural and artificial landscapes. Vegetatio 126: 5–26.

Weeda, E. J., R. van der Meijden & P. A. Bakker, 1990. Rode Lijst van de in Nederland verdwenen en bedreigde planten over de periode 1980–1990. Gorteria 16: 2–26.

Young, P., 1996. The 'new' science of wetland restoration. Envir. Sci. Technol. 30: 292A–296A.

Zandee, M., 1981. Contribution to the biosystematics of two cyanogenic polyploid complexes (*Juncus articulatus* and *Holcus mollus*). Ph.D. thesis, Univ. of Leiden.

Zeeman, W. P. C., 1986. Application in land, nature and water management; the Reitma case study. Proc. & Inf. CHO-TNO 34: 117–126.

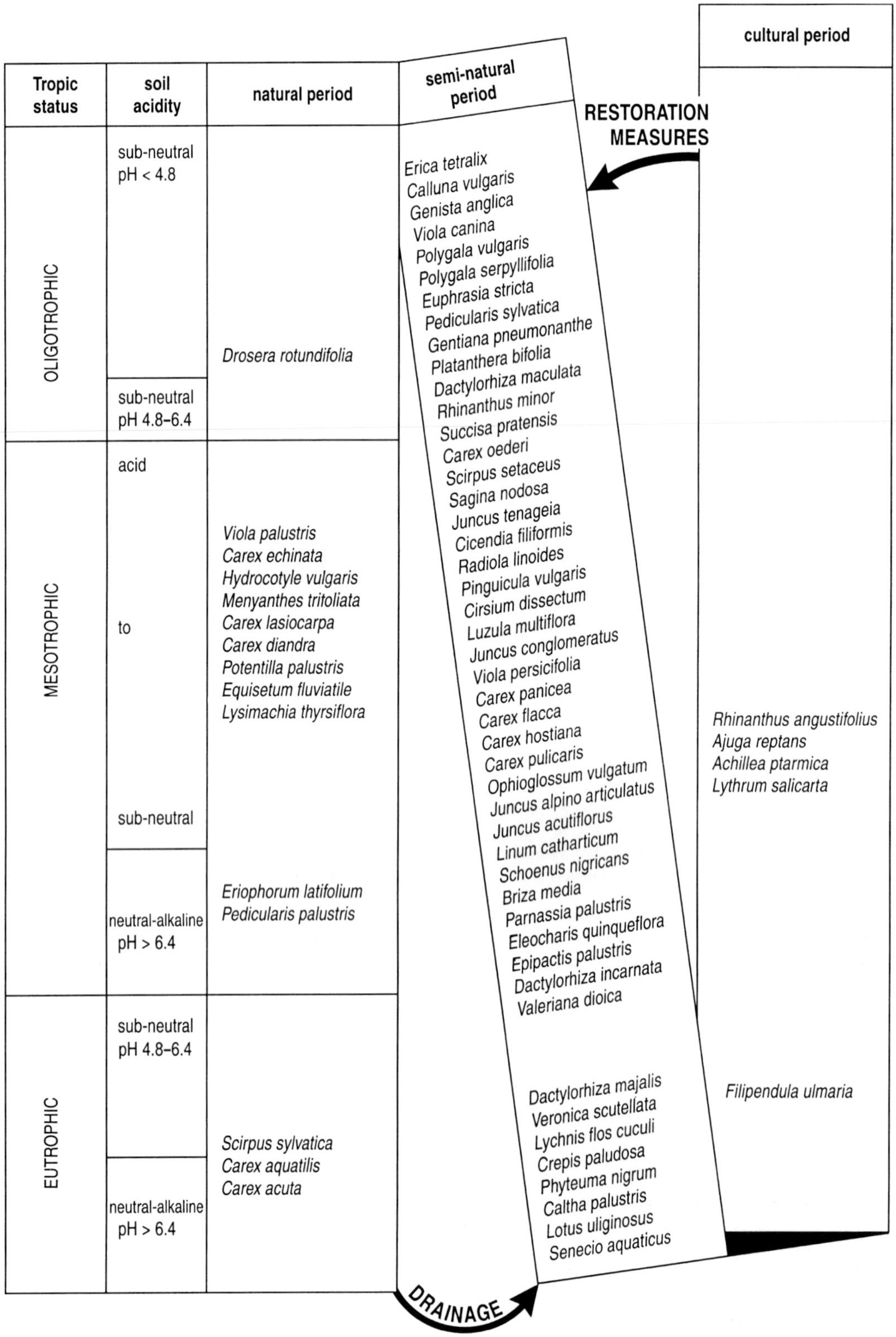

Appendix 1. Target species of natural, semi-natural and cultural landscapes which have been used to judge whether restoration measures in brook valley meadows have been successful.

Hydrobiologia **478:** 171–180, 2002.
P.H. Nienhuis & R.D. Gulati (eds), Ecological Restoration of Aquatic and Semi-Aquatic Ecosystems in the Netherlands (NW Europe).

Restoration of aquatic macrophyte vegetation in acidified and eutrophicated shallow soft water wetlands in the Netherlands

J.G.M. Roelofs[1], E. Brouwer[1] & R. Bobbink[2]
[1]*Department of Aquatic Ecology and Environmental Biology, Research Group Environmental Biology, University of Nijmegen, Toernooiveld 1, 6525 ED Nijmegen, The Netherlands*
Tel.: +31-24-3652340. Fax: +31-24-3652134. E-mail: J.Roelofs@sci.kun.nl
[2]*Landscape Ecology, Faculty of Biology, Utrecht University, P.O. Box 800.84, Utrecht, The Netherlands*

Key words: soft water, wetlands, acidification, eutrophication, carbon dioxide, alkalinisation, restoration

Abstract

Soft water lakes possess a highly characteristic vegetation adapted to limitation of carbon. Based upon hydrology, vegetation and geographic distribution, boreal and Atlantic lake types can be distinguished. Reducing the input of nutrients or liming, or both, the stream or its catchment is generally sufficient to restore typical soft water vegetation of boreal soft water lakes. The vegetation of Atlantic soft water lakes is subject to many anthropogenic degradation processes. Therefore, spontaneous recovery in the near future is not expected and restoration is urgently required. Removal of nutrient-rich, anoxic, organic sediments is a prerequisite for restoration of these lakes. In acidified or acid-sensitive lakes, additional measures against acidification are required. Controlled supply of calcareous, nutrient-poor water is much better than direct liming. The effects of these restoration measures strongly depend on the detrimental effects of processes such as atmospheric deposition, drainage, catchment acidification, eutrophication and reduced colonisation rates.

Introduction

The macrophyte vegetation of shallow soft water lakes and wetlands is very different from that of both acidic and alkaline waters. Very characteristic are macrophytes with an extensive root system and short leaves in a rosette, e.g. *Isoetes lacustris* (quillwort). These isoetids often form dense stands on the lake sediment. The highest macrophyte diversity in these lakes or wetlands can be found in the Atlantic regions of Europe, including the Netherlands. Furthermore, soft water macrophytes occur scattered throughout the world in mountainous areas, especially *Isoetes* species (Keeley et al., 1983; Gacia et al., 1994). This characteristic vegetation is adapted to the restricted availability of nutrients and carbon dioxide. An important problem for submerged plants is that the diffusion rate of carbon dioxide in water is 10^5 lower than in the air. Bio-available carbon can be present in the water layer as carbon dioxide (CO_2) or bicarbonate (HCO_3^-). Bicarbonate is the main contributor to the alkalinity, or to the acid-neutralising capacity (ANC) of lakes. In alkaline lakes, vegetation is dominated by plants that can also use HCO_3^- from the surrounding water as a carbon source if this is present in high concentrations. In the Netherlands, we find shallow soft water lakes and pools on the Pleistocene deposits in the south-eastern part and in some coastal dunes. Because of the rainwater surplus, most other nutrient concentrations, like nitrogen and carbon (HCO_3^- and CO_2) were originally low. In acidic lakes, the carbon equilibrium shifts towards dominance of CO_2, enabling the growth of CO_2-utilising, acid-resistant macrophytes, such as *Sphagnum* species.

Generally, the HCO_3^- concentration in soft water lakes is below 200 μM, which is insufficient to sustain net photosynthesis (a.o. Sand-Jensen, 1983; Roelofs et al., 1984; Maberly, 1985a; Wetzel et al., 1985; Madsen et al., 1996). Most soft water macrophytes have developed morphological adaptations for acquiring CO_2. Isoetid macrophytes take up CO_2 from the sediments through their extensive root system, where

CO_2 levels are 10–100 times higher compared with the water layer (Roelofs et al., 1984). Nymphaeid species have floating leaves, which enable them to use CO_2 from the atmosphere. Several small species, like *Elatine hexandra* (Lapierre) DC, obtain their CO_2 from the water layer immediately above the sediment, where concentrations are generally higher (Maberley, 1985b). Larger species filling the water column, like *Myriophyllum alterniflorum* L., often possess finely divided leafs which efficiently take up CO_2 (Madsen & Sand-Jensen, 1994). As many soft water macrophytes grow close to the sediment, they depend on clear water. In the water layer of most soft water lakes and of wetlands the availability of nitrogen and phosphorus is low (NO_3 and NH_4 <5 μM, ortho-PO_4 <0.5 μM; Seddon, 1972; Roelofs, 1983). This leads to characteristic algal communities, occurring in very low densities. Nutrients are primarily obtained from the sediment and nitrate is the dominant source of nitrogen for most soft water macrophytes (Schuurkes et al., 1986). The presence of isoetid species contributes to the maintenance of oligotrophic conditions in soft water lakes. The extensive root system aerates the upper sediment layer via radial oxygen loss (Pedersen et al., 1995). Aeration of the rooting zone stimulates nitrification of ammonium and subsequent denitrification occurs in deeper, anoxic layers (Risgaard-Petersen & Jensen, 1997). Furthermore, the oxygenated upper sediment layer acts as a barrier for phosphate input from the sediment to the water layer (Andersen & Olsen, 1994; Christensen & Wigand, 1998). The oxygen release via the roots is less prominent in other rooting soft water macrophytes, such as some *Myriophyllum* and *Batrachium* species (Flessa, 1994).

Because of the shallow nature of Dutch soft water lakes and moorland pools, partial or total desiccation can occur in very dry summers. These waters are fed by precipitation, directly entering the lake or reaching it via superficial, local groundwater or surface water inflows and in general these systems have sandy or peaty sediments. Water losses occur as a result of infiltration in the sediment and evaporation. Sediment characteristics, water depth and ANC are the principal determinants of the vegetation in all soft water lakes and wetlands.

Macrophyte communities of the Atlantic lake type and littoral communities of the boreal lake type are more influenced by fluctuations such as water level, succession to terrestrial communities, climatic extremes, grazing and peat formation than permanently submerged communities. Due to limited nutrient availability and the dynamic circumstances (wave action, ice, water level fluctuation), the dominance of large helophytes along soft water lakes is restricted to sheltered places with moderate sediment development.

As a consequence of their shallow nature and occurrence in densely populated areas, many causes for the decline of Atlantic soft water lakes exist. An overview is given in chapters 2 and 3, where the chances for spontaneous recovery from acidification and eutrophication are explored. An understanding of degradation processes is of vital importance in lake restoration projects, while examples of natural recovery can reveal something of the recovering capacity of these systems. The decline of soft water lake vegetation is expected to continue in the near future and, therefore, restoration measures are urgently required.

In this paper, two types of degraded soft water lakes are distinguished: eutrophicated, well buffered lakes and eutrophicated, acidified or acid-sensitive lakes. The effect of the removal of recently accumulated organic matter has been studied in both lake types. Furthermore, the effect of additional measures against acidification in acidified or acid-sensitive lakes, such as addition of powdered limestone and the supply of alkaline, nutrient poor water, have been studied. In the final section, the effectiveness of rehabilitation measures against the negative effects of eutrophication and acidification is summarised.

Decline of vegetation of soft water lakes and wetlands

Acidification and eutrophication are widely accepted as important causal factors for the decline of soft water lake vegetation in Western Europe. In the 20th century, over 90% of the habitat of Atlantic soft water macrophytes in the lakes in the Netherlands have locally disappeared as a consequence of acidification or eutrophication (Arts, 1990). In many soft water lakes, precipitation of atmospheric nitrogen compounds above the critical level of 5–10 kg N ha^{-1} yr^{-1} has caused acidification and the accumulation of ammonium (Schuurkes et al., 1987; Bobbink & Roelofs, 1995).

However, many additional causes for decline can be found. (Arts & Leuven, 1988). Most soft water lakes originated under completely natural conditions, particularly through large scale erosion by ice, wind and water. The same conditions also strongly inhibited further succession. With the coming of man, these natural dynamics were gradually controlled. However, many soft water lakes appeared or persisted as a result of other anthropogenic activities such as peat excavation, cattle grazing, reed-cutting and the supply of alkaline water in acidic lakes (Arts, 1990). Nowadays, most of these activities have ceased, whilst the natural dynamics have been further affected by drainage and habitat fragmentation. This resulted in a swift succession towards terrestrial plant communities in many lakes. Isolated systems returned to their original acid state or were acidified as a consequence of reduced seepage and atmospheric deposition.

Acidification

Large scale acidification of soils and lakes occurs in areas with high deposition of acid or acidifying compounds, especially SO_x, NO_x and NH_x. Apart from direct acidification of the water layer, atmospheric deposition causes an acceleration of the weathering of the catchment soil, higher concentrations of CO_2, cations and sulphate and a lower pH of the drainage water (Reuss et al., 1987). Drainage water from soils with a relatively high cation exchange capacity that receive large amounts of acid can sometimes become alkalised (Kilham, 1982). Prolonged acid deposition leads eventually to cation depletion of the catchment soil and further acidification of the drainage water.

The influx of acid water causes acidification of soft water lakes and mobilisation of metals. Nymphaeid species occur along a wide pH gradient, most isoetid species can survive acidification down to pH 4.5 (Roelofs, 1983). Other soft water macrophytes are seldom found in waters with pH <5 (Catling et al., 1985; Rørslett, 1991; Maessen et al., 1992). Elevated levels of metals like aluminium, which occur after acidification of the water layer, have little or no effect on soft water macrophytes (Maessen et al., 1992; Albers & Camardese, 1993). However, Yan et al. (1988) reported a negative correlation of the occurrence of soft water macrophytes and extremely high Cu and Ni levels. Aluminium can precipitate with phosphate in acid environments, which leads to oligotrophication (Jansson et al., 1986). In many west European Atlantic soft water lakes, acidification has been caused by ammonium deposition. After acidification of the water layer below pH 4.5, nitrification ceases, ammonium accumulates and CO_2 levels increase (Roelofs, 1986; Rudd et al., 1988). This inhibits the growth of most soft water macrophytes, which use nitrate as their dominant nitrogen source, and stimulates the growth of ammonium utilising, acid-tolerant species (Roelofs, 1983; Schuurkes et al., 1986). The growth of *Sphagnum* and *Juncus bulbosus* is especially stimulated by elevated levels of both CO_2 and NH_4^+ (Paffen & Roelofs, 1991; Roelofs et al., 1994, Lucassen et al., 1999). After cation depletion of the catchment soil, acidification can continue to values between pH 3.5 and 4.5. This results in the disappearance of all characteristic soft water macrophytes. Furthermore, an organic layer develops on the sediment due to increased plant growth and reduced decomposition under acid circumstances. Macrophyte growth eventually becomes limited to some *J. bulbosus*, *Sphagnum* and nymphaeids.

Eutrophication

Eutrophication can be caused by the inflow of polluted water or by atmospheric deposition of eutrophicating substances, primarily nitrogen compounds. Local sources of pollution can be agricultural runoff, industrial and domestic wastewater or the supply of eutrophic water from nearby water bodies. An additional source of eutrophication can be the afforestation of the lakes' immediate surroundings. Trees contribute considerably to the accumulation of organic matter (leaves) on the sediment and accelerate eutrophication and succession on the shore by inhibiting wind- and wave action and by promoting the deposition of atmospheric substances. In shallow lakes, sediment enrichment and concomitant expansion of nymphaeids and helophytes occur throughout the lake. These lead eventually to the disappearance of all soft water macrophytes and subsequent transition into terrestrial vegetation.

The supply of allochthonous surface water often leads to alkalinisation and an increase in the amount of inorganic carbon present as bicarbonate (HCO_3^-) in the water layer. This enables the growth of macrophytes using bicarbonate from the water layer as carbon source. Particularly when ANC increases above 1 meq l^{-1}, soft water macrophytes are replaced by species of alkaline water, such as *Myriophyllum spicatum* L. and *Elodea canadensis* Michx. An increased ANC also accelerates the decomposition of organic matter (Kok & van de Laar, 1990). In originally poorly buffered lakes, large amounts of accumulated organic matter thus become more biodegradable (Brouwer et al., 1999). Accelerated decomposition causes an increased recycling of N, P and C and the availability of P increases with increasing ANC (Roelofs, 1991).

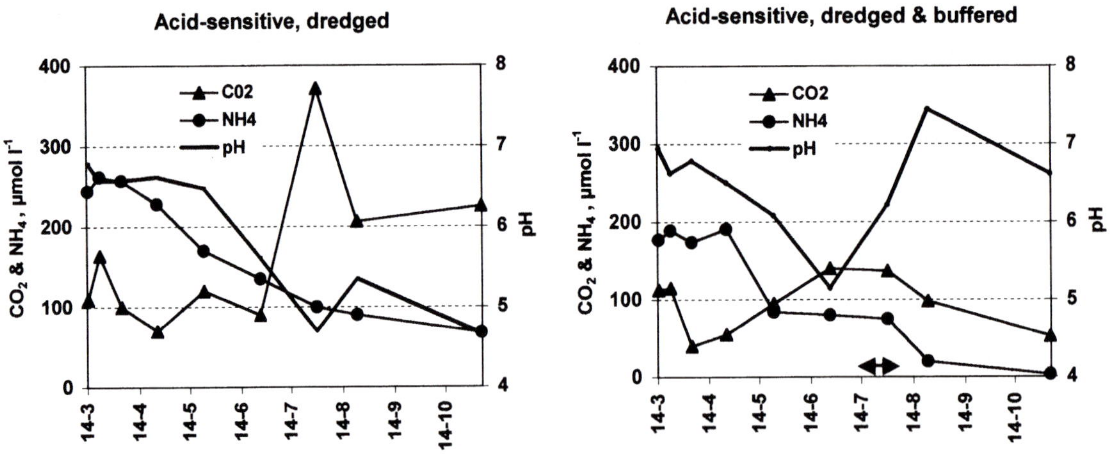

Figure 1. Example of the effect of groundwater supply in acidified or acid-sensitive, dredged soft water lakes. After draining and dredging of Lake Goorven and Lake Van Esschenven at the end of 1995, the lakes filled with rain and local water. In July 1996, re-acidification of Lake Goorven was counteracted by the supply of groundwater (arrow mark). In this year, soft water macrophytes became abundant in Lake Goorven and almost disappeared from Lake Van Esschenven.

the effect of alkalinisation on acidified seepage lakes (Garrison et al., 1996). The pH of Lake Max was 5.1 after acidification. By pumping up alkaline groundwater, the pH could be manipulated to any desired level between 5.1 and 6.7. Groundwater-supply increased nitrification rates and caused shifts in phytoplankton, zooplankton and fish communities. Some eutrophication occurred due to the relatively high phosphate load in the groundwater.

In three acidified and eutrophicated shallow soft water lakes in the Netherlands, small amounts of alkaline, nutrient-poor groundwater were extracted from deeper aquifers and were supplied after removal of the sapropelium layer by dredging (Roelofs et al., 1996; Brouwer & Roelofs, 1998). The supply of groundwater was ceased when the alkalinity reached a level of 200 μeq l^{-1}. During the supply, pH rose from 4 to 7, because of a rapid decline of ammonium and carbon dioxide concentrations in the water layer (see Fig. 1). Within a few weeks *J. bulbosus* plants turned brown and the growth of *Littorella uniflora* was stimulated. Extensive recolonisation by soft water macrophytes was reported from all these three lakes, although recolonisation was somewhat inhibited by the increased turbidity as a consequence of accelerated decomposition. Ground- or surface water was supplied repeatedly on annual basis to counteract the acid substances that were supplied from the still acid sediment, from inflow of acid groundwater and from acid atmospheric deposition. Twelve years after sediment removal and controlled periodical supply of small amounts of alkaline, nutrient-poor surface water in the acid-sensitive Lake Beuven, almost no re-eutrophication or acidification has been observed. The vegetation is now dominated by isoetid soft water macrophytes. Reacidification in acid-sensitive lakes with a non-acidified sediment, like Lake Beuven, occurs at a far slower rate, leading to a more stable pH. Therefore, it is expected that the large, annual pH fluctuations that occur in the acidified lakes with groundwater supply will decrease when the sediment has been sufficiently loaded with cations. Nevertheless, the largest populations of the isoetid species *Littorella uniflora* and *Lobelia dortmanna* in the Netherlands exist in Lake Beuven and in the Bergvennen lake series, both rehabilitated sites with a calcareous water supply. A great variety of soft water macrophytes can profit particularly from this type of restoration measure (see Fig. 2).

In nitrogen saturated soft water lakes, removal of the accumulated organic matter is a prerequisite for restoration and additional measures against acidification are required. Liming soft water lakes has many deleterious side-effects and does not lead to recovery of characteristic macrophytes. Supply of calcareous, nutrient poor water is a more effective measure against acidification and leads to recovery of soft water macrophytes.

Reduction of nutrient loads

There is an extensive literature on the restoration of eutrophicated, alkaline lakes. A prerequisite for all lake rehabilitation projects is the reduction of the external nutrient load to pre-eutrophication levels. In many cases, this condition has not been met and re-eutrophication occurs at various rates. Until recently, however, virtually nothing was known about the restoration of aquatic plant communities in eutrophicated shallow soft water lakes (Maurizi & Poillon, 1992) and major gaps still exist in our knowledge on this subject. In regions with Atlantic soft water lakes and wetlands, it is more difficult to meet the conditions for restoration of soft water ecosystems. Eutrophication and acidification sources are more numerous and drainage and habitat fragmentation are not easily reversed. In many shallow Atlantic soft water lakes, the anoxic layer of organic matter present on the sediment makes dredging an essential prerequisite of restoration in these systems.

After reduction of the external nutrient load, the transparency of shallow lakes increases, thereby enabling the swift colonisation of, especially, characean communities (Simons et al., 1994; Brouwer et al., 1999; van den Berg et al., 1999). Several factors can stimulate this colonisation process. In eutrophicated, alkaline lakes, biomanipulation and alum-treatments (Ozimek et al., 1990) have sometimes been applied successfully. Also, periodic emergence of the sediment of shallow lakes stimulates germination and colonisation. In Lake Banen, many soft water macrophytes returned after the reduction of the external alkalinity and nutrient load and periodic desiccation of parts of the lake (Brouwer et al., 1999).

However, most angiosperms do not easily colonise anoxic sediments. This sediment remains an important nutrient source within the lake and causes algal blooms (Moss et al., 1996). During prolonged periods with a high water table, aquatic macrophytes die-off as a consequence of the formation of reduced toxic substances in the sediment and light deprivation. Therefore, removal of these reduced layers is necessary before restoring eutrophicated soft water lakes.

Sediment removal from eutrophicated lakes

Removal of accumulated sapropelium strongly stimulated the germination of soft water macrophytes under experimental conditions in shallow soft water lakes in the Netherlands (Bellemakers et al., 1996). Furthermore, an important source of nutrients is removed from the system. After reduction of the external nutrient load and sediment removal in Lake Beuven, a stable community of soft water macrophytes was re-

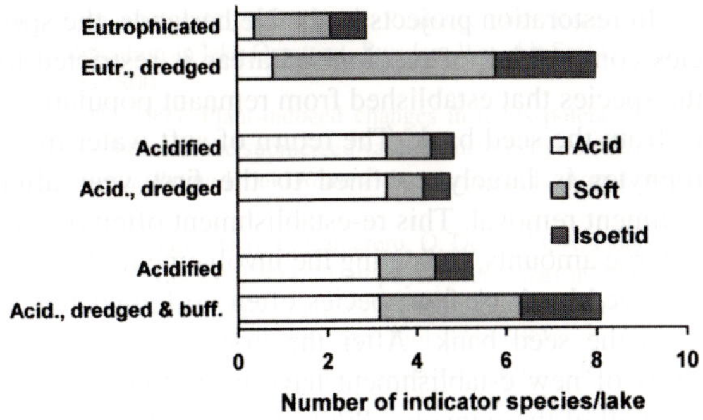

Figure 2. Average number of indicator species present in five eutrophicated and 12 acidified or acid-sensitive Dutch soft water lakes, before and 5 years after sediment removal. Nutrient poor, alkaline water was added to six of the acidified or acid-sensitive lakes. Indicator species of acid water (*Juncus bulbosus* and the mosses *Sphagnum cuspidatum* Hoffm., *Sphagnum denticulatum* Brid. and *Drepanocladus fluitans* (Hedw.) Warnst.) and soft water (isoetids versus other characteristic soft water macrophytes) were distinguished.

established (Brouwer & Roelofs 2001). Similar results were obtained in dune pools and Lake Broekse wielen in the Netherlands (Brouwer et al., 1996; Roelofs et al., 1996). If the eutrophicated sediment layer is not removed completely, macrophyte establishment is seriously impaired (Moss et al., 1996). In Lake Banen, partial removal of the sediment was followed by re-establishment of soft water macrophytes. However, especially during periods with a high water table, these macrophytes partly disappeared again (Brouwer, unpublished data).

In Atlantic soft water lakes, the condition of the sediment is a crucial factor in lake restoration. Eutrophicated, anoxic sediment strongly inhibits the improvement of water quality. Therefore, dredging and reduction of nutrient input are the most important restoration measures in these systems. Furthermore, recovery of the original level of alkalinity is of great importance in restoring the original vegetation.

Restoring soft water lakes in a fragmented environment

Especially in densely populated areas such as the Netherlands, a large reduction in the number and size of soft water habitats has occurred. Furthermore, many (constructed) connecting waterways are also strongly affected or have disappeared totally. The remaining Atlantic soft water lakes have therefore become very isolated. The reduced dispersal possibilities between soft water lakes lead to decreased (re-)colonisation rates.

Maurizi, S. & F. Poillon, 1992. Restoration of Aquatic Ecosystems. National Acadamy Press, Washington D.C.: 552 pp.

Moss, B., J. Stansfield, K. Irvine, M. Perrows & G. Phillips, 1996. Progressive restoration of a shallow lake: a 12-year experiment in isolation, sediment removal and biomanipulation. J. appl. Ecol. 33: 71–86.

Ozimek, T., R. D. Gulati & E. van Donk, 1990. Can macrophytes be useful in biomanipulation of lakes? The Lake Zwemlust example. Hydrobiologia 200/201: 399–407.

Paffen, B. G. P. & J. G. M. Roelofs, 1991. Impact of carbon dioxide and ammonium on the growth of submerged *Sphagnum cuspidatum*. Aquat. Bot. 40: 61–71.

Pedersen, O. & K. Sand-Jensen, 1995. Diel pulses of O_2 and CO_2 in sandy lake sediments inhabited by *Lobelia dortmanna*. Ecology 76: 1536–1545.

Pietsch, W. H. O., 1996. Recolonization and development of vegetation on mine spoils following brown coal mining in Lusatia. Wat. Air Soil Pollut. 91: 1–15.

Rasmussen, L., H. Hultberg & B. J. Cosby, 1995. Experimental studies and modelling of enhanced acidification and recovery. Wat. Air Soil Pollut. 85: 77–88.

Reuss, J. O., B. J. Cosby & R. F. Wright, 1987. Chemical processes governing soil and water acidification. Nature 329: 27–31.

Risgaard-Petersen, N. & K. Jensen, 1997. Nitrification and denitrification in the rhizosphere of the aquatic macrophyte *Lobelia dortmanna* L. Limnol. Oceanogr. 42: 529–537.

Roelofs, J. G. M., 1983. Impact of acidification and eutrophication on macrophyte communities in soft waters in the Netherlands. I. Field observations. Aquat. Bot. 17: 139–155.

Roelofs, J. G. M., 1986. The effect of airborne sulphur and nitrogen deposition on aquatic and terrestrial heathland vegetation. Experientia 42: 372–377.

Roelofs, J. G. M., 1991. Inlet of alkaline river water into peaty lowlands: effects on water quality and *Stratiotes aloides* L. stands. Aquat. Bot. 39: 267–293.

Roelofs, J. G. M., J. A. A. R. Schuurkes & A. J. M. Smits, 1984. Impact of acidification and eutrophication on macrophyte communities in soft waters. II. Experimental studies. Aquat. Bot. 18: 389–411.

Roelofs, J. G. M., T. E. Brandrud & A. J. P. Smolders, 1994. Massive expansion of *Juncus bulbosus* L. after liming of acidified SW Norwegian lakes. Aquat. Bot. 48: 187–202.

Roelofs, J. G. M., R. Bobbink, E. Brouwer & M. C. C. de Graaf, 1996. Restoration ecology of aquatic and terrestrial vegetation on non-calcareous sandy soils in The Netherlands. Acta Bot. Neerl. 45: 517–541.

Rørslett, B., 1991. Principal determinants of aquatic macrophyte richness in Northern European lakes. Aquat. Bot. 39: 173–193.

Rudd, J. W. M., C. A. Kelly, D. W. Schindler & M. A. Turner, 1988. Disruption of the nitrogen cycle in acidified lakes. Science 140: 1515–1517.

Sand-Jensen, K., 1983. Photosynthetic carbon sources of stream macrophytes. J. exp. Bot. 34: 198–210.

Schindler, D. W., 1974. Eutrophication and recovery in experimental lakes: implications for lake management. Science 184: 897–899.

Schindler, D. W., 1986. The significance of in-lake production of alkalinity. Wat. Air Soil Pollut. 30: 931–944.

Schnoor, J. L. & W. Stumm, 1986. The role of chemical weathering in the neutralization of acidic deposition. Schweiz. Z. Hydrol. 48: 171–195.

Schuurkes, J. A. A. R., C. J. Kok & C. Den Hartog, 1986. Ammonium and nitrate uptake by aquatic plants from poorly buffered and acidified waters. Aquat. Bot. 24: 131–146.

Schuurkes, J. A. A. R., M. A. Elbers, J. J. F. Gudden & J. G. M. Roelofs, 1987. Effects of simulated ammonium sulphate and sulphuric acid rain on acidification, water quality and flora of small-scale soft water ecosystems. Aquat. Bot. 28: 199–226.

Seddon, B., 1972. Aquatic macrophytes as limnological indicators. Freshwat. Biol. 2: 107–130.

Simons, J., M. Ohm, R. Daalder, P. Boers & W. Rip, 1994. Restoration of Botshol (The Netherlands) by reduction of external nutrient load: recovery of a characean community, dominated by *Chara connivens*. Hydrobiologia 275/276: 243–253.

Van Dam, H., 1997. Vennen herstellen zich gedeeltelijk van verzuring. H_2O 30: 366–370.

Van den Berg, M., M. Scheffer, E. H. van Nes & H. Coops, 1999. Dynamics and stability of *Chara* sp. and *Potamogeton pectinatus* in a shallow lake changing in eutrophication level. Hydrobiologia 409: 335–342.

Wetzel, R. G., E. S. Brammer, K. Lindström & C. Forsberg, 1985. Photosynthesis of submersed macrophytes in acidified lakes. II. Carbon limitation and utilization of benthic CO_2 sources. Aquat. Bot. 22: 107–120.

Yan, N. D., G. E. Miller, I. Wile & G. C. Hitchin, 1988. Richness of aquatic macrophyte floras of soft water lakes of differing pH and trace metal content in Ontaria, Canada. Aquat. Bot. 23: 27–40.

Hydrobiologia **478:** 181–203, 2002.
P.H. Nienhuis & R.D. Gulati (eds), Ecological Restoration of Aquatic and Semi-Aquatic Ecosystems in the Netherlands (NW Europe).

Restoration of coastal dune slacks in the Netherlands

A.P. Grootjans[1], H.W.T. Geelen[2], A.J.M. Jansen[3] & E.J. Lammerts[4]
[1]*Department of Plant Biology, University of Groningen, P.O. Box 14, 9750 AA Haren, The Netherlands*
E-mail: A.P.Grootjans@biol.rug.nl
[2]*Amsterdam Water Supply, P.O. Box 8169, 1005 AD Amsterdam, The Netherlands*
[3]*WMO, Water Company Overijsel, P.O. Box 10.005, 8000 GA Zwolle, The Netherlands*
[4]*Ministry of Agriculture, Nature Management and Fisheries, P.O. Box 30027, 9700 RM Groningen, The Netherlands*

Key words: conservation, dune slacks, hydrology, stable state, succession, public water supply, restoration

Abstract

In order to stop the continuous decline of typical dune slack communities along the Dutch coast, restoration projects have been carried out since 1952. Restoration measures consisted of re-introducing traditional management techniques in dune slacks, such as mowing, grazing and sod removal, or constructing artificial dune slacks to compensate for lost biodiversity elsewhere. An analysis of successful and unsuccessful projects showed that constructing new dune slacks was not very successful for maintaining new populations of endangered dune slack species, since such projects were often carried out in areas where seed banks were depleted, while hydrological conditions and seed dispersal mechanisms were sub-optimal. The construction of sand dikes to prevent sea intrusion in large beach plains was, unintentionally, a temporary success for the establishment of many Red List species, although such measures often disrupted natural dune slack formation. Successful sites were all characterised by a regular discharge of calcareous groundwater provided by local or regional hydrological systems, where not very long ago populations of typical dune slack plants were present. Under such conditions, sod removal was a successful measure to create pioneer stages which were relatively stable, due to a very slow accumulation of organic matter in the topsoil. It is argued that new and more flexible coastal defence strategies can provide new opportunities for natural and relatively stable pioneer stages of dune slack formation, suitable for the long term preservation of endangered dune slack species.

Introduction

Coastal dune areas represent one of the last nutrient-poor landscapes in The Netherlands (Fig. 1) and are characterised by pronounced geomorphological and hydrological gradients, thus creating good conditions for a high biodiversity of plants and animals. Most of the species featuring on the Dutch Red List of endangered plant species now have their main distribution in the coastal area, but few are real endemic species (van der Maarel & van der Maarel-Versluys, 1996). Most species are restricted to the dune areas because practically all their main land populations have become extinct due to severe drainage, acidification and eutrophication. Pioneer communities, in particular, are rich in Red List species. The last four decades, however, have shown a steep decline in typical dune slack vegetation, especially along the Holland coast, due to intensive use of that area for the production of drinking water. This decline has triggered numerous activities to restore the species-rich dune slack stages with many Red List species. Sometimes such activities were successful, but more often success was limited. An interesting phenomenon is, though, that some human activities in the framework of coastal defence, unintentionally triggered the development of pioneer dune slack stages with many Red List species.

On order to provide dune managers with a good blue-print for successful dune slack restoration we have studied some 50 projects and selected 12 that had been monitored in a systematic way for at least six years. In the present paper, we will evaluate the

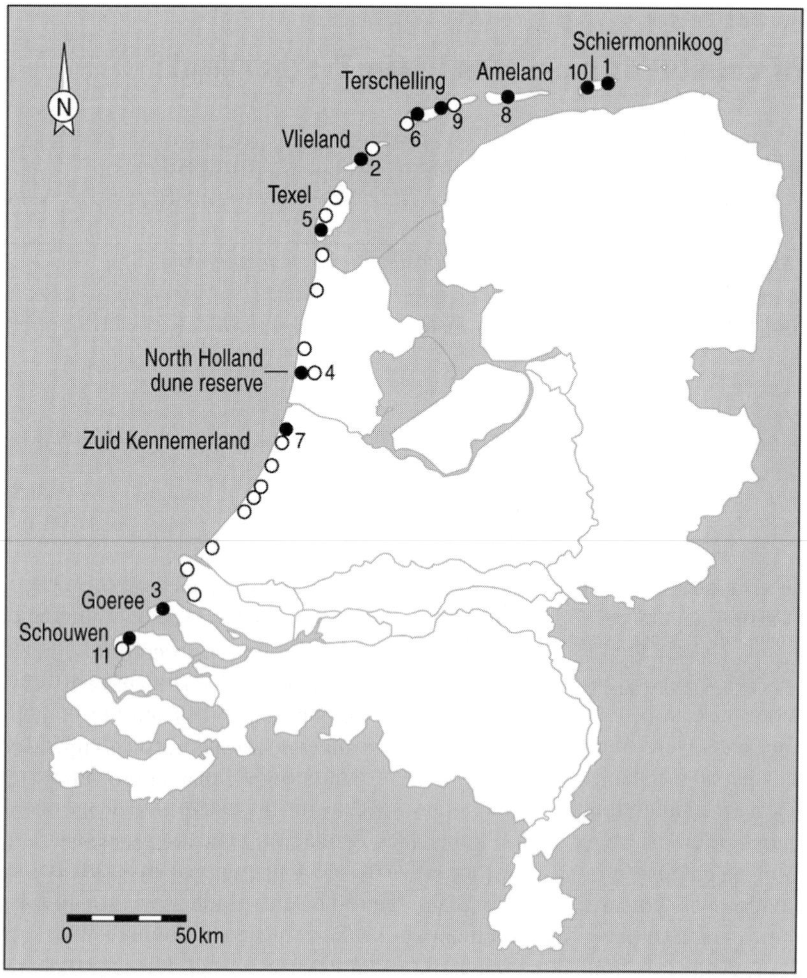

Figure 1. Dune areas in The Netherlands with position of the research areas. 1 = Beach Plain Schiermonnikoog, 2 = Kroon's Polder Vlieland, 3 = Middelduinen Goeree, 4 = Reggers Sandervlak, 5 = Moksloot Texel, 6 = Hunenplak Terschelling, 7 = Vogelmeer Kennemerduinen, 8 = Lange Duinen Ameland, 9 = Koegelwieck Terschelling, 10 = Kapenglop Schiermonnikoog, 11 = Vroongronden Schouwen.

successes and failures of these restoration projects and use the number of re-established Red List species to judge their success. The analysis is preceded by a short review on the structure and functioning of dune slacks, the ecological and hydrological requirements of dune slack species, and various (human) disturbances in dune slack functioning. Measures to restore acidified or eutrophicated dune slacks will be shortly discussed and successful and less successful projects assessed. Finally, the state of the art of dune slack restoration will be summarised in a conceptual model.

Structure and functioning of dune slacks

Site factors and plant adaptations

Dune slacks are low-lying areas within the coastal dunes where high water tables prevail during winter and spring (van der Laan, 1979; Boorman et al., 1997; Grootjans et al., 1998). During the summer period the water table may drop 50–100 cm below the surface. Young dune slacks are very nutrient poor but show a large variability with respect to species composition, depending on their substrate characteristics and hydrological regime. Figure 2 shows the results of a canonical correspondence analysis (CCA), relating the occurrence of plant communities from 51 sites on the Dutch Wadden Sea islands with measured en-

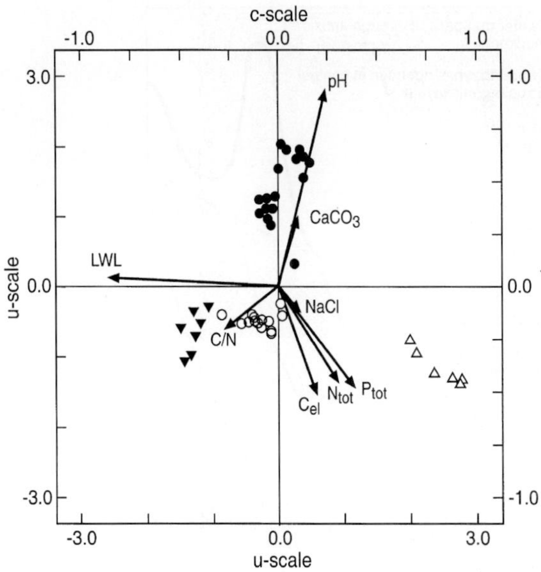

Figure 2. Results of Canonical correspondence analysis (CCA) showing the relationships between measured environmental factors (0–10 cm) and species composition in dune slacks on the Dutch Wadden Sea islands (from Lammerts & Grootjans, 1999). LWL = mean lowest water level, C_{el} = carbon content, N_{tot} = total nitrogen content, P_{tot} = total phosphorus content.

vironmental factors (Lammerts & Grootjans, 1998). The length of the arrows corresponds with the magnitude of greatest variation in the total data set. In this analysis, the mean lowest water level and the pH are the factors explaining most of the variation in species composition. The occurrence of the typical dune slack community is clearly related to sites with a relatively high pH and intermediate soil moisture conditions. Late successional stages, such as eutrophic reeds and small sedge marshes occur at less calcareous sites with higher soil nutrient stocks (N-total, P-total). Due to nutrient-poor conditions the vegetation cover of pioneer communities is low. Young slacks consist of a mixture of different life and growth forms: annuals, biennials, perennials, young shrubs and trees (Crawford & Wishart, 1966; Ranwell, 1972). Older slacks may be completely covered with shrubs and trees when natural succession has taken place. Nutrient stocks are high, both in the vegetation and in the shallow organic soil layer. The Dutch coast has an Atlantic climate where precipitation dominates over evaporation. This leads to prominent decalcification processes in the top layer and to rapid acidification in areas with a low initial lime content, as is the case in the Wadden Sea area (less than 2% $CaCO_3$).

Most typical dune slack species have to cope with hydrological very dynamic conditions. Anoxia tolerance is one of the dominant factors responsible for a stable vegetation within the dune slack (Studer-Ehrensberger et al., 1993). A very critical factor is the height and the length of the winter flooding. Plant species with well-developed aerenchyma such as *Schoenus nigricans, Littorella uniflora* can counteract the deprivation of oxygen from the soil by high levels of radial oxygen loss (ROL). They can survive as mature plants when flooded as long as a small part of the shoot can keep in contact with the atmosphere (Armstrong, 1975). Anoxic conditions, therefore, may prevent the establishment of late-successional species without ROL capabilities since they enhance the supply of reduced iron and manganese and also sulphide. Sulphide in particular can be very harmful for plant species if it remains in a reduced state. In addition to anoxia, dune slack species have to cope with nutrient-poor conditions (Schat et al., 1984). Many species, such as *Littorella uniflora, Centaurium pulchellum*, and *Radiola linoides*, have a very low nutrient demand because they are very small. Others, such as *Schoenus nigricans*, are long-lived tussocks forming species with a diameter of up to 35 cm and a height of over 70 cm. *Schoenus* has a very efficient recycling of nitrogen and phosphorus from senescent to juvenile leaf and shoot tissue (Ernst & van der Ham, 1988). In this way, they can capture much nutrients in the tussock making them unavailable for fast growing herbaceous species and later successional grasses such as *Calamagrostis epigejos*.

Types of dune slacks

Two types of dune slacks can be distinguished on the basis of their geomorphological history; primary and secondary slacks (Boorman et al., 1997). Primary slacks originate from sandy beaches, which have been partially or fully cut off from the influence of the sea. The soil in young primary slacks can be buffered by any combination of brackish surface water, $CaCO_3$-rich sand, and calcareous groundwater. Secondary slacks result from blowouts. They are usually calcareous and may or may not be supplied by calcareous groundwater. In both cases, the vegetation may be very similar since infiltration of rainwater also leads to base-rich conditions, since $CaCO_3$ is dissolved in the rooting zone almost immediately. Secondary, *in situ*, carbonate deposition has also been observed in dune slacks. It occurs when exfiltrating

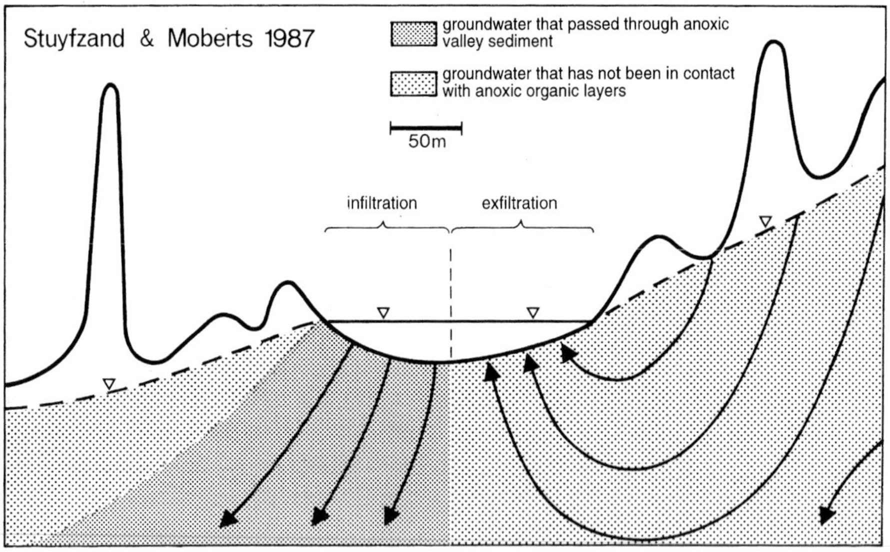

Figure 3. Groundwater flow pattern in a cross section through a dune slack with high water levels in the upper part of the dune area and low water levels close to the sea. Groundwater exfiltrates at the upgradient side of the slack and infiltrates at the opposite site, but only when the slack is flooded (after Stuyfzand & Moberts, 1987).

Figure 4. Vegetation development in a wet dune slack starting as sparsely vegetated pioneer stages (a and b), proceeding to a stage with many Red List species (c), until tall grasses and shrubs (d) take over (drawing by Rob Beentjes).

groundwater, (over) saturated with respect to calcite, enters the slack. Loss of CO_2 from this water results in carbonate precipitation at the soil–air interface (Krauskopf, 1982; Chafetz, 1994), thus counteracting soil acidification in a very significant way. Such carbonate precipitation can only occur when water tables remain very high during summer (c. 40 cm below the surface) and capillary rise of calcareous groundwater is possible.

Slacks in decalcified areas can harbour basiphilous vegetation types only if they are supplied by calcareous groundwater from surrounding dune areas. The best examples are situated at the low-lying periphery of the dune system, where most of the groundwater of the main hydrological system discharges. However, seepage slacks can also be found close to the top of the main hydrological system when thick clay or peat layers prevent infiltration to deeper layers and give rise to local groundwater flow towards adjoining dune slacks. Such slacks function as 'flow-through lakes' with groundwater discharge in one part of the slack and infiltration of surface water in another (Fig. 3). This hydrological mechanism can only function during the time that the slack is flooded and when high water tables in the surroundings occur (Stuyfzand & Moberts, 1987; Grootjans et al., 1996). Under such conditions, calcareous groundwater from deeper layers can flow towards the upgradient parts of the slack. The influx of calcareous groundwater stimulates mineralisation of organic matter and consequently the accumulation of organic matter is lower here than at the infiltration sites (Sival & Grootjans, 1996).

Succession

Natural succession in dune slacks can be roughly divided into 4 phases (Fig. 4): a pioneer phase (phase 1) in which small pioneer species establish on an almost bare soil, which is usually covered with a thin layer of green algae and laminated microbial mats (van Gemerden, 1993; Grootjans et al., 1997), (phase 2) colonisation of phanerogams adapted to very low nutrient availability, (phase 3) development of a moss layer of pleurocarpic bryophytes and establishment of typical dune slack species, and on older phase (phase 4) in which rapid accumulation of organic matter and increase of tall grasses and shrubs appears, which leads to the decline of typical dune slack species. The microbial mat in phase 1 stabilises the sandy substrate (Pluis & de Winder, 1990). The Cyanobacteria in the microbial mats can fix nitrogen (Stal et al., 1994) and may, therefore, assist in the colonisation by phanerogams. The dense layer of pleurocarpic bryophytes in phase 3 promotes the rapid built up of organic material and the establishment of tall grass and willow species. These eventually form a dense vegetation cover which prevent light penetration through the dense canopy, leaving little space for small dune slack species. Red list species are most abundant in phase 2 and 3. Orchids only occur in these early successional stages, where they receive nutrients from soil mycorrhizas (Smith, 1966). Juvenile orchids in particular are almost totally dependent on nutrients from mycorrhizas.

Succession along most of the Holland and Zeeland coast differs from that on the Wadden sea islands, because the initial lime content is higher (Fig. 1). Decalcification in the topsoil, therefore, begins in early stages of succession on the Wadden Sea islands. Furthermore, primary dune slacks on the Wadden Sea islands are often flooded by brackish surface water, which is a rare phenomenon along the Holland and Delta coasts.

Stability of pioneer stages

Large differences in the stability of the pioneer phase have been observed in dune slack succession (Lammerts & Grootjans, 1995). Micro-organisms and algae may contribute to the stability of the pioneer stage of dune slacks. They can form so-called microbial mats (van Gemerden, 1993) which are most abundant in tidal flats and salt marshes, but occur also in pioneer stages of dune slacks. Recent research suggests that in some fresh water dune slacks microbial mats may assist in retarding the natural vegetation succession (Grootjans et al., 1997), thus extending the life span of pioneer stages. The shift from pioneer stage to more mature stages usually takes place between 20 and 30 years (van der Maarel et al., 1985). In some dune slacks, however, pioneer stages may last for at least 30–60 years (Sival & Grootjans, 1996: Peterson, 2000). The succession rate and soil development in dune slacks is largely controlled by the productivity of the ecosystem, decomposition of organic matter and recycling of nutrients within the ecosystem (Koerselman, 1992; Verhoeven et al., 1996). The low productivity during the first years of vegetation development in dune slacks is the main cause of the low content of organic matter and the slow succession rate. The high hydrological dynamics may stimulate a rapid mineralisation of organic matter, especially under alkaline conditions. Tussock forming species as *Schoenus nigricans* can modify their environment by retention of nutrients in living and dead biomass, resulting in spatial aggregation of nutrients.

Fertilisation experiments in different chrono-sequential stages, originating from sod cutting at different times intervals, showed that late successional grass species, such as *Calamagrostis epigejos* and *Agrostis stolonifera*, are already present in pioneer stages, but that their growth was limited by a low availability of both nitrogen *and* phosphorus (Ernst et al., 1996; Lammerts et al., 1999). The low productive pioneer species, in particular *Juncus* and *Carex* species, were only nitrogen limited (Lammerts & Grootjans, 1997), most likely because they have a very low phosphorus demand (Willis, 1963; van Beckhoven, 1995). *Schoenus nigricans* showed no response at all to either nitrogen or phosphorus additions. This implies that, as long as phosphorus limits the growth of tall grasses, a basiphilous pioneer vegetation can persist for quite some time, even when nitrogen availability increases. Buffer mechanisms that keep the soil pH above 6, appear to be crucial for maintaining a low phosphorus availability. Adding lime in the fertilisation experiments had the same effect on biomass production as the effects of co-limitation by nitrogen and phosphorus.

Atmospheric input of nitrogen may accelerate the accumulation of organic matter in the topsoil considerably, because the growth of most pioneer and mid-successional species is N-limited and, therefore, responsive to additional supply of nitrogen.

Summarising, the hydrological regime of a dune slack is essential for a good functioning of the dune slack ecosystem. Factors that stabilise the longevity of pioneer stages comprising many Red List species are always associated with a regular supply of groundwater.

Disturbances in ecosystem functioning

Large scale disturbances of dune slack environment along the Holland coast in particular started around 1853 when the vast stock of fresh dune water became a major source of drinking water production for the large cities. Large dune areas actually became drinking water catchments. The exploitation of dune water resulted in a large scale lowering of the water table by 2–3 m on average (Bakker & Stuyfzand, 1993). At the same time, large parts of the dune area were saved as a landscape in a time of rapid industrialisation and rapid urbanisation in this densely populated area.

Factors that contributed to the dramatic decline of wet dune slacks along the Holland coast during the second half of the century were lowering of the water levels in the adjacent polder areas, reclamation for agricultural use and afforestation with pine plantations (see Janssen & Salman, 1995 for an overview). In a later stage, a new drinking water production technique was developed: infiltration of surface water from the rivers Rhine and Meuse. This input of polluted river water led to increased water tables in the dune slacks, but at the same time promoted eutrophication (van Dijk et al., 1985; van Dijk & Grootjans, 1991). In some areas, desiccated dune slacks were used as infiltration ponds, in others new infiltration ponds were excavated in the dune area.

In less affected dune areas, as the Wadden Sea coast, 70 years of dune fixation, in which every spot of bare soil was covered with branches or hay, resulted in an almost complete stop in natural dune formation. This has led to ongoing succession, which was accelerated by atmospheric deposition from industrial areas. The total amount of atmospheric N-deposition has increased between 1930 and 1980 from c. 10 kg N ha^{-1} yr^{-1} to c.25 kg N ha^{-1} yr^{-1} (Stuyfzand, 1993) and has stabilised between 25 and 35 kg N ha^{-1} yr^{-1}in the late 1990s (ten Harkel & van der Meulen, 1996; van Wijnen, 1999).

Decreased grazing by cattle and rabbits also enhanced grass encroachment (Veer, 1997) and the development of woodland. A positive feedback mechanism exists between increased biomass production and decreased groundwater availability. Tall vegetation types, such as shrubs and forests intercept more nutrients from atmospheric deposition than relatively open and short vegetation types, which leads to increased growth and a higher evapotranspiration. The result is an increased drop in water tables during the summer and consequently in a decreased discharge of groundwater in the dune slacks (Stuyfzand, 1993). If the supply of groundwater decreases, shrubs and tall grass species invade the site and pioneer communities lose the competition due to increased availability of nutrients.

A future threat to wet dune ecosystems could be caused by predicted sea level rise due to overall changes in climatic conditions. Sea level rise as a consequence of global warming could either lead to a drop or a rise in the phreatic water level, depending on erosion and accretion rates along the coastline (Noest, 1991).

Restoration of dune slacks; why and how?

Nature conservancy tries to stop the decline of pioneer communities by initiating restoration projects. The aim is to restore or stabilise pioneer stages of dune slack succession in order to provide endangered plant and animal species with habitats where their required site conditions are sustained for a long time. Dune slacks harbour many endangered fen species which have become almost extinct in mires on the mainland in most of Western Europe. Dune slacks, therefore, are listed high on the international conservation agenda. Restoration of dune slacks is relatively easy compared to restoration of mires. The restoration of fen ecosystems, for instance, usually requires a complete reconstruction of large hydrological systems (Pfadenhauer & Grootjans, 1999). In young dune slacks, the pool of stored nutrients is still small compared to the large peat deposits in fens. Consequently dune slack release less nutrients to the environment when changes occur in the hydrology.

The aim, therefore, is to create nutrient-poor, open and wet conditions that remain more or less stable for some decades. Most nature conservation organisations realise that the best option for continuous rejuvenation of dune slack ecosystems is to allow natural dune forming processes to take their course (Janssen, 1995). Encouraging these natural processes is, however, often blocked by established forms of land use in dune areas: sea defence, tourism, housing, forestry, production of drinking water etc.

The following management techniques have been applied in Dutch dune slacks to combat the negative effects of acidification, eutrophication and desiccation: (i) mowing, (ii) grazing, (iii) rewetting (iv) sod-cutting, and (v) creating new slacks.

Mowing

Mowing is a traditional management technique, which when applied without addition of fertiliser or manure, prevents grasses, willows and tree species to dominate the vegetation, thus favouring many non-competitive Red List species. The production of dead biomass is less because part of the biomass is harvested. This leads to decreased nutrient cycling within the ecosystem and might even lead to a shift in the type of nutrient limitation (Koerselman et al., 1995; Verhoeven et al., 1996). Mowing does not prevent the accumulation of organic matter in the top layer and it does not prevent acidification in decalcified soils. Mowing, therefore, is most efficient in sustaining basiphilous pioneer stages in areas where the soil is well buffered against acidification and where nutrient cycling is still low; in calcareous dune soils and in soils with a strong discharge of calcareous groundwater. Mowing early in the season can be very harmful, because it affects reallocation of nutrients in *Schoenus* tussocks and prevents seed set (Ernst, 1991).

Grazing

Grazing is often applied in dry dunes to combat grass and shrub encroachment. In wet dune slacks, cattle and horses may damage young pioneer stages by trampling and by polluting surface water. The effect of grazing by rabbits is most pronounced in early stages of succession. They may eat 40–50% of the shoot biomass during the first three years after sod cutting (unpublished data from van der Veen). Accumulation of organic matter is not prevented by grazing, but perhaps retarded, due to a lower input of litter (Kooijman & de Haan, 1995). Species-richness is locally stimulated by cattle grazing and some dominant species, such as *Calamagrostis epigejos, Salix repens* and *Carex arenaria* decrease in cover (van Dijk, 1992; Annema & Jansen, 1998; de Bonte et al., 1999). In general, grazing promotes low structured vegetation types and creates a more fine-grained vegetation pattern. In Moksloot area cattle reduced tall reeds and tall sedges including rare species such as *Ranunculus lingua*, leaving a monoculture of *Iris pseudacorus*

Rewetting

At present, adaptations in the hydrological system are applied frequently to rise water tables in dune areas. The extraction of natural dune water is further reduced. In the National Park of 'Zuid Kennemerland', for instance, existing groundwater extraction (9 Mm^3 a year) was eliminated. A large production canal of 2.5 km has been eliminated in the dunes near Amsterdam, leading to a spectacular rise in water tables of 2 m. Rewetting was combined with a restoration project of the whole dune landscape. (Geelen et al., 1995). In the near future, the elimination of groundwater extraction will lead to further rewetting of hundreds of hectares of desiccated dune slacks. In the Meyendel area, a production field with surface infiltration was recently removed to restore the hydrological system. This termination of production fields became possible due to the development of new production techniques,

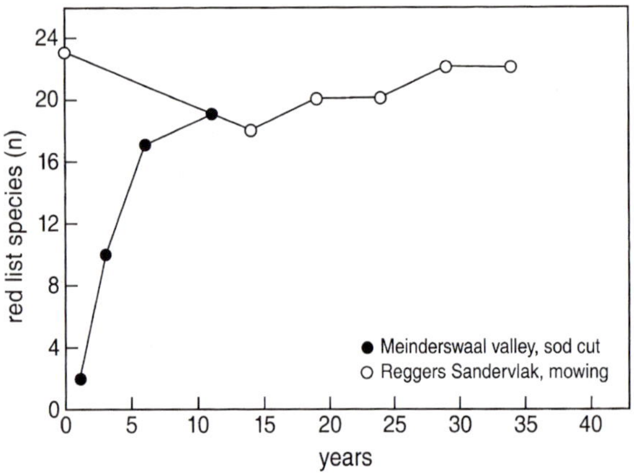

Figure 6. Changes in the number of Red List species during succession in the dune slacks with a calcareous soil (>2% $CaCO_3$) Reggers Sandervlak' and 'Meinderswaal' valley. Restoration measures were mowing and sod cutting, respectively.

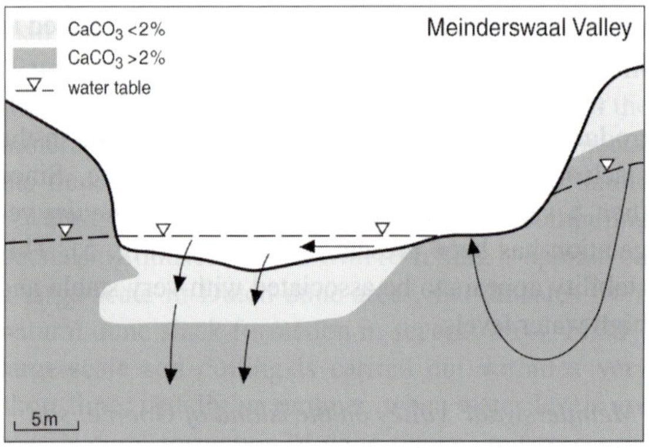

Figure 7. Hydrological system of the Meinderswaal valley (changed after Sival et al., 1997). Due to higher water levels in the upstream area calcareous groundwater is discharging in the dune slack at the right end side. This water proceeds as surface water and infiltrates again at the left end side.

but c. 1–5 cm of the organic soil remained. In 1982, sheep grazing was re-introduced in most of the dune area (c. 120 ha), including the Meinderswaal valley. In 1990, a small herd of Charolais cattle replaced the sheep, since the sheep only grazed the dry parts of the dunes, leaving the slacks untouched. Cows, however, successfully prevented a rapid spread of *Phragmites australis*, by eating the protein-rich tops. *Salix repens* and *Juncus subnodulosus*, however, were not grazed by either cattle or sheep, and gradually increased in cover.

In 1989, the vegetation of the upper part of the valley consisted of a pioneer community, in which *Anagallis tenella* established in 1990. This species was last recorded on the island in 1920. Typical dune slack species soon re-established since 1990 in both parts of the slack. The number of Red List species has increased steeply since 1990 (Fig. 6). *Schoenus nigricans, Dactylorhiza incarnata* and *Eleocharis quinqueflora* are now as abundant as described by Wevers (1920), but also new species established populations. *Parnassia palustris* and *Epipactis palustris*, however, did not return, probably due to lack of seeds.

The successful restoration of the Meinderswaal valley is most likely associated with the presence of an intact local hydrological system, supplying the slack with unpolluted calcareous groundwater (Sival et al., 1998). During winter, most of the valley is inundated. Infiltrated precipitation water passes through calcareous sands above a thin layer of very fine sand and seeps up along the southern edge of the valley (Fig. 7). Here, thin calcareous crusts are present on the soil surface.

'Reggers Sandervlak' North Holland Dune Reserve; hydrological measures and mowing

Reggers Sandervlak is a large secondary dune slack in the North Holland Dune Reserve situated some 150–300 m from the sea. The slack was formed about 250–300 years ago and still has a calcareous top layer due to a very high original lime content. The water tables in the slack had dropped considerably (0.2–0.5 m) due to extraction of drinking water, coastal retreat, lowering of the water tables in polder areas and by increased evapotranspiration by pine plantations. The slack became wetter after a decrease in groundwater extraction since 1976. Due to these changes in the hydrology the slack had lost many Red List species: *Dactylorhiza incarnata, Orchis morio, Gymnadenia conopsea* and *Cicendia filiformis* disappeared completely while others remained in small numbers. In the early 1980s *Hippohaë rhamnoïdes, Salix repens, Rubus caesius* and *Calamagrostis epigejos* dominated the vegetation. To prevent a further loss of Red List species a mowing regime was introduced and about 3 ha was mown every year in autumn. This mowing regime was rather successful and the populations of Red List species increased (Fig. 6). Even new red list species appeared, such as orchid species as *Dactylorhiza majalis* and *Dactylorhiza maculata*, which spread relatively easy. As a restoration technique mowing in this dune slack was efficient because most species, although threatened, were still present and the calcareous soil in this slack is well buffered against acidification.

'Moksloot' area on the island of Texel: hydrological measures and sod cutting

The Moksloot area, consisting of c. 250 ha of dunes and dune slacks, is situated at the south-western part of the island of Texel. The first vegetation descriptions of the area were made in c.1850 by Holkema (1870) who found many basiphilous pioneer and marsh communities in the largest slacks. Since then a retreat of the west-coast, the digging of a ditch (Moksloot) in 1880, drainage in nearby polder areas and, since 1956, extraction of groundwater in the slacks for drinking water purposes led to much drier conditions. Eutrophic reed beds with *Phragmites australis* and *Carex disticha* dominated the lower parts of the slacks, while the higher parts were covered by tall grasses such as *Calamagrostis epigejos* or willow shrubs (*Salix repens*). Some areas were mown in August to preserve species-rich grasslands with *Anagallis tenella, Ophioglossum vulgatum, Schoenus nigricans* and *Epipactis palustris*. After the termination of groundwater extraction in 1993, it was decided to restore the original species-rich dune slacks and top soil removal was applied in 16 slacks with a total area of over 35 ha. The depth of sod cutting varied between 10 and 40 cm.

In 1995, a small herd of Highland cattle and Exmar ponies was introduced to prevent a rapid growth of the vegetation in the sod cut areas. Only one small sand blown valley was fenced to prevent intensive trampling by cattle and horses. The vegetation development was monitored in several slacks from 1994 onwards (Grootjans et al., 2001).

Surprisingly, the species richness measured in the monitoring plots one year after sod cutting was approximately the same as in plots situated in adjoining sites that had not been sod cut. During the five years of monitoring, the species richness showed a steady increase (Fig. 8). The large slacks flooded by surface water started with a much higher diversity than the smaller and more isolated slacks directly after sod cutting. Almost 90% of the species was already present in 1994 in the largest slack. Species that contributed to the species richness belonged to plant communities ranging from dry dunes to eutrophic marshes. In some plots even species appeared which have their optimum in brackish sand beaches, which were not present in the study area before sod cutting. This rapid increase in species richness is unexpected, particularly in the isolated and fenced slack E, since no mature vegetation was left, no surface water could reach the area and grazing by cattle was prevented. Apparently other dispersal agents, such as dispersal by wind and birds or rabbits, are active as well.

The number of Red List species was very low at the start of the restoration measures, but increased rapidly in most slacks after 3–4 years (Fig. 9). In 1998, the number of Red list species and total species diversity was higher in the larger slacks with regular flooding by surface water compared to the more isolated slacks.

Apart from flooding frequencies, the hydrological regime favours the establishment of several Red List species in yet another way. Species, such as *Anagallis tenella, Schoenus nigricans* and *Potamogeton coloratus* showed a clear preference to those parts of the slack that were influenced by iron-rich groundwater. The groundwater flow pattern during early spring and its calcium concentration in the soil profile along an east west transect through three parallel dune slacks is shown in Figure 10. The central slack F receives groundwater from the adjoining slack C along it eastern shores, where pronounced precipitation of iron-hydroxides can be observed in early spring. This groundwater proceeds as surface water to infiltrate at the western shoreline and flows to the next exfiltration area of the low-lying slack (G).

These examples show that, despite the loss of most of the seed bank (c. 94% of the seed bank was lost), seed dispersal mechanisms were very effective in the area. Most of the species of the local species pool were still present nearby. Flooding in winter and early spring transported large numbers of seeds and plant fragments to the sod cut areas. Furthermore, the Moksloot area was visited by many ducks, gulls and geese, which used the sand flats as resting places. These birds may well have contributed to the rapid dispersal of many plant species that were rare or even completely absent from the area and originated from a regional species pool. The hydrological conditions had improved considerably for typical dune slacks species, due to the termination of the groundwater extraction and the rewetting of dune slacks situated upstream in the hydrological gradient. Calcareous groundwater and surface water could re-enter most slacks in spring and this has led to a higher pH in summer in most slacks (Grootjans et al., 2001).

Less successful projects; which ones and why

Because most restoration projects in the dune area have only started recently, it is difficult to judge whether they will be successful in the future. In some

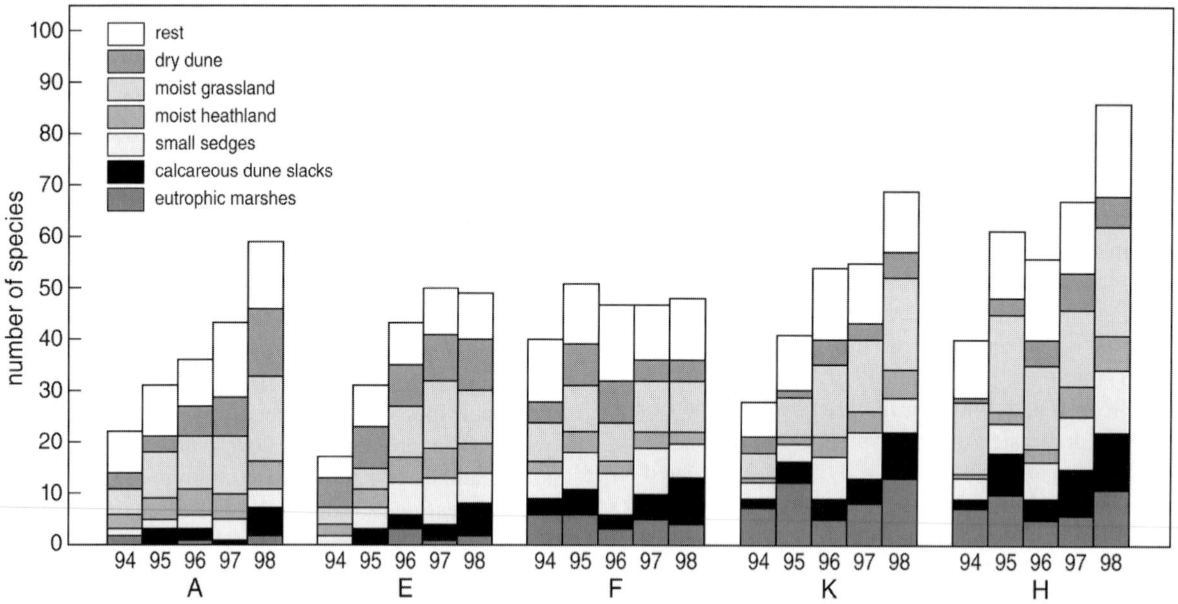

Figure 8. Changes in species-richness over 5 years, observed in 5 sod-cut dune slacks in the Moksloot area on the island of Texel (from: Grootjans et al., 2001). The changes in species-richness were measured in permanent plots of 10 × 10 m. A = isolated acidic slack, E = isolated acidic secondary slack, F = isolated calcareous through-flow slack, K and H = primary slacks flooded by surface water.

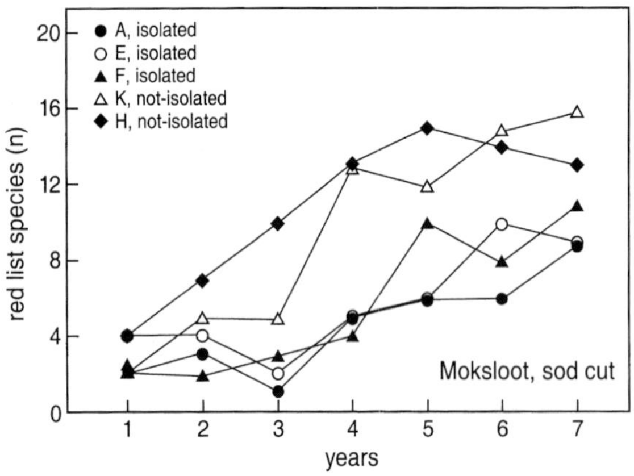

Figure 9. Changes in the number of Red List species during succession in sod-cut dune slacks in the Moksloot area, Texel (see also Fig. 8). All slacks were partly sod cut in 1993. The number of Red List species refer to the sod cut parts only (from Grootjans et al., 2001).

projects many target species established new populations shortly after the start of the project, but soon disappeared again. In other projects, almost none of the target species appeared during the first 5 years, but this might be due to a lack of seeds and such projects may become successful later. We will, therefore, classify projects as less successful if only a few of the target species established after 5 years, or when most of the target species showed a rapid decline shortly after establishment.

In the following, we will discuss some less successful projects along the Dutch coast.

'Van Hunenplak' on the island of Terschelling: creating a new dune slack

The dune slack Dodemanskisten on Terschelling (see map Fig. 1) was a natural dune slack which was once famous for its wealth of Red List species, mostly pioneer species of rather acid to neutral habitats (Holkema, 1870). These pioneer communities have existed for many decades, probably due to intensive sand blowing which prevented rapid succession (Westhoff & van Oosten, 1991). Planting of pine forests and drainage in the surroundings led to a loss of most species around 1940. Nowadays the slack is an eutrophic duck pond. To compensate for these losses a new pond had been excavated (van Hunenplak) nearby in a decalcified dune complex in 1950. This was the first example of a constructed dune slack in the history of the Netherlands. Although some target species established populations during the first 5 years (Fig. 11; *Pilularia globulifera, Cicendia filiformis, Juncus pygmaeus*, and *Littorella uniflora*), practically all species disappeared within 10 years (Sýkora, 1978). Wet heathland spe-

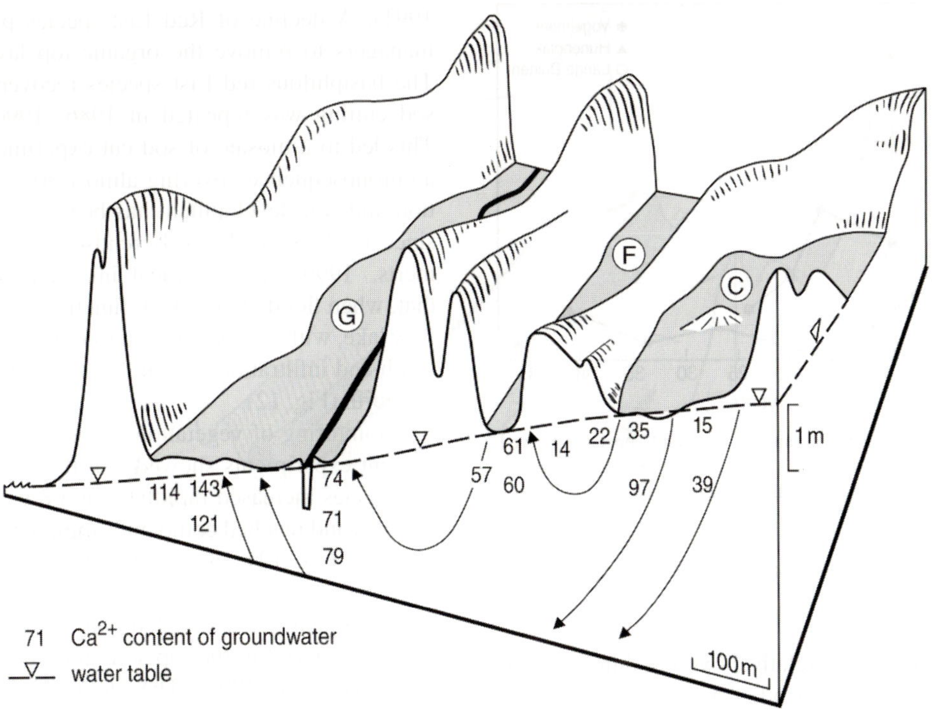

Figure 10. Hydrological system of the Moksloot area on the island of Texel (changed after Grootjans et al., 2001). The flow pattern of the groundwater has been derived from the Ca^{2+}-content of the groundwater in the soil profile. Dune slack C functions as an infiltration area and supplies dune slack F with calcareous groundwater, which infiltrates again at the western part of the slack to surface again at the eastern part of slack G. The centre of slack g is supplied with slightly brackish groundwater from deeper layers.

cies moved in quickly due to rapid acidification of the top soil. At present the pond is completely surrounded by birch wood and natural processes to counteract acidification, such as sand blowing, have become impossible.

'Vogelmeer' in the National Park Zuid Kennemerland: creating a new slack

The dune slack Vogelmeer (bird lake; see map Fig. 1) was constructed between 1951 and 1955 to compensate for the almost total loss of wet dune slack vegetation due to groundwater extraction for the public water supply and by afforestation (Londo, 1971). The vegetation development in most of the slack started in 1952. Between 1956 and 1968, vegetation development was monitored in great detail by Londo (1971). The dune slack is situated in relatively calcareous dunes along the western part of the Dutch coast (lime content 4–9%). With respect to species richness the project was successful. Between 1956 and 1959, the total number of species increased, but between 1959 and 1968 the species richness remained nearly constant at c. 170 species. The number of Red List species, however, showed a slow increase during the first 15 years (Fig. 11) and started to decline shortly afterwards, due to growth of tall grasses, sedges and shrubs. The aim to regenerate the plant community Junco baltici-Schoenetum nigricantis (Schaminée et al., 1995) failed due to decreasing water tables in the surroundings. Some characteristic species (*Dactylorhiza incarnata, Epipactis palustris* and *Schoenus nigricans)* established small populations, but they all soon disappeared. Londo (1971) compared his project with other constructed dune slacks and found that deep lakes with narrow shores were very unsuccessful in developing typical dune slack vegetation with many Red List species. Newly constructed dune slacks should resemble natural dune slacks with shallow pools and large shores. The new Vogelmeer did have such features, but yet was only moderately successful compared to natural dune slacks elsewhere. Londo concluded that dispersal problems could play a role in the early stages of succession, but the failure to develop a vegetation type with many Red List species most likely was due to unfavourable environmental conditions.

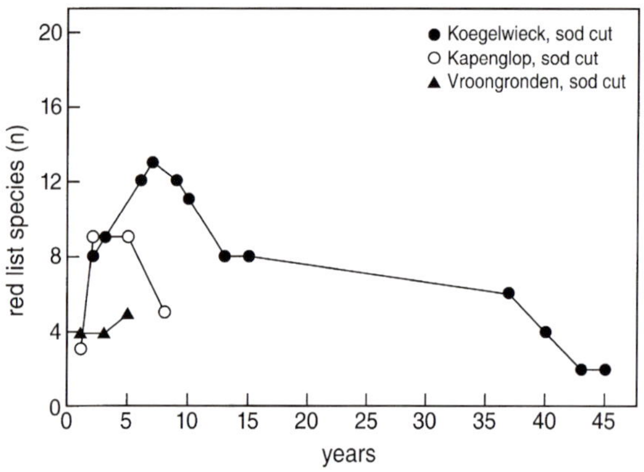

Figure 13. Changes in the number of Red List species observed in three dune slacks with a disturbed hydrological system (Koegelwieck, Terschelling; Kapenglop, Schiermonnikoog, and Vroongronden, Schouwen). The changes in the dune slack Koegelwieck were derived from sod cut experiments forming a chronosequence.

part of the production field to the western part of the island. Most Red list species appeared within 5 years. The prospects of this new restoration attempt are better since accumulation of the organic matter in the top soil is much slower in the newly sod-cut sites. At present, further improvement of the hydrological system is prevented by low water levels in the local infiltration areas of the slack. Higher water levels, however, would flood a tennis court during the winter. More adequate planning in the future could solve these problems.

'Vroongronden' on the island of Schouwen; sod cutting

A large sod-cutting project was carried out in 1994 in the dunes of the Delta coast on the island of Schouwen to restore typical dune slack vegetation types that were lost during the last three decades, due to lowered groundwater tables and increased acidification. The gently sloping dune area included a series of small dune slacks, which were flooded during wet periods and where relic populations of *Littorella uniflora, Apium inundatum, Eleocharis quinqueflora, Radiola linoides* and *Anagallis minima* had remained in only one small slack. Six other Red List species became extinct in the period between 1960 and 1985. After sod cutting *Littorella uniflora, Carex oederi, Anagallis minima* immediately established small populations in all sod-cut slacks (Fig. 14). *Radiola linoides* appeared two years later. Five years after carrying out the restoration measures, none of the other Red List species re-appeared. Detailed hydrological research showed that some of the slacks still function as 'through flow lakes'; during very wet periods precipitation water infiltrates in the whole area, which causes a pronounced raise in water tables, with highest water levels in the centre of the dune area. All dune slacks are flooded but the central ones have slightly higher (absolute) water tables compared to the slacks situated down-slope (Fig. 15). Since the slacks are not connected, a temporary cascade of water levels exists during very wet periods. Under such conditions, the up-slope slacks provide the down-slope slacks with calcareous groundwater. The groundwater has become calcareous passing through the shallow calcareous soil layers. These dune soils are relatively young and decalcification has occurred only in the top soil layers. Weather conditions from year to year have a pronounced effect on this hydrological mechanisms. But drainage ditches on both sides of the hydrological gradient have a much stronger impact. In this case, the ditch preventing flooding at the adjoining road, also prevents winter water tables to raise sufficiently to generate regular groundwater flow to the dune slacks in the centre of the dune area. Drainage in the down-slope camping site prevents regular groundwater discharge in the slacks at the lower end of the dune area. Interference with this very delicate hydrological system is probably the reason why conditions in the slacks are not optimal for Red List species. Considering the fact that only species with relic populations within the area established new populations after sod cutting, we may also assume that either seed banks of many other species were depleted or that dispersal mechanisms were ineffective.

Evaluation and restoration theory

Evaluation of case studies

Based on the case studies mentioned above, the following patterns of the restoration process in dune slacks can be observed (Fig. 16): (1) Unsuccessful projects include restoration projects where measures have been carried out in unsuitable sites; seed banks were depleted and dispersal mechanisms were not effective. Most of these projects were carried out to compensate for losses in existing nature reserves. (2) Temporary success, followed by rapid decrease of typical dune slack species can be observed in areas where seed banks were still present and dispersal mechanisms were effective, but where soil conditions and

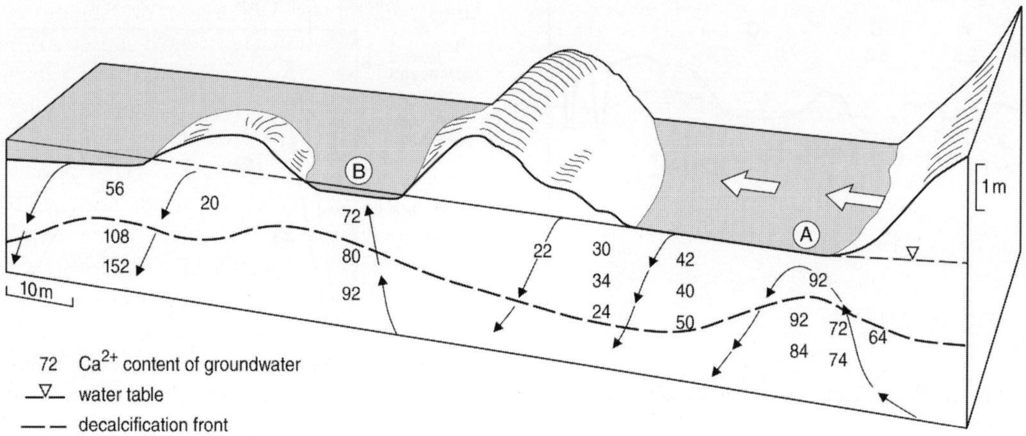

Figure 14. Hydrological system of the dune slack Kapenglop on the island of Schiermonnikoog (changed after Sival et al., 1997). The flow pattern of the groundwater has been interpreted from the calcium content of the groundwater in the soil profile and the position of the decalcification front. Calcareous groundwater enters the slack underneath the eastern slope (point A), where the decalcification front is situated high in the soil profile. The groundwater mixes with calcium-poor precipitation water at the soil surface and infiltrates again to deeper layers. In the central part of the transect (point B), groundwater discharges when the water level in the eastern slack has dropped below the surface.

hydrological regimes were sub-optimal. Examples are fresh water slacks surrounded by decalcified dune areas, where many changes in the hydrological regime had occurred. (3) Successful reconstruction of dune slacks in areas with hydrological systems that were little influenced by man and where dispersal mechanisms were effective. Pioneer stages, however, were relatively short lived. A good example is the Koegelwieck on the island of Terschelling. (4) Very successful projects where many typical dune slack species established in large numbers and persisted for many decades. These examples include the large sandy beaches on the Wadden Sea islands, which were enclosed by artificial dikes of drift sand. Local and regional species pools were available and dispersion mechanisms were still effective. Successful projects also include some examples in which a traditional management was re-installed. Good examples are Meinderswaal valley and Reggers Sandervlak situated in the calcareous dune areas of the mainland coast.

Analysing the restoration process further, we can distinguish three different environmental conditions influencing the stability and recovery process of young dune slack ecosystems. These conditions are most favourable when the soil is nutrient-poor and well-buffered (pH > 6), prolonged flooding with base-rich groundwater occurs and when the soil remains moist during dry summers. The latter condition is usually only fulfilled when the slack receives an additional flow of calcareous groundwater from adjoining dune areas. Under such conditions the accumulation of organic matter is slow (Fig. 17a). Nutrient-poor conditions and flooding in spring prevent the establishment of very productive species, while base-rich conditions stimulate the mineralisation of organic matter production of the pioneer species, keeping the nutrient stocks low. The best results are obtained in large beach plains that have been partly embanked by artificial sand dikes. Vegetation development (stage 1) starts under near natural conditions here. Usually the typical dune slack species can easily colonise such areas since small populations are practically always present in older stages nearby, and dispersal mechanism (wind, water, animals) appear to be very effective. If a well developed soil seed bank is present (stage 1a) the pioneer stages are rich in Red List species immediately after the restoration measures. Typical dune slack communities (stage 2) are maintained for a long time due to regular sand blowing which often starts the succession anew. Seepage slacks with a regular supply of anaerobic calcareous groundwater can also maintain pioneer stages for many decades, even when the dunes are fixed and no sand blowing occurs. It is still unclear which mechanisms are responsible for such 'stable states'. Late successional stages with grasses and shrubs may also harbour many Red List species (stage 3), but soon change into tall and very species-poor stands (stage 4). Accumulated nutrient stocks in the soil compartment appears to act as a threshold here

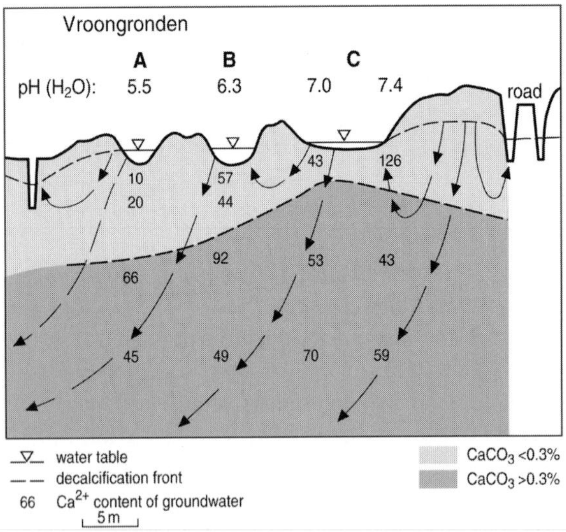

Figure 15. Hydrological system of the Vroongronden on the island of Schouwen with groundwater flow pattern derived from calcium contents of the groundwater and the position of the decalcification front in the soil profile. Also indicated is the pH of the top soil (0–10 cm) in various slacks. The infiltration area is situated close to a road, which is flanked by two drainage ditches. Precipitation passes through the calcareous sand deposits at c. 50 cm below the surface. Part of the calcareous groundwater enters slack C the eastern slope, where very high values of calcium were measured in the topsoil, possibly as a result of dissolution of secondary carbonate deposits. The decalcification front is situated high in the profile underneath the whole slack. The groundwater mixes with calcium-poor precipitation water at the soil surface and infiltrates again to deeper layers. Slack B is also supplied with calcareous groundwater from slack C. In slack A infiltration of precipitation water prevails.

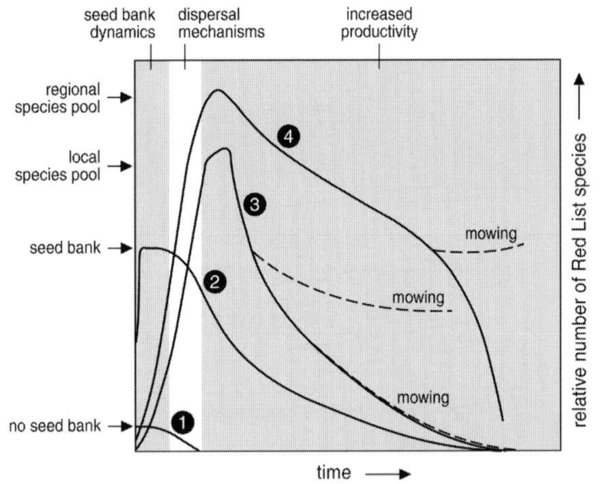

Figure 16. Conceptual model of occurrence of endangered dune slack species (Red List species) after restoration measures have been carried out. 1 = unsuccessful projects where measures have been carried out in unsuitable sites and where seed banks are depleted. 2= temporary success, followed by rapid decrease of target species is encountered in slacks where environmental conditions are unfavourable, but where seed banks are still present. 3 = successful, but short-lived, reconstruction of pioneer vegetation with many Red List species. Dispersal mechanisms are effective, but environmental conditions are sub-optimal. Mowing may sometimes retard a rapid spread of later successional species and a rapid decline in Red List species. 4 = very successful projects where many typical dune slack species establish in large numbers and persist for many decades. Natural processes retard the succession towards late successional stages. A mowing regime may stabilise the pioneer stage even longer.

(compare Hobbs & Norton, 1996), preventing further establishment of Red List species. The soil seed bank underneath such late successional stages may be still rich in Red List species (4a), but when time passes, the soil seed bank becomes depleted.

Less favourable conditions exist when a rapid acidification of the top layer occurs during succession. The result is a much faster accumulation of organic matter (Fig. 17b) and a general increase in productivity of the ecosystem (Gerlach et al., 1994). Pioneer stages after restoration measures, such as sod cutting, usually do not last long (10–15 years). If seed banks have been depleted and dispersal mechanisms are ineffective, the results of restoration measures in such areas are often very disappointing.

When the availability of nutrients in newly created slacks is high during the first successional stages, or when natural slacks have been polluted, fast growing plants species, such as *Phragmites australis, Calamagrostis epigejos,* or even *Urtica dioica* (van Dijk & Grootjans, 1993) can establish almost immediately and effectively store nutrients in living and dead material. The period for establishment of slow-growing pioneer species is often so short that conditions for them have become unfavourable even before most of the species have reached the area (Fig. 17c).

Restoration prospects and gaps in knowledge

Failures of restoration projects are often caused by not implementing existing knowledge or by selecting sites that cannot be expected to produce good results. Sometimes restoration measures are carried out in areas that are available, not because they are the best suitable sites for restoration of species-rich dune slacks. Existing ecological and hydrological knowledge is often good enough to estimate the success of planned restoration projects, but such knowledge is not always used, because of the costs of implementing such knowledge.

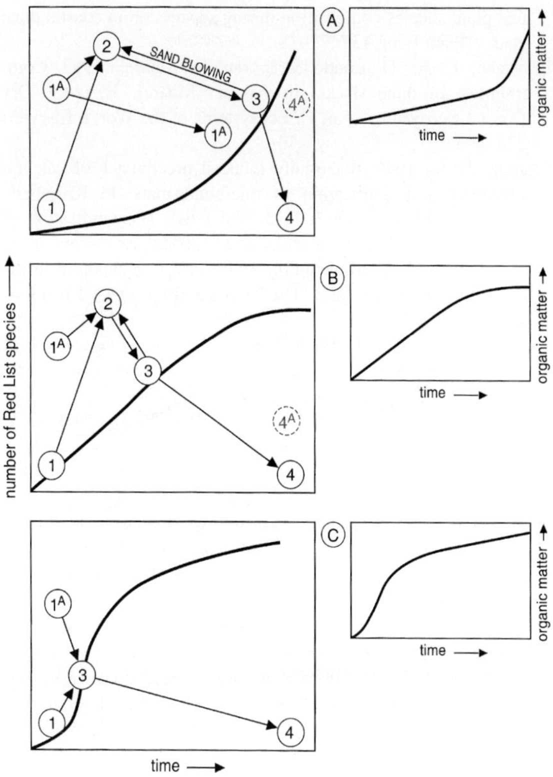

Figure 17. The rate of organic matter accumulation appears to be a threshold for a successful restoration of typical dune slack vegetation with many Red List species. When the soil is nutrient-poor, well buffered, and fed by base-rich ground water, the accumulation of organic matter is slow (Fig. 17a). Typical dune slack communities (stage 2 = target community) develop from species-rich pioneer stages (stage 1a), if a well-developed soil seed bank is available at the site immediately after restoration. If dispersal mechanisms are effective the target community may also develop from species-poor pioneer stages. Once established the target community can be maintained for a long time by regular sand blowing, which prevents a rapid accumulation of organic matter. Late successional stages with grasses and shrubs may also harbour many Red List species (stage 3), but soon change into tall and very species-poor stands (stage 4). Restoration of the target community is relatively easy, since many target species have survived in the seed bank or as relic populations in the late successional stages. Target communities with many Red List species are very short-lived when the accumulation of organic matter in soil or vegetation is rapid (Fig. 17b). The target community cannot develop if the accumulation of organic matter is very fast due to unfavourable environmental conditions. Some of the Red List species can occur temporarily in pioneer or later successional stages, but they will soon disappear. Restoration will not succeed when late-successional stages have existed for a long time and soil seed banks of the target community have been depleted.

When restoration goals are only formulated on the species level, it does not really matter in which type of ecosystems their populations are being restored. Many endangered dune slack species grow well in a large variety of dune slack types (Lammerts et al., 1999). Some species occur in large numbers in artificially created primary dune slacks, the same species can also be maintained in old secondary slacks on calcareous substrates by a mowing regime. Restoration projects are considered successful if large numbers of endangered Red List species have established new populations or if relic population increase in numbers. Such goals are formulated by most nature conservation organisations and they have broad political backing. Politicians, however, want quick results and creating new pioneer stages in existing dune slacks where relic populations are still present is often very successful during a short period. Such type of restoration often triggers comments as "gardening with wild species in natural mosaics" (Allan & Hoekstra, 1992), and the Dutch nature conservancy is well trained in such activities. The range of natural variability of ecological key parameters is well known in the case of dune slack ecosystems (Lammerts et al., 1999); it can be easily assessed whether such key parameters are outside the limits of required conditions, and it is well known which sites are best suited for optimal results. Such societal desires may, however, change (Allen, 1994) and the concept of 'wilderness' may become more popular. In the latter concept the method (i.e. no management measures) is often better specified than the goal, and this may conflict with a more traditional goal of preserving endangered species and landscapes. For recovery of natural ecosystems along the coast, a landscape-ecological approach (Naveh, 1994) is required. In the case of dune slacks, we need to specify natural references, identify the natural processes that form and sustain these ecosystems. Particularly important is to identify ecological and geomorphological mechanisms that create (semi-) stable states in pioneer communities, in which endangered species can survive for a long time without human interference. In our case, a reference could be the natural ecosystem formed by dynamic processes by wind and water. We have some information on what such references look like, but we have little information on dispersal strategies of endangered plant species, and how they establish populations in an increasingly unstable coastal area, e.g. connected to sea-level rise (Noest, 1991). For the Netherlands, a sea-level rise of 50–60 cm is expected over the next 100 years (van den Bergh & Nijkamp, 1998), which will no doubt lead to land losses in coastal areas that are not protected by firm dikes. However, the new government policy of 'dynamic preservation' of the Dutch coast also offers new possibilities to enhance nature conservation values in newly formed dune areas. This coastal defence policy primarily aims at ensuring safety against flooding and sustainable preservation of values and interests attached to dunes and beaches (de Ruig, 1998). A method to carry out this policy is 'sand nourishment' in places where sand is most needed. In a country that spends nearly 30 million USD a year on sand nourishment (de Ruig, 1998), the statement 'restoring the natural ecosystem' is not necessarily a vague statement, because such activities could stimulate the natural formation of dunes, dune slacks and even salt marshes. Creating a natural dune slack ecosystem is not an unachievable goal (Hobbs & Norton, 1996) in the Netherlands. Besides knowledge on ecological and geomorphological processes creating or sustaining such ecosystems, we need to formulate societal strategies to prevent human interventions in areas where these ecosystems can develop.

Acknowledgements

Part of the research was supported by the Ministry of Agriculture, Nature Management and Fisheries (Survival Plan of Dutch Nature: OBN). Rienk Slinks, Martin Annema and Kees Bruin are kindly acknowledged for supplying unpublished data on occurrence of Red List species in dune slacks. We thank Prof. Eddy Van der Maarel for commenting on an earlier version of the manuscript, and Dirk Visser for preparing the figures.

References

Allen, T. F. H. & T. W. Hoekstra, 1992. Towards a Unified Ecology. Columbia Press, New York.

Annema, M. & A. J. M. Jansen, 1998. Restoration of the Vroongronden area Middelduinen and Oostduinen on Goeree. Stratiotes 17: 20–60 (in Dutch with English summary).

Armstrong, W., 1975. Waterlogged soils. In Etherington, J. R. (ed.), Environment and Plant Growth. Wiley, London: 181–218.

Bakker, T. W. M., 1990. The geohydrology of coastal dunes. In Bakker, T. W. N., P. D. Jungerius & J. A. Klijn (eds), Dunes of the European Coasts, Geomorphology, Hydrology, Soils. Catena supplement 18, Catena Verlag, Cremlingen: 109–119.

Bakker, T. W. N. & P. J. Stuyfzand, 1993. Nature conservation and extraction of drinking water in coastal dunes. In Vos, C. C. & P. Opdam (eds), Landscape Ecology of a Stressed Environment. Chapman & Hall: 224–260.

Bekker, R. M., E. J. Lammerts, A. Schutter & A. P. Grootjans, 1999. Vegetation development in dune slacks: the role of persistent seed banks. J. Veg. Sci. 10: 45–54.

Berendse, F., E. J. Lammerts & H. Olff, 1998. Soil organic matter accumulation and its implication for nitrogen mineralisation and plant species composition during succession in coastal dune slacks. Plant Ecol. 137: 71–78.

Boorman, L. A., G. Londo & E. van der Maarel, 1997. Communities of dune slacks. In van der Maarel, E. (ed.), Dry Coastal Ecosystems, part C. Ecosystems of the World. Elsevier: 275–293.

Chafetz, H. S., 1994. Bacterially induced precipitates of calcium carbonate and lithification of microbial mats. In Krumbein, W. E., D. M. Paterson & L. J. Stal (eds), Biostabilisation of Sediments. BIS Verlag, Oldenburg: 149–163.

Colijn, C. J. & A. C. Binnendijk, 1998. Physical planning in the coastal region of Zeeland, The Netherlands. J. coastal Cons. 4: 135–142.

Crawford, R. M. M. & D. Wishart, 1966. A multivariate analysis of the development of dune slack vegetation in relation to coastal accretion at Tentsmuir Fife. J. Ecol. 54: 729–744.

De Bonte, A. J., A. Boosten, H. G. J. M. van der Hagen & K. V. Sýkora, 1999. Vegetation development influenced by grazing in the coastal dunes near The Hague, The Netherlands. J. coastal. Cons. 5: 59–68.

De Raeve, F., 1989. Sand dune vegetation and management dynamics. In Van der Meulen, M. J., P. D. Jungerius & J. Visser, (eds), Perspectives in Coastal Dune Management. SPB Academic Publishing: 99–109.

De Ruig, J. H. M., 1998. Coastline management in The Netherlands. J. coastal Cons. 4: 127–134.

De Vries, V., 1961. Vegetatiestudie op de westpunt van Vlieland. Ph.D. thesis, Univ. of Amsterdam.

Ernst, W. H. O., 1991. Ökophysiologie von Pflanzen in Küstendunen Europas in einem Gradiënten von der Nordsee zum Mittelmeer. Ber. Der Reinhold Tüxen-Gesell. 3: 451–465.

Ernst, W. H. O. & N. F. van der Ham, 1988. Population structure and rejuvenation potential of *Schoenus nigricans* in coastal wet dune slacks. Acta bot. neerl. 37: 451–465.

Ernst, W. H. O., Q. L. Slings & H. J. M. Nelisssen, 1996. Pedogenesis in coastal wet dune slacks after sod-cutting in relation to revegetation. Plant Soil 180: 219–230.

Geelen, L. H. W., E. F. H. Cousin & C. F. Schoon, 1995. Regeneration of dune slacks in the Amsterdam Waterwork dunes. In Healy, M. G. & J. P. Doody (eds), Directions in European Coastal Management. Samara Publ.: 525–532.

Grootjans, A. P., F. P. Sival & P. J. Stuyfzand, 1996. Hydrogeochemical analysis of a degraded dune slack. Vegetatio 126: 27–38.

Grootjans, A. P., W. H. O. Ernst & P. J. Stuyfzand, 1998. European dune slacks: strong interactions between vegetation, pedogenesis and hydrology. Trends Ecol. Evol. 13: 96–100.

Grootjans, A. P., P. S. Hartog, L. F. M. Fresco & H. Esselink, 1991. Succession and fluctuation in a wet dune slack in relation to hydrological changes. J. Veg. Sci. 2: 545–554.

Grootjans, A. P., F. P. van den Ende & A. F. Walsweer, 1997. The role of microbial mats during primary succession in calcareous dune slacks: an experimental approach, J. coastal. Cons. 3: 95–102.

Grootjans, A. P., F. H. Everts, K. Bruin & L. F. M. Fresco, 2001. Restoration of wet dune slacks on the Dutch Wadden Sea island of Texel; recolonisation after large scale sod-cutting. Rest. Ecol. 9: 137–146.

Hobbs, R. J. & D. A. Norton, 1996. Towards a conceptual framework for restoration ecology. Rest. Ecol. 4: 93–110.

Holkema, F., 1870. De plantengroei der Nederlandsche Noordzee-eilanden. Ph.D. thesis, Univ. of Groningen.

Janssen, M. P., 1995. Coastal management: restoration of natural processes in fore dunes. In Healy, M. G. & J. P. Doody (eds),

Directions in European Coastal Management. Samara Publ.: 195–198.

Janssen, M. P. & A. H. P. M. Salman, 1995. A national strategy for dune conservation in The Netherlands. In Salman, A. H. P. M., H. Berends & M. Bonazountas (eds), Coastal Management and Habitat Conservation. EUCC, Leiden: 215–230.

Jones. P. S. & J. R. Etherington,1989. Ecological and physiological studies of sand dune slack vegetation, Kenfig pool and dunes local nature reserve, Mid-Glamorgan, Wales, UK. In van der Meulen, F., P. D. Jungerius & J. Visser (eds), Perspectives in Coastal Dune Management. SPB Academic Publishing: 297–303.

Koerselman, W., 1992. The nature of nutrient limitation in Dutch dune slacks. In Carter, R. W. G., T. G. F. Curtis & M. J. Sheehy-Skeffington (eds), Coastal Dunes. Ballgame: 189–199.

Koerselman, W. & J. T. A. Verhoeven, 1995. Eutrophication of fen ecosystems; external and internal nutrient sources and restoration strategies. In Wheeler, B. D., S. C. Shaw, W. J. Foot & R. A. Roberts (eds), Restoration of Temperate Wetlands. John Wiley: 91–112.

Kooijman, A. M. & M. W. A. De Haan, 1995. Grazing as a measure against grass encroachment in Dutch dry dune grassland: effects on vegetation and soil. J. coastal Cons. 1: 127–134.

Krauskopf, K. B.,1982. Introduction to Geochemistry. McGraw-Hill, Auckland.

Lammerts, E. J. & A. P. Grootjans, 1996. Nutrient deficiency in dune slack pioneer vegetation: a review. J. coastal Cons. 3: 87–94.

Lammerts, E. J. & A. P. Grootjans, 1998. Key environmental variables determining the occurrence and life span of basiphilous dune slack vegetation. Acta bot. neerl. 47: 369–392.

Lammerts, E. J., A. P. Grootjans, P. J. Stuyfzand & F. P. Sival, 1995. Endangered dune slack gastronomers in need of mineral water. In Salman, A. H. P. M., H. Berends & M. Bonazountas (eds), Coastal Management and Habitat Conservation. EUCC, Leiden: 355–369.

Lammerts, E. J., D. M. Pegtel, A. P. Grootjans & A. van der Veen, 1999. Nutrient limitation and vegetation change in a coastal dune slack. J. Veg. Sci. 10: 11–122.

Londo, G., 1971. Pattern and process in dune slack vegetation along an excavated lake in the Kennemer dunes (the Netherlands). Thesis, Univ. of Nijmegen (in Dutch with English summary).

Naveh, Z., 1994. From biodiversity to ecodiversity: a landscape-ecological approach to conservation and restoration. Rest. Ecol. 2: 180–189.

Noest, V., 1991. Simulated impact of sea level rise on phreatic level and vegetation of dune slacks in the Voorne dune area (the Netherlands). Landsc. Ecol. 6: 89–97.

Olff, H., J. Huisman & B. F. Van Tooren, 1993. Species dynamics and nutrient accumulation during early primary succession in coastal sand dunes. J. Ecol. 81: 693–706.

Parikh, A. & N. Gale, 1998. Vegetation monitoring of created dune swale wetlands, Vandenberg Air Force Base, California. Rest. Ecol. 6: 83–93.

Petersen, J., 2000. Die Dünentalvegetation der Wattenmeer-Inseln in der südlichen Nordsee. Ein pflanzensoziologische und ökologische Vergleichsuntersuchung unter Berücksichtigung von Nutzung und Naturschutz. Dissertation Univ. Hannover, Husum Verlag (with English summary).

Pfadenhauer, J. & A. P. Grootjans, 1999. Wetland restoration in Central Europe: aims and methods. Appl. Veg. Sci. 2: 95–106.

Pluis, J. L. A. & B. De Winder, 1990. Natural stabilisation. In Bakker, T. W. M., P. D. Jungerius & P. A. Klijn (eds), Dunes of the European coasts. Catena supplement 18: 195–208.

Ranwell, D. S., 1972. Ecology of Salt Marshes and Sand Dunes. Chapman and Hall.

Schaminée, J. H. J., A. H. F. Stortelder & V. Westhoff, 1995. De vegetatie van Nederland, part 2. Opulus Press, Uppsala/Leiden.

Schat, H., A. H. Bos & M. Scholten, 1984. The mineral nutrition of some therophytes from oligotrophic dune slack soils. Acta Oecol., Oecol. Plant. 5: 119–131.

Sival, F. P., 1996. Mesotrophic basiphilous communities affected by changes in soil properties in two dune slack chronosequences. Acta bot. neerl. 45: 95–106.

Sival, F. P. & A. P. Grootjans, 1996. Dynamics of seasonal bicarbonate supply in a wet dune slack: effects on organic matter, nitrogen pool and vegetation succession. Vegetatio 126: 39–50.

Sival, F. P., H. J. Mücher & S. P. J. Van Delft, 1998. Carbonate accumulation affected by hydrological conditions and their relevance for dune slack vegetation. J. coastal Cons. 4: 91–100.

Sival, F. P, A. P. Grootjans, P. J. Stuyfzand & T. Verschoore de la Houssaye, 1997. Variation in groundwater composition and decalcification depth in a dune slack: effects on basiphilous vegetation. J. coastal Cons. 3: 79–86.

Smith, S. E., 1966. Physiology and ecology of orchid mycorrhizal fungi with reference to seedling nutrition. New Phytol. 65: 488–499.

Stal, L. J., M. Villbrandt & B. de Winder, 1994. Nitrogen fixation in microbial mats. In Krumbein, W. E., D. M. Paterson & L. J. Stal (eds), Biostabilisation of Sediments. BIS Verlag, Oldenburg, FRG: 384–399.

Studer-Ehrensberger, K., C. Studer & R. M. M. Crawford, 1993. Competition at community bounderies: mechanisms of vegetation structure in a dune slack complex. Funct. Ecol. 7: 156–168.

Stuyfzand, P. J., 1993. Hydrochemistry and hydrology of the Coastal Dune area of the western Netherlands. Ph.D. thesis, Free Univ. of Amsterdam.

Stuyfzand, P. J. & F. M. L. Moberts, 1987. De bijzondere hydrologie van kwelplassen in duinen met kunstmatige infiltratie. H2O 20: 52–57.

Ten Harkel, M. J. & F. Van der Meulen, 1996. Impact of grazing and atmospheric deposition on the vegetation of dry coastal dune grasslands. J. Veg. Sci. 7: 445–452.

Van Beckhoven, K., 1995. Rewetting of coastal dune slacks: effects on plant growth and soil processes, Ph.D. thesis, Free Univ. of Amsterdam.

Van den Bergh, J. & P. Nijkamp, 1998. Economical aspects of global change impact and response strategies in the coastal zone of The Netherlands. J. coastal Cons. 4: 161–168.

Van der Laan, D., 1979. Spatial and temporal changes in the vegetation of dune slacks in relation to the groundwater regime. Vegetatio 39: 43–51

Van der Maarel, E. & M. Van der Maarel-Versluys, 1996. Distribution and conservation status of littoral vascular plant species along the European coasts. J. coastal Cons. 2: 73–92.

Van der Maarel, E., R. G. A. Boot, D. Van Dorp & J. Rijntjes, 1985. Vegetation succession on the dunes near Oostvoorne, The Netherlands: a comparison of the vegetation in 1959 and 1980. Vegetatio 58: 137–187.

Van der Veen, A., A. P. Grootjans, J. de Jong & J. Rozema, 1997. Reconstruction of an interrupted beach plain succession using a GIS. J. coastal Cons. 3: 71–78.

Van Dijk, H. W. J., 1992. Grazing domestic livestock in Dutch coastal dunes: experiments, experiences and perspectives. In Carter, R. W. G., T. G. F. Curtis & M. J. Sheehy-Skeffington (eds), Coastal Dunes. Balkema: 235–250.

Van Dijk, H. W. J. & A. P. Grootjans, 1993. Wet dune slacks: decline and new opportunities. Hydrobiologia 265: 281–304.

Van Dijk, H. W. J., M. A. W. Noordervliet & W. T. de Groot, 1985. Nutrient supply of herbaceous bank vegetation in Dutch coastal dunes; the importance of nutrient mobilisation in relation to (artificial) infiltration. Acta bot. neerl. 34: 301–319.

Van Gemerden, H.,1993. Microbial mats: a joint venture. Mar. Geol. 113: 3–25.

Van Wijnen, H., 1999. Nitrogen dynamics and vegetation succession in salt marshes. Ph.D. thesis, Univ. of Groningen.

Verhoeven, J. T. A., W. Koerselman & A. F. M. Meuleman, 1996. Nitrogen- or phosphorus-limited growth in herbaceous, wet vegetation: relations with atmospheric inputs and management regimes. Trends Ecol. Evol. 11: 494–496.

Weeda, E. J., R. van der Meijden & P. A. Bakker, 1990. Rode Lijst van de in Nederland verdwenen en bedreigde planten over de periode 1980–1990. Gorteria 16: 2–26.

Westhoff, V., 1947. The vegetation of dunes and salt-marshes on the Dutch islands of Terschelling, Vlieland and Texel. Ph.D. thesis, Univ. of Utrecht.

Westhoff, V., 1989. Dunes and dune management along the North Sea coasts. In van der Meulen F., P. D. Jungerius & J.Visser (eds), Perspectives in Coastal Dune Management. SPB Academic Publishing: 41–51.

Westhoff, V. & M. van Oosten, 1991. De plantengroei van de Waddeneilanden. KNNV, Utrecht.

Willis, A. J., 1963. Braunton Burrows: the effects on the vegetation of the addition of mineral nutrients to the dune soils. J. Ecol. 51: 353–374.

Appendix 1. Red list species used in the evaluation of restoration projects in Dutch dune slacks

Plant community	**Red List species**	**Plant community**	**Red List species**
Potamion graminae	*Potamogeton gramineus*	Caricion davallianae	*Parnassia palustris*
	Potamogeton coloratus		*Juncus arcticus*
	Potamogeton polygonifolius		*Carex oederi*
Littorellion uniflorae	*Littorella uniflora*		*Anagallis tenella*
	Echinodorus ranunculoides		*Dactylorhiza incarnata*
Hydrocotylo-Baldellion	*Eleocharis multicaulis*		*Eleocharis quinqueflora*
	Deschampsia setacea		*Epipactis palustris*
	Pilularia globulifera		*Equisetum variegatum*
	Apium inundatum		*Gentianella amarella*
Saginion maritimae	*Centaurium littorale*		*Gentianella campestris*
	Sagina nodosa		*Herminium monorchis*
	Gnafalium luteo-album		*Liparis loeselii*
Nanocyperion flavescentis	*Juncus capitatus*		*Schoenus nigricans*
	Cicendia filiformis		*Pedicularis palustris*
	Anagallis minima		*Gymnodenia conopsea*
	Radiola linoides		*Linum catharticum*
	Juncus pygmeus	Nardo-Galion saxatilis	*Antennaria dioica*
	Hypericum humifusum		*Botrychium lunaria*
Armerion maritimae	*Oenanthe lachenalli*		*Platanthera bifolia*
	Odontites vernus		*Polygala vulgaris*
	Centaurium pulchellum		*Carex caryophyllea*
	Scirpus rufus		*Hypericum pulchrum*
Junco-Molinion	*Cirsium dissectum*		*Pedicularis sylvatica*
	Carex pulicaris		*Gentiana pneumonanthe*
Ericion tetralicis	*Lycopodium inundatum*	Other communities	*Cladium mariscus*
	Dactylorhiza maculata		*Dactylorhiza majalis*
Empetrion nigri	*Genista tinctoria*		*Scirpus cariciformis*
	Pyrola rotundifolia		*Rhinanthus minor*
	Pyrola minor		*Anthyllis vulneraria*
			Briza media

threatened by extinction or, in case of some species and populations, already extinct. The European sturgeon (*Acipenser sturio*, Linnaeus, 1758) which inhabits the Northern European coasts and rivers, is extinct in large parts of its range primarily due to overfishing (Lelek, 1987; Rochard et al., 1990; Anonymous, 1994; Billard, 1995). In north-western Europe, the sturgeon is only present in the river system of the Gironde–Garonne, Dordogne (France). Since the mid 1980s, a restocking programme was initiated in this river system (Castelnaud & Trouvery, 1984). It is unlikely that stocks from the Gironde system will recolonize the river Rhine system.

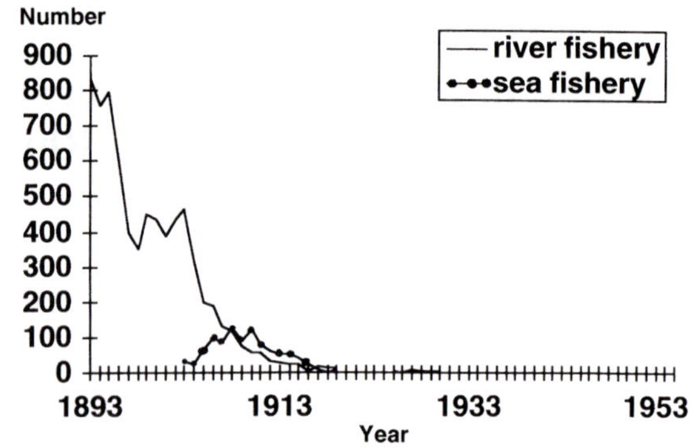

Figure 1. Number of sturgeon (*Acipenser sturio*) caught in the Lower Rhine and North Sea; data from Fisheries Inspection (Anonymous, 1920) and Verhey (1961).

Historical distribution

In this paper, sturgeon is only discussed in relation to the river Rhine system. From the Iron Age (700–300 BC) and Roman times (57 BC–406 AD), it is known that the sturgeon inhabited Dutch waters. This was demonstrated by Prummel (1989), who excavated fragments of a dermal plate and of spina pectoralis on the former island of Voorne-Putten, dating back to the Iron Age. Moreover, Brinkhuizen (1989) described remains of a 1-m long sturgeon found near Velsen (North Sea coast), and early Middle Ages (600–1200) sturgeon remains are known from the area near Oldenburg (Germany) (Prummel, 1991). Kinzelbach (1987) carried out a reconstruction of the now extinct population of the sturgeon in the river Rhine-system. He based his study on records from the 15th century. Sturgeon were common in the Delta and in the Lower Rhine until about 1910, becoming extinct as a breeding species in 1942. Kinzelbach (1987) described in detail the sturgeon distribution higher up in the river systems and stated that the species was more common higher up the river than was formerly assumed. Nevertheless, about two thirds of the sturgeon population stayed in the Lower Rhine and Delta and about one third swam up to the Middle and Upper Rhine (up to 850 km upstream) and their tributaries. It is estimated that about 3000–4000 sturgeons lived in the Rhine from perhaps 1440 until 1900. The stock was larger in the middle and at the end of the 16th century.

It is unknown where sturgeon nurseries existed in the river Rhine. It is assumed that spawning took place in the deeper parts of the river, mainly between May and August. Minimum water depth requirements for spawning are 3 m, although spawning can occur in depths up to 20 m (Jakob, 1996). The so-called 'pits' in which sturgeons were observed (e.g. Mohr, 1952) in the river Eider, north Germany are in fact refugia for fish to rest for a while during the spawning process. Deelder & Huussen (1973) and Lobregt & van Os (1977), indicate that sturgeon spawn in very shallow water, however, on what grounds this statement is based is unclear. Spawning of sturgeon has never been observed in Dutch waters. The smallest sturgeons mentioned in the literature in comparable German waters (Eider, Elbe-river system) were 40–60 cm, or at least 1–2 years old. Hence, there has been no indication of a spawning area at all (Hoek, 1910).

After spawning the adults migrate rapidly to the sea, where they remain before returning to spawn again. The juveniles leave the freshwater area and move into the estuary. They may migrate for a while into the sea and then return to the estuary. The fact that sturgeon males require about 8 years to reach sexual maturity, and females even 14 years, makes them especially vulnerable to fishing. Verhey (1949, 1961, 1963) analysed the sturgeon fishery of the Lower Rhine, with an emphasis on the period from 1900 to 1931. In the year 1749, 8.999 sturgeons were caught in the Lower Rhine and landed at Dordrecht fish market. Between 1824 and 1852, an average of 3000 sturgeons per year were caught. Complete reliable fishery statistics of sturgeon date from the end of the 19th century. Figure 1 is based on data collected by the Dutch Fishery-inspectorate (Anonymous, 1920), augmented with Verhey's data. Also added to this figure are catch data of sturgeon by the North Sea fishing fleet (1903–1919) (Anonymous, 1920). After a strong decline of river caught sturgeon, sea catches exceeded the river catches during a short period, but after 1960 no river catches of sturgeon in the Lower Rhine were reported. Sturgeons, approximately 1 m in length, were occasionally caught in the North Sea or washed ashore on the beaches (Nijssen & de Groot, 1987).

After 1992, however, juvenile sturgeons have been reported from the Lower Rhine and adjacent waters (Volz & de Groot, 1992). The first individual (8-2-1992) from the river Merwede (Rhine-estuary) proved to be a juvenile *Acipenser sturio*; the body was mutilated, perhaps by a propeller of a ship. Since then, occasional small sturgeons have been reported (e.g. Hartgers et al., 1998). After careful examination, they proved to be hybrids, with many features of the Russian sturgeon (*A. gueldenstaedti*, Brandt, 1833) and stellate sturgeon (*A. stellatus*, Pallas, 1811). The reported revival of the sturgeon in the Rhine can not be taken seriously, as the specimen collected are likely originating from accidental releases of aquarium fish.

Causes of decline

What are the causes of the decline and ultimate extinction of the sturgeon in the Rhine? Kinzelbach (1987), assuming that the fisheries took about 30% of the sturgeons in the Rhine, calculated that around the turn of the 19th century, the stock of the Middle- and Upper Rhine was 100–1000 sturgeons. In the Lower Rhine and Delta, the stock was about 2400 sturgeons. He concluded that the decline of the sturgeon started in the mid 19th century and that the species became extinct in the thirties of the 20th century. The most likely cause was the overfishing, and the destruction of spawning grounds, due to canalisation and the construction of dams and weirs. For a restoration endeavour the following requirements are essential: a gravel bottom (1.7–6.5 cm gravel), with a water velocity of 1 m s^1 and a water temperature of 15–20 °C, needed for spawning. Oxygen concentrations of $\geq$6 mg l^{-1}, water depths of 3–20 m, and a spawning area per female of about 300 m^2 are also required for successful reproduction.

Increased fishing pressure proved to be as destructive as the deterioration of the spawning areas. The river fisheries caught far too many adult specimens, the sea fisheries finished off the immature and remaining adults (Steinert, 1951; Mohr, 1952). Birstein (1993) and Shagaeva et al. (1993) described serious changes in the structure of gonads of sturgeon and their larvae in the Volga due to water pollution. As for most of the world's sturgeons, the combination of river pollution, river regulation, and overfishing brought about the eventual collapse of a once abundant population (reviewed in Barannikova, 1987).

Table 1. Dutch coregonid catches 1910–1939 (data in the archive of RIVO-DLO, IJmuiden)

Year	Landings (kg)	Year	Landings (kg)
1910	2769	1925	65
1911	3586	1926	597
1912	1665[a]	1927	534
1913	1241[a]	1928	115
1914	3170[a]	1929	140
1915	14 682[a]	1930	24
1916	3298	1931	109
1917	4737	1932	139
1918	1175	1933	129
1919	1644	1934	53
1920	720	1935	1
1921	972	1936	2
1922	130	1937	46
1923	25	1938	0
1924	690	1939	3

[a] Underestimate.

Coregonids

Depending upon the taxonomic system selected, either two or three anadromous coregonid species inhabit, or have inhabited the Rhine: *Coregonus lavaretus* (Linnaeus, 1758) (whitefish); *C. oxyrinchus* (Linnaeus, 1758) (houting); and *C. albula* (Linnaeus, 1758) (vendace). Vendace have been caught only a few times in the Rhine (the last time in 1927), but it is most likely that the fish were simply incorrectly identified when caught. At the turn of the century, whitefish were caught in all large rivers such as the Rhine, IJssel and Meuse, houting was very common in the Wadden Sea (Redeke,1934). At present only a small original population of houting still exists in the Danish Wadden Sea, with a proven spawning ground in the river Vidå near Tønder. Since the discovery of the existence of this spawning ground, houting has been released in the Baltic Sea, in German and Danish waters, and also in the German river Eider (Grøn, 1987; Grøn et al., 1988; Jäger, 1999; Scheffel, 1999).

Coregonid fishing occurs from August to November. Prior to 1910, catch data are difficult to trace because fish were sold directly by the fishermen at local markets. However, it was possible to obtain data for 1910–1939 (Table 1). Although the fishery was never large, the obvious decline between 1916 and 1920 is remarkable. In 1917, the catch was still around

Table 2. Overview of releases of fry and juvenile coregonids in The Netherlands. (Reuter, 1966; Nijssen & de Groot, 1987)

Year	Numbers	Age	Location
1907	50 000	fry	Meuse/IJssel
1908	400 000	4 weeks	Meuse/IJssel
1922	60 000	6 weeks	?
1923	22 000	fry	Meuse
1924	–	–	–
1925	500 000	fry	Meuse/IJssel
1926	2121 000	4 weeks	Meuse/IJssel
1927	2442 000	fry	Meuse/IJssel
	1000	6 months	Reeuwijkse plassen
	1000	6 months	Amstel
1928	50 000	3–4 weeks	Meuse
	175 000	3–4 weeks	Meuse
	400 000	fry	Gelderse IJssel
	1625 000	1 year	Meuse
	900 000	1 year	Gelderse IJssel
	9220	1 year	Meuse
1929	500 000	fry	Meuse
	500 000	fry	IJssel
	150 000	fry	Meuse
	749	1 year	Meuse/Reeuwijkse plassen
1930	5150	1 year	Meuse
1931	1000 000	fry	Meuse
1937	50 000	fry	Reeuwijkse plassen
	50 000	fry	Bergumermeer
	50 000	fry	Binnen Maas
	50 000	fry	Alkmaardermeer
	50 000	fry	Beulakerwijde
	700 000	fry	Lake IJsselmeer
	1800	6 months	Reeuwijkse plassen/Binnen
1938	failed	–	Meuse/Alkmaardermeer
1939	–	–	–

5000 kg: the decline was therefore far quicker than that observed for salmon (de Groot, 1992). The dramatic downward trend and near-extinction of the species was reason for great concern by government authorities. Artificial rearing was tried, for the first time in 1907, and a restocking programme with eggs, 1-month and 1-year-old whitefish, imported from Peipus Lake, Russia, was initiated. In Dutch inland waters, a total of nearly 119 million larval or juvenile whitefish were released between 1907 and 1939 (Table 2); however, there were no increases in abundance.

There is a growing concern in Germany for the preservation of coregonids. Non-migratory stock in German alpine lakes are threatened by pollution,

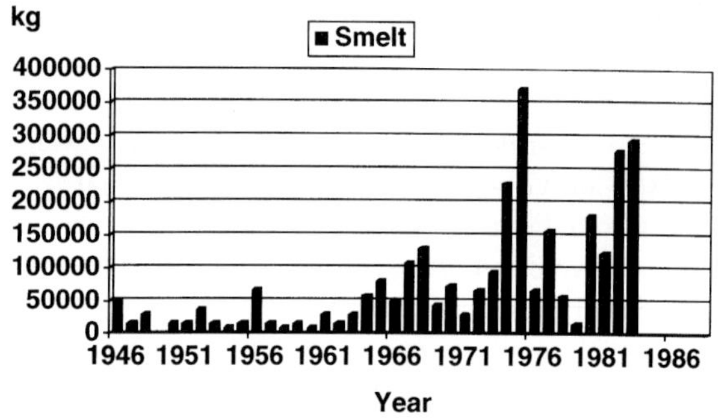

Figure 2. Smelt (*Osmerus eperlanus*) catches in Hollands Diep and Haringvliet between 1946 and 1988.

hence, extensive annual restocking operations are carried out in large lakes created by barrages (so called 'Talsperre'). Coregonids are released mainly in the States of Nordrhein-Westfalia and Hessen. It is assumed that the present increase in the Rhine, as well as in the Dutch Lake IJsselmeer, must originate from these Danish/German restocking attempts (de Groot & Nijssen, 1997; Hartgers et al., 1998).

Coregonid populations have not recovered in Dutch inland waters. Migratory forms of coregonids are unable to enter the inland waters owing to the disappearance of suitable tidal inlets. The houting spawned in the Netherlands, very likely in the northern province of Groningen, adjacent to the Wadden Sea, up to the beginning of the 20th century. However, the rivers and waterways of Groningen have been changed from 1870 to 1977, and the tidal movement still observed in the harbour of the town of Groningen disappeared. Channels were deepened, the surrounding land was better irrigated and exploited, and sluice systems were built to prevent the inlet of seawater during flood tide. At present, plans to regulate the water level in Lake Lauwersmeer, a former inlet of the Waddensea, will definitely close off the entry of the houting into inland waters in the northern part of the Netherlands. This blockade between the small rivers and the sea is exacerbated owing to subsidence of the land mass caused by the exploitation of natural gas, disturbing the natural run off of water from land to sea. All these factors make the return of houting for spawning in Dutch inland rivers a remote possibility.

(*Osmerus eperlanus* Linnaeus, 1758)

Smelt (*Osmerus eperlanus*) in Dutch waters are present in both migratory (anadromous) and land-

Table 3. Lake IJsselmeer smelt catches 1980–1998 (data Fisheries Directorate, The Hague)

	Landings (kg)
1980	157 000
1981	160 000
1982	2309 000
1983	1235 000
1984	1305 000
1985	1357 000
1986	1881 000
1987	2317 000
1988	3319 000
1989	1982 000
1990	1481 000
1991	2504 000
1992	1517 000
1993	586 000
1994	1402 000
1995	2345 000
1996	856 000
1997	1032 000
1998	327 000

Table 4. Smelt catches in Holland's Diep and Haringvliet 1914–1919 (data Fisheries Directorate, The Hague)

Year	Total (kg)
1914	140 771
1915	109 366
1916	225 955
1917	101 097
1918	176 055
1919	146 275

locked (non-migratory) forms. The migratory form is known in the estuaries, the Wadden Sea and the lower reaches of the large rivers. Migratory smelt have essentially vanished from Dutch coastal waters (Fig. 3) owing to the construction of the large Delta works project, the Haringvlietdam, a safety scheme against flooding, completed in 1970, and virtually sealing off the estuary from the sea. The non-migratory form is found in large quantities in the Lake IJsselmeer (Table 3), the Frisian lakes and in the provinces of North and South Holland (Table 4, Fig. 2).

Migratory smelt reach a length of up to 30 cm in 8 years (Mohr, 1941). Non-migratory smelt of the IJssel Lake, a large freshwater body, reach lengths of 19

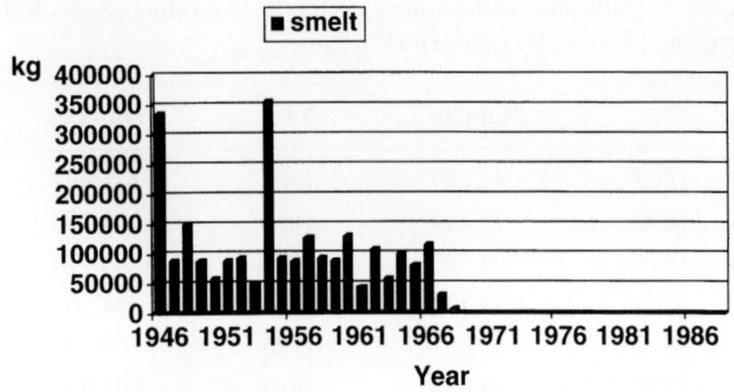

Figure 3. Smelt (*Osmerus eperlanus*) catches in the Dutch Wadden Sea between 1946 and 1988.

cm in their fourth year (first year: 6 cm, second year: 12 cm; third year: 14 cm), but these large individuals are, however, uncommon (Redeke, 1907, 1914, 1922; Havinga, 1928, 1954). Redeke (1914) indicated that migratory smelt spawned in the lower reaches of the Rhine and in the freshwater tidal areas of Hollands Diep, between Moerdijk-bridges and Willemstad. The migratory smelt was always rare near Arnhem and Nijmegen in the upper Rhine reaches of the Netherlands (van den Ende, 1847). The non-migratory smelt spawn in both lakes and rivers from February to April when water temperatures are between 4 and 12 °C. The eggs are ellipsoidal and average fecundity is 40 000 eggs. The eggs adhere to stones, gravel, or water plants, but never to silt bottoms. Average egg density is 6 eggs cm^{-2}, but sometimes eggs are also found in layers. The larvae hatch between temperatures of 18–20 °C.

One reason for the decline of the river Rhine smelt is the building of extensive coastal defence constructions, which prevent migration of fish to their spawning habitats. An additional cause of the decline is the deposition of enormous amounts of river borne silt, following the closure of the Haringvliet estuary (Anonymous, 1987). Silt-covered and heavily polluted river bottoms are unsuitable for egg survival.

Allis shad (*Alosa alosa* Linnaeus, 1758) and twaite shad (*A. fallax* Lacépède, 1803)

The allis shad and the twaite shad both belong to the riverine Clupeidae, and supported considerable commercial fisheries in the past. Historically, the catches of the allis shad were rather important, but the twaite shad has gained some importance concurrently with the disappearance of allis shad.

Table 5. Allis shad (*Alosa alosa*) landings at Kralingseveer, Rotterdam (1869–1894) (after Hoek, 1894)

Year	Number	Year	Number
1869	42 217	1882	141 542
1870	79 184	1883	103 746
1871	94 786	1884	84 170
1872	79 322	1885	184 209
1873	69 851	1886	179 439
1874	116 033	1887	167 966
1875	85 181	1888	148 846
1876	71 431	1889	128 837
1877	67 495	1890	123 233
1878	91 998	1891	53 568
1879	91 232	1892	43 915
1880	65 707	1893	34 289
1881	122 398	1894	35 500

Allis shad

In the past, adult fish were caught by the Dutch river fisheries between the third week of March and the first week of June. In good years, the catch made up about 20% of the annual income for the local fishermen, and salmon comprised the remaining 80%. The decline in allis shad over the years is striking (Fig. 4, Table 5). Landings averaged 207 423 (1881–1890), 54 685 (1891–1900), 39 701 (1901–1910), 1249 (1911–1920) and finally 13 (1931–1936) individuals (Redeke, 1938). Nationally as well as internationally, protective fishery measures were difficult to agree upon. The reasons were simple: restricting allis shad fishing from March to June would also have had repercussions on salmon fishing. Hence, the main reason for the sharp decline in stocks was probably overfishing.

The allis shad entered rivers to spawn when the water was about 11–12 °C. Males arrived in small schools a few weeks earlier in the season than females. Allis shad was commonly caught in the Rhine from March to June, and the spawning period was May to June. Actual spawning never took place in Dutch waters, but rather in the higher reaches of the Rhine, near Koblenz; the Mosel, near Trier; and the Neckar, between Neckarsteinach and Hirschhorn. Spawning took place mainly at night in the upper water column, and over a gravely bottom. During spawning the fishes could be heard splashing through the surface waters, a sound described by the late Roman Ausonius in his poem ‘Moselia’. The best known description of this spawning was given by Vincent (1894). Parental fish leave the river after spawning. The eggs are fertilized in mid-water and then sink to the bottom, where they enter the interstices of the gravel. A female produces about 50 000 eggs per kg body weight. Depending on the water temperature, larvae emerge within 4–8 days (22–24 °C). Larvae and juveniles are transported downriver into the tidal zone. Vertical diurnal activity, rhythmically synchronized with water ebb and flow, permits them to remain in the estuary for over 1 year.

River alterations (e.g. deepening, barrages, etc.) have negatively effected the spawning areas of allis shad. Nevertheless, Lelek (1987) stated that “poor water quality is probably chiefly responsible for the decline and extinction in many European rivers”. This seems less true for the Rhine, where overfishing and destruction of spawning habitat can be considered as the main factors leading to extinction. Pollution combined with physical difficulties for anadromous fish species to enter into the Rhine-Scheldt estuary will certainly hamper future reintroduction efforts. In the past the lower reaches of the river Rhine acted as a nursery. Due to the construction of large flood defence works (Delta works), juvenile fish of both allis and twaite shad can no longer thrive throughout the year in the lower river reaches, because the complex system of vertical migration combined with the tidal movement for species maintenance no longer exists.

The theory presented by Redeke (1938), based on earlier observations of Hoek (1894, 1899) that the hybridisation of allis and twaite shad played an important role in their disappearance, is supported by recent electrophoretic work (Boisneau et al., 1992); this would explain the disappearance of the allis shad from the relatively clean Rhône river (Rameye et al., 1976). As far as the river Rhine population is concerned, even if this process had occurred, the effect would have been insignificant because the spawning areas of both species were geographically well separated.

Twaite shad

Twaite shad spawned under conditions similar to those of the allis shad. However, the twaite shad stayed in the freshwater tidal zone, spawning mainly in the Dutch part of the Rhine river, whereas the allis shad moved further upstream. Because the twaite shad had a shorter distance to move upstream, the entry period and the length of the stay of the adult fish in the area were much shorter than that of the allis shad. April and May were the months of river

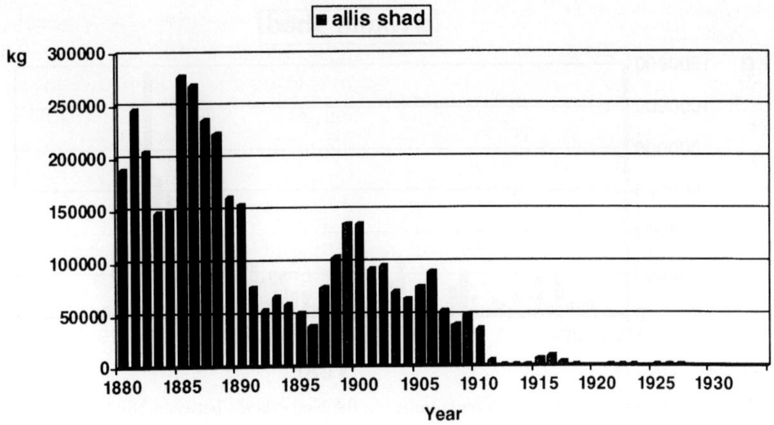

Figure 4. Allis shad (*Alosa alosa*) landings in the Netherlands between 1880 and 1934.

entry and spawning, with actual spawning occurring over 3 weeks. Well known spawning areas were in the Merwede (near Woudrichem) and in the Bergse Maas (near Genderen), but actual spawning sites were never identified. After spawning, the adult twaite shad left the river but the juveniles remained in the lower reaches together with young allis shad. The fish reached maturity within 3 years (Hoek, 1899).

As twaite shad were considered inferior during the days when allis shad were plentiful, no catch data were kept. Only when allis shad catches were declining did this fish become of interest, leading to a fishery between 1933 and 1944 (Figs 5 and 6). Records in early years certainly underestimate true catches; however, it is obvious that catches peaked in 1934–1939. Although the statistics during World War II (1940–1945) underestimate catches, the decline in landings are obvious. In 1950, catches increased to some extent, but since 1966 when Hollands Diep and Haringvliet were virtually cut off from the sea by the Delta works, catches dropped to zero in the lower Rhine river reaches. Twaite shad are currently found in very small numbers in several locations close to the coast of the Netherlands. Observations in Lake IJsselmeer do not coincide with the spawning period, and justified the assumption that the fish enter Lake IJsselmeer unintentionally or only for a short period of time (Hartgers et al., 1998).

Decline of the twaite shad in the Rhine was probably caused by several factors, not only overfishing. River regulation works destroyed several spawning habitats, and pollution and silting up may also have played a role in the stock depletion. Closing off of the river, thereby changing the freshwater tidal system in the estuary into a river with a one-way flow, has likely been the fatal blow for the population. However, the introduction of a new sluicing regime in the Haringvliet (Delta works) is being reconsidered, to allow part of the former tidal movement (Anonyous, 1998), and this may lead to a partial recovery of the twaite shad population in the south-west of the Netherlands.

Sea trout (*Salmo trutta* Linnaeus, 1758)

The life cycle of the sea trout resembles that of other anadromous fish such as the salmon. However, an important difference is that both sexually mature and immature trout ascend the river together in autumn on their return from the sea (Shearer, 1955). Trout can reproduce several times over their life-span, with the eggs being laid in gravel beds in the upper reaches of the river (Stuart, 1953; Frost & Brown, 1967). Sea trout hybridise with brown trout, and are rarely found in rivers with currents in excess of 60 cm s^{-1}.

In contrast to salmon, sea trout generally remain closer to the coast; most of the trout leaving Dutch waters head northwards (Svärdson & Fagerström, 1982) and stay within 100–350 km of the coast. Compared with salmon catches, the numbers of sea trout caught in the river Rhine were relatively small and had an average length of 50 cm and weighed approximately 1 kg.

The fact that a significant proportion of sea trout that are landed, never appear in the fisheries statistics is not new. As early as 1920, officials noted that many of the sea trout caught were not sold at market and therefore not included in the catch statistics. Nevertheless, using information from Hoek (1893) and catch figures from the Fisheries Inspectorate, which

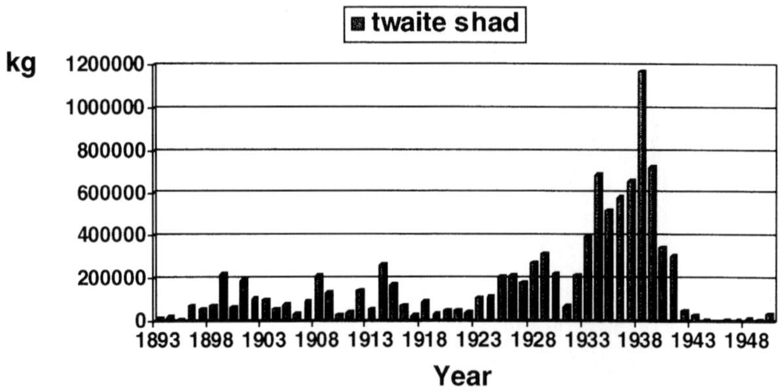

Figure 5. Twaite shad (*Alosa fallax*) landings in the Netherlands between 1893 and 1950.

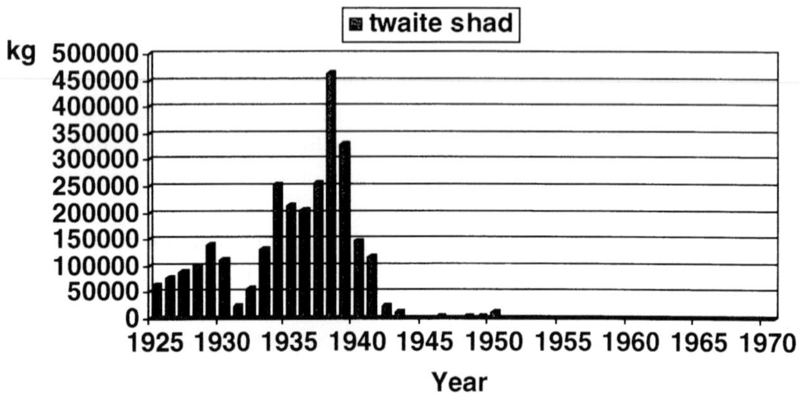

Figure 6. Twaite shad (*Alosa fallax*) landings in the Netherlands (Hollands Diep and Haringvliet) between 1925 and 1970.

later became known as the Fisheries Division and the Department of Fisheries, estimates can be made of the number of trout landed between 1886 and 1986 (Fig. 7). Annual trout catches over this period varied considerably, but no evidence can be found to suggest that there has been a systematic decline in numbers (de Groot, 1989b). Reports from professional fishermen are too infrequent to allow firm conclusions to be drawn about trout stocks, but do provide an indication of the continuing presence of trout in the river.

The action that has most affected the sea trout and other migratory fish in the Rhine is the closure of the sea inlets in the south-west of the Netherlands. The presence of sea trout in the IJsselmeer could be due to trout entering via the rivers Rhine and IJssel. However, it is also plausible that sea trout enter via the sluices in the Barrier Dam at Kornwerderzand and Den Oever. The fact that trout are mainly caught in the lake and not in the rivers is not sufficient justification for assuming that trout are only present in large numbers in Lake IJsselmeer and that this serves as a catchment area for trout. Thorpe's (1990) hypothesis about the archetypal life history strategy comprising specific physiological and morphological adaptations of trout to the freshwater environment, places the occurrence of sea trout in the Rhine and Meuse in a different light. Sightings of silver-coloured trout in Nordrhein-Westfalen or in the Meuse can no longer be taken as proof that trout have swum upstream from the sea. These fish could equally well originate from brown trout populations and have been forced to leave their environment because of insufficient food supplies or inadequate conditions. Similarly, the presence of sea trout in the lower reaches of the rivers Meuse, Lek, Merwede or in the Rhine-Meuse-Scheldt estuary can not be viewed as conclusive evidence that these fish are intending to ascend the river. In contrast to the behaviour of the salmon, young sea trout often undertake what are known as 'dummy runs'. Sexually immature trout that inhabit coastal waters often accompany their sexually mature counterparts a short distance up the river before returning to the sea. Although it is generally agreed that trout enter Dutch rivers from the sea, it is difficult to establish to what extent this occurs. The

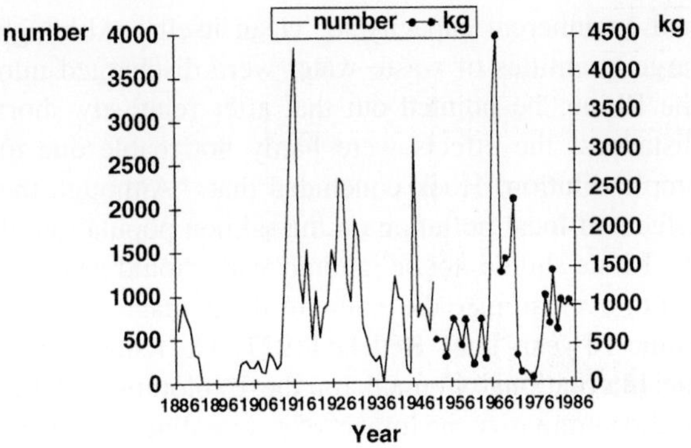

Figure 7. Catches of sea trout (*Salmo trutta*) over the period 1886–1986; data not available for 1892–1901 (source RIVO-data).

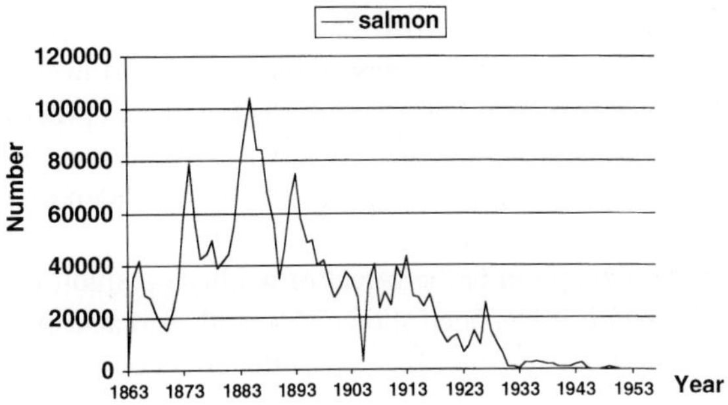

Figure 8. Dutch salmon (*Salmo salar*) catches 1863–1957 (source RIVO-data).

irregular catch statistics merely provide an indication of such movements.

Salmon (*Salmo salar* Linnaeus, 1758)

The history of the Dutch fishing industry can be traced back to the oldest inhabitants of the Netherlands, although it is difficult to specify exactly when salmon fishing began. The first written evidence to suggest that salmon were being caught in reasonable numbers dates back to 1100. However, proper statistics describing the number of salmon caught in the Netherlands at that time are not available. Similarly, few details are to be found about domestic consumption or export. Due to the lack of data, it is impossible to give precise estimates of the number of salmon caught in the Dutch, German, French and Swiss sections of the Rhine basin on an annual basis. The most detailed historical records (1863–1957) that were kept concern the salmon caught in the Dutch part of the Rhine (Fig. 8).

The end of the 1920s witnessed a major decline in salmon catches in the Rhine. In subsequent years, the numbers of salmon landed in the Dutch section was stabilised at about 1000–2000 fish per year, up to 1944. However, these figures would probably have been higher, had organised salmon fishing not ceased in Dutch waters in 1933. To rehabilitate the Rhine salmon population, experiments began as early as 1861 releasing salmon fry which had been reared under controlled conditions. These activities assumed greater importance after the ratification of the Salmon Convention in 1886. Enormous numbers of fry and parr were released into the Rhine at the end of the 19th and the beginning of the 20th centuries. Restocking operations in Switzerland and Germany between 1879 and 1912, to which the Netherlands contributed, involved about 160 million young salmon. In addition, about 13 million young salmon were released into the Rhine in the Netherlands over the period 1861–1897. Throughout the period that serious restocking operations were carried out, most observers were convinced of the positive effect of this policy. It was felt that without such action the decline in salmon numbers would have been even more rapid. However, in 1947, a report on the effect of restocking operations on salmon catches concluded that no clear statistical evidence could be found to support such a link from the available data (de Groot, 1989a, 1992).

A number of factors is hold responsible for the disappearance of the salmon from the Dutch coastal area and the large rivers. De Jonge & de Jong (2002) and Nienhuis et al. (2002) devote an extensive discussion to the anthropogenic disturbances that have led the deterioration of surface waters in the Netherlands. In the following paragraphs, a brief survey will be given of the main factors behind the decline of the once important salmon fishery in the Netherlands.

Canalisation and flood control

The Rhine is basically a glacier stream that receives an important supply of water during the summer months. As a consequence, the difference between seasonal high and low water levels in the Rhine is less marked than in a river fed by rainwater, such as the Meuse. This feature makes the Rhine particularly attractive to shipping. The move towards greater industrialization in the 19th century led to deepening of the navigation channel and, where necessary, the construction of weirs and locks. The fact that canalisation of the Rhine proceeded at a more rapid pace than, for instance, that of the Meuse, reinforced the natural superiority of the Rhine as a major inland shipping route. It is not only the presence of locks and weirs along the Rhine that are preventing the return of salmon. The condition of the many streams and rivulets that traditionally formed the spawning grounds for these fish has also changed by human intervention (van de Ven, 1976). Moreover, simply restoring a few tributaries of the Rhine – which is suggested by certain parties as a potential solution – would not guarantee the return of the salmon on a permanent basis, as a variety of habitats will be required.

At present, the Nieuwe Waterweg offers the only open link with the sea in the Netherlands. However, the intensity of shipping along this route could discourage salmon migration. Since the construction of the 'Afsluitdijk' (Barrier Dam), which closed off the Zuiderzee, salmon have been hampered from using their once traditional route to re-enter the Rhine via the river IJssel. Sluices are not forming an absolute barrier, and a few salmon entered the freshwater via the sluices in the Afsluitdijk and those of the North Sea Canal (IJmuiden) (Larsson, 1984). This was recently demonstrated by the first record of an Arctic charr (*Salvelinus alpinus* Linnaeus, 1758) in the Netherlands. The fish was caught (6-7-1999) in a fykenet near Breezand. The total length was 26.7 cm, the weight: 90 g.

Sand and gravel extraction

Sand and gravel extraction activities in Dutch rivers are not thought to have had a serious impact on salmon stocks, mainly because salmon only pass through the Netherlands, either on their way to the sea as smolts or kelts, or in ascending the river as grilse, 2-sea-winter, or multi-sea-winter salmon. Sand and gravel extraction operations in the smaller German rivers and streams will have had a much more dramatic effect on the salmon population since these areas were historically used as spawning grounds. The increase in silt levels in the river and subsequent sedimentation on the spawning grounds make these spots unsuitable for reproduction purposes.

Waste water discharges

As early as the beginning of this century, discharges of polluted waste water were identified as one of the possible reasons for the decline in the salmon population in the Rhine (Hoek, 1916b). Hoek was convinced that the Rhine, as a large fast-flowing river, would have had an inherent capacity to 'clean itself'. Although large quantities of waste water were discharged into the Rhine, he pointed out that after relatively short distances, the effects were hardy noticeable due to ample dilution. Hoek concluded that: "Although the effects of local pollution on the salmon population of the Rhine should not be ignored, care should be taken not to overestimate the extent of the damage caused". Some 10 years later, Redeke (1927) was requested by the International Council for the Exploration of the Sea to summarize the impact of river pollution on fisheries. In general, it can be said that by 1927 domestic and industrial waste waters were considered to pose a far greater threat to the fish population than had been thought possible less than 25 years before. The number of fish kills had increased dramatically. However, it was often difficult or even impossible to give precise explanations for such events, in spite of the fact that a large number of potential causes could be identified (van Brummelen, 1989).

An exception is the effect of acidity on smolt development. It has been shown that acid (and acid plus aluminium) exposures that have little effect on survival and growth of parr or smolts in freshwater, completely destroy the function of the gill in saltwater (Staurnes et al., 1996). The smolt inhabit the main stream of rivers and estuaries, which are more heavily impacted by pollutants then high elevation tributaries where fry and parr live. Pollutants may also, intervene with the development of olfactory imprinting (Bardach & Case, 1965; Holl, 1965; Bardach, et al., 1965, 1967; Holl et al., 1970).

The impact of the fishing industry on salmon stocks

With the advantage of hindsight, it can clearly be said that the fishing industry has had a negative effect on salmon stocks in the Rhine. However, in the final analysis, economic forces dictate that the fishing industry often can not fish a particular species to extinction. The loss of income as the size of the catch declines can bring such activities to a natural end, as was the case with the Rhine. The Dutch salmon fishing industry effectively ceased to exist after 1933 and that of the Germans after 1950.

Feasibility of restocking operations for salmon

At present, salmon catches in the entire North Atlantic are declining. (1970 – 10 735 ton; 1980 – 10 127 ton; 1990 – 4943 ton; 1995 – 3625 ton; 1996 – 3132 ton; 1997 – 2377 ton; 1998 – 2401 ton; ICES, 1999). Many fisheries, both commercial and recreational, have already been closed. Fishing seasons are shortened, and more and more anglers are returning their caught fish to the river. The situation is quite serious. This is particularly true at the southern limit of the salmon distribution range in Portugal, Spain and the United States e.g. the American Government has decided to list the wild salmon in eight rivers as 'endangered' under the Endangered Species Act (Information NASCO, Edinburgh). The effects of Global Warming become manifest, just as with the cod (*Gadus morhua* Linnaeus, 1758) in the North Sea. Even though a number of excellent salmon rivers are still to be found in countries around or near the North Sea, salmon catches (rod catches) in these rivers are on the decline as well. Salmon culture on the contrary flourishes. In 1998, the production of Norway was 360 536 tons and in Scotland 110 917 tons. If attempts to halt the decline are successful, the purity of the original species will have been affected by interbreeding with non-indigenous or cultivated salmon (Gross, 1998) Experience from France and Spain has shown that eggs recovered from indigenous fish are more suitable for rearing and eventual release than imported eggs from Scotland and Norway. It is, therefore, clear that restocking the tributaries of the Rhine will be a slow and arduous process.

Migration behaviour of stocked fish may also be impaired. It is not clear whether certain smells that are characteristic of the spawning grounds are still discernible to salmon that have passed through rivers rich in detergents (Holl, 1965; Holl et al., 1970). A further complication is that none of the rivers in the vicinity of the Rhine can serve as alternative entry points for salmon returning from sea. Rivers such as the Ems, Weser and Elbe, which once had relatively small salmon populations, now can not be regarded as true salmon rivers. The latter also applies to the Meuse and river Scheldt. Consequently, factors such as straying and infiltration, which play an important role in populating remote rivers, are likely to be of little significance in the Rhine basin.

It should be possible, however, for river systems along the main stream of the Rhine to be recolonized by salmon, and hence to increase the size of the spawning grounds. It is, therefore, essential that a wealth of spawning streams be made available for salmon in the Rhine. However, creating the right conditions is no guarantee of success. Salmon interbreed with trout (Refstie & Gjedrem, 1975), and this may endanger restocking programmes. In 1992, 43% of the salmonid fry in the German river Sieg (Rhine – river system), where a restocking programme was conducted, consisted of hybrids of salmon and trout (Schreiber et al., 1994). A natural hesitation exists to kill the trout in a river system before carrying out restocking with salmon. However this would be the best solution, as trout also prey on salmon. Only when a salmon stock has established itself, trout should be allowed to return.

The limited success of establishing spawning stocks in German tributaries of the river Rhine system by simply releasing fry obtained from all over Europe (Norway, Scotland) might be explained by the work by Youngson & McLaren (1998), who concluded that restocking a river with naturally spawned salmonid ova resulted in a far better survival ratio then using fry from hatchery reared salmonids. If the Rhine is to be rehabilitated as a salmon river, it is essential that the water quality of the river should meet the requirements for salmon rivers as laid down in the relevant EC Directives. Finally, it can be concluded that establishing a natural population of salmon in the Rhine without outside help will certainly not be realized within the foreseeable future (10–20 years). After all, the Rhine is still one of the most polluted rivers in Europe and it suffers from the added disadvantages that it contains a large number of locks and weirs that restrict fish access to traditional habitats.

To conclude

At the end of the 19th century, the once important freshwater fisheries, mostly those on the large rivers forming the Rhine-Meuse estuary, were declining rapidly. The decline in catches of salmon and allis shad became the reason for great concern, but so did the declining catches of far less important species such as sea trout, and the coregonids (houting and whitefish), sturgeon, twaite shad and smelt. The reason for the decline of these species must be found in the rapidly changing abiotic conditions resulting from river regulation works, the construction of numerous barrages and sluice systems to improve shipping on the rivers, dredging away the gravel of the spawning beds, increasing turbidity, and the impact of various chemical pollutants, and changes in oxygen content and temperature. In the early days of the 20th century, it was thought that mass releases of fertilized eggs or juvenile stages would counteract the negative effects of river development. This was unsuccessful in all cases for the salmonids. Restocking the waters with eggs of the allis shad was abandoned after a few experiments. It was never tried to restock the sturgeon in the river Rhine. Hybridisation might also explain anadromous fish disappearance from clean rivers, for example in France. The mixing of salmon escaped from hatcheries with wild salmon, will lead to genetic, behavioural and ecological interactions. As a rule, these interactions are likely to be negative in their effect on the viability of wild populations (Youngson & Verspoor, 1998).

To place much emphasis on recent observations of an odd salmon, houting, whitefish is, in fact, misleading. Many of the declines of anadromous species mentioned above can not be attributed to a single factors, such as the construction of dams, pollution, dewatering of streams or sand and gravel extraction. It is better not to single out a culprit as the reason of decline, as most factors do not act singly, but rather in concert.

References

Anonymous, 1920. Jaarverslag Visscherij inspectie 1919 (4): 77–81 (in Dutch).

Anonymous, 1987. Vervuilde baggerspecie is groot aandachtsgebied van overheid. Land en Water 27: 60–61 (in Dutch).

Anonymous, 1994. Sturgeons, urgent action is needed. IUCN-Vorschlag für die 9 Vertragstaaten konferenz zum Washingtoner Artenschutzübereinkomen – Bundesamt für Naturschutz, 9 August 1994, Bonn: 15pp.

Anonymous, 1998. Vierde Nota Waterhuishouding Regeringsbeslissing Min. Verkeer en Waterstaat, Den Haag: 49–59 (in Dutch).

Barannikova, I. A., 1987. Review of sturgeon farming in the Soviet Union. Vopr. Ichthiol. 27 (5): 735–746.

Bardach, J. E. & J. Case, 1965. Sensory capabilities of the modified fins of squirrel hake (*Urophycis chuss*) and searobins (*Prionotus carolinus* and *P. evolans*). Copeia 1965: 194–206.

Bardach, J. E., M. Fujiya & A. Holl, 1965. Detergents: effects on the chemical senses of the fish *Ictalurus natalis* (le Sueur). Science 148: 1605–1607.

Bardach, J. E., M. Fujiya & A. Holl, 1967. Investigations of external chemoreceptors of fishes. In Hayaski, T. (ed.), Olfaction and Taste. 11th. Proc. Sec. Int. Symp. Tokyo, Sept. 1965. Pergamon Press, Oxford: 647–665.

Billard, R., 1995. Elements sur la biologie des esturgeons. Pêche Marit. 74 (1): 33-47.

Birstein, V. J., 1993. Sturgeons and paddlefishes: threatened fishes in need of conservation. Conserv. Biol. 7: 773–787.

Boisneau, P., C. Mennesson-Boisneau & R. Guyomard, 1992. Electrophoretic identity between allis shad, *Alosa alosa* (L.) and twaite shad, *A. fallax* (Lacépède). J. Fish Biol. 40: 731–738.

Brinkhuizen, D. C., 1989. Ichthyo-archeologisch onderzoek: methoden en toepassing aan de hand van Romeins vismateriaal uit Velsen (Nederland), PhD Thesis University of Groningen: 312 pp. (in Dutch).

Castelnaud, G. & M. Trouvery, 1984. Premiers résultats de trois années de marquage de l'esturgeon *Acipenser sturio* dans le bas estuaire de la Gironde. CEMAGREF/AGEDRA, Etude No. 18–Ser. Esturgeon 2: 1–25.

De Groot, S. J., 1989a. Literature survey into the possibility of restocking the River Rhine and its tributaries with Atlantic Salmon (*Salmo salar*). Publs. Reps. Ecol. Rehab. Rivers Rhine and Meuse 11: 1–56.

De Groot, S. J., 1989b. Literature survey into the possibility of restocking the River Rhine and its tributaries with sea trout (*Salmo trutta trutta*). Reps. Ecol. Rehab. Rivers Rhine and Meuse 12: 1–14.

De Groot, S. J., 1992. Decline and fall of the salmon fisheries in the Netherlands: is restocking the Rhine a reality. Aquaculture. Fish Managem. 23: 253–264.

De Groot, S. J. & H. Nijssen, 1997. The North Sea houting, *Coregonus oxyrinchus*, back in the Netherlands (Pisces, Salmoniformes, Salmonidae). Bull. Zool. Mus. Univ. Amsterdam 16: 21–24.

De Jonge, V. N. & D. J. de Jong, 2002. Ecological restoration in coastal areas in the Netherlands: concepts, dilemmas and some examples. Hydrobiologia 478/Dev. Hydrobiol. 166: 7–28.

Deelder, C. D. L. & A. H. Huussen Jr., 1973. Opmerkingen betreffende de kuilvisserij op de voormalige Zuiderzee, voornamelijk in de zestiende eeuw. Holland 5(5): 221–242 (in Dutch).

Frost, W. E. & M. E. Brown, 1967. The Trout, New Naturalist, vol. 21, London, Collins: 286 pp.

Grøn, P., 1987. Saving the North Sea houting. In: Proc. 5th Int. Wadden Sea Symp. Sept. 29th–Oct. 3rd 1986, Esbjerg, Denmark: 146–159.

Grøn, P., B. Møller. M. B. Nielsen, M. Ejbye-Ernst & L. K. Larsen, 1988. Snaeblen – en truet fiskeart. Ribe Amtsrad, Sonder jylands Amtsrad, Tonder, Ribe: 31 pp.

Gross, M. R., 1998. One species with two biologies: Atlantic salmon (*Salmo salar*) in the wild and in aquaculture. Can. J. Fish. aquat. Sci. 55 (supl. 1): 131–144.

Hartgers, E. M., A. D. Buijse & W. Dekker, 1998. Salmonids and other migratory fish in Lake IJsselmeer Publs. Reps. Ecol. Rehab. Rivers Rhine and Meuse 76: 1–41.

Havinga, B., 1928. De spiering in de Zuiderzee. Meded. Zuiderzee Comm. 1: 1–18 (in Dutch).

Havinga, B., 1954. Vissen. In: de Beaufort, L. F. (ed.), Veranderingen in de Flora en Fauna van de Zuiderzee (thans IJsselmeer) na afsluiting in 1932. Den Helder, Neth. Zool. Soc.: 253–267 (in Dutch).

Hoek, P. P. C., 1893. Über die Meerforelle des Rheingebietes. Z. Fisch. 1: 3–20.

Hoek, P. P. C., 1894. De elft op onze rivieren. Meded. Vissch. 1: 49–54, 65–69, 81–86 (in Dutch).

Hoek, P. P. C., 1899. Neuere Lachs- und Maifish studien. Tijdschr. Ned. Dierk. Ver. 2 Ser. 6: 156–242.

Hoek, P. P. C., 1910. De achteruitgang der steurvischerij onzer beneden-rivieren. Meded. Vissch. 17: 103–109, 118–124 (in Dutch).

Hoek, P. P. C., 1916a. Rapport over de beteekenis der zalmvisscherÿ op de Maas. In Versl. Staatscomm. Zalmvraagstuk 2: 159–172.

Hoek, P. P. C., 1916b. Rapport over den Rÿn als zalmrivier. In versl. Staatscomm. Zalmvraagstuk 2: 173–238.

Holl, A., 1965. Vital staining by trypan blue: its selectivity for olfactory receptor cells of the brown trout bullhead, *lctalurus natalis*. Stain Technol. 40: 269–273.

Holl, A., E. Schulte & W. Meinel, 1970. Funktionelle Morphologie des Geruchsorgans und Histologie der Kopfanhänge der Nasenmuräne , *Rhinomuraena ambonesis* (Teleostei, Anguilliformes). Helgoländer. wiss. Meeresunters. 21: 103–125.

ICES, 1999. Report of the Working Group on North Atlantic Salmon. ICES CM 1999/ACFM 14: 1–288.

Jäger, T., 1999. Die Wiedereinbürgerung des Nordseeschnäpels. In: Der Nordseeschnäpel. Offenbach, Verband Deutscher Sportfischer e.V: 3–11.

Jakob, E., 1996. Das Potential des Unteren Niederrheins ab Laich- und Bruthabitat des europaïschen Stör *Acipenser sturio*. Das Land NRW – Landesanstalt für Ökologie, Bodenordnung und Forsten/Landesamt für Agrarordnung NRW, Dezernate für Fischerei, Albaum, Köln: 45 pp.

Kinzelbach, R., 1987. Das ehemalige Vorkommen des Störs, *Acipenser sturio* (Linnaeus, 1758), im Einzugsgebiet des Rheins (Chondrostei: Acipenseridae). Z. angew. Zool. 74: 167–200.

Larsson, P. O., 1984. Remote straying of salmon (*Salmo salar* L.) from the Swedish west coast and possible effects on sea ranching operations. Aquaculture 38: 83–87.

Lelek, A., 1987. The Freshwater Fishes of Europe – Vol. 9. Threatened Fishes of Europe, AULA-Verlag, Wiesbaden: 42–45, 58–61.

Lobregt, P. & J. van Os, 1977. De laatste riviervissers, Heerewaarden: 131 pp. (in Dutch).

Mohr, E., 1941. Stint (*Osmerus artedi*). In Demoll, R. & Maier (eds), Hndb. Binnenfisch. Mitteleuropas: Vol. 3 (7a): 1–11, 504–526.

Mohr, E., 1952. Der Stör: Die neue Brehm-Bücherei. 84. Akad. Verlagsgesellschaft Geest und Portig K.G, Leipzig: 66 pp.

Nienhuis, P. H., A. D. Buijse, R. S. E. W. Leuven, A. J. M. Smits, R. J. W. de Nooij & E. M. Samborska, 2002. Ecological rehabilitation of the lowland basin of the river Rhine (NW Europe). Hydrobiologia 478/Dev. Hydrobiol. 166: 53–72.

Nijssen, H. & S. J. de Groot, 1987. De vissen van Nederland, Kon. Ned. Natuurh. Ver., Publ. 42: 224 pp. (p. 67–68) (in Dutch).

Prummel, W., 1989. Iron Age animal husbandry, hunting, fowling and fishing on Voorne-Putten (the Netherlands) Palaeohistoria 31: 235–265.

Prummel, W., 1991. Kapitel 12- Haus- und Wildtiere, in Stargard/Oldenburg. In Müller-Wille, M. (ed.), Ein slawischer Herrschersitz des frühen Mittelalters in Ostholstein. Karl Wachholtz Verlag, Neumünster: 299–306.

Rameye, L., A. Kiener, C. P. Spillmann & J. Biousse, 1976. Aspects de la biologie de l'alose du Rhône. Pêche et difficultes croissantes de ses migrations. Bull. Fr. Piscic. 49: 50–76.

Redeke, H. C., 1907. De spiering (*Osmerus eperlanus*). In Redeke, H. C. (ed.), Rapport over onderzoekingen betreffende de visscherij in de Zuiderzee ingesteld in de jaren 1905 en 1906. Min. Landbouw, Nijverheid en Hand's. Gravenhage: 79–81 (in Dutch).

Redeke, H. C., 1914. De spiering en de spieringvisscherij. Meded. Vissch. 21: 51–54, 70–74, 83–85 (in Dutch).

Redeke, H. C., 1922. Visschen. In Flora en Fauna der Zuiderzee. Den Helder, Nederlandse Dierkundige Vereeniging: 426–442 (in Dutch).

Redeke, H. C., 1927. River pollution and fisheries. Rapp. P.-v. Reun. Cons. perm. int. Explor. Mer 43: 1–50.

Redeke, H. C., 1934. Über den Rheinschnäpel, *Coregonus oxyrhynchus* L. Verh. int. Ver. Limnol. 6: 352–357.

Redeke, H. C., 1938. Uber den Bastard Clupea alosa finta Hoek. Arch. Néerld. Zool. Suppl. 3: 148–158.

Refstie, T. & T. Gjedrem, 1975. Hybrids between Salmonidae species. Hatchability and growth rate in the freshwater period. Aquaculture 6: 333–342.

Reuter, J. H., 1996. Houdt de houting. Visserij-Nieuws 19: 206–210 (in Dutch).

Rochard, E, G. Castelnaud & M. Lepage, 1990. Sturgeons (Pisces: Acipenseridae); threats and prospects. J. Fish Biol. 37 (Supplement A): 123–132.

Scheffel, H. J., 1999. Biologie des Nordseeschnäpels. In: Der Nordseeschnäpel. Offenbach, Verband Deutscher Sportfischer e.V.: 12–37.

Schreiber, A., M. Schenk, J. Lehmann & F. J. Stürenberg, 1994. Genetische Untersuchungen an Meerforellen- und Lachswildfängen aus dem Rheinsystem in Nordrhein-Westfalen. Fischer und Teichwirt. 2: 52–53.

Shagaeva, V. G., M. P. Nikolskaya, N. V. Akimova, K. P. Markov & N. G. Nikolskaya, 1993. A study of the early ontogeny of Volga sturgeon (Acipenseridae) subjected to human activity, J. Ichthyol. 33(6): 123–141.

Shearer, W. M., 1955. 'Homing Instinct' in Sea Trout. Nature 176: 171–172.

Staurnes, M., F. Kroglund & B. O. Rosseland, 1996. Water quality requirement of Atlantic salmon (*Salmo salar*) in water undergoing acidification or liming in Norway. Wat. Air Soil Poll. 85: 347–352.

Steinert, H., 1951. Kaviar in der Fischkinderklinik? Orion 6 (18): 713–716.

Stuart, T. A., 1953. Spawning migration, reproduction and young stages of Loch Trout (*Salmo trutta* L.) Sci. Invest. Freshw. Fish. Scot. 5: 1–39.

Svárdson, G. & A. Fagerström, 1982. Adoptive differences in the long-distance migration of some trout (*Salmo trutta* L.) stocks. lnst. Freshw. Res. Drottningholm Rep. 60: 51–80.

Thorpe, J. E., 1990. Sea trout: an archetypal life history strategy for *Salmo trutta* L. In Piggins, D. J. & W. M. Shearer (eds), Sea Trout in Scotland. Symp. Scot. Mar. Biol. Ass. Oban, Scotland, June 1987: 1–4.

Van Brummelen, T. C., 1989. Chemicals affecting the spawning migration of anadromous fish by causing avoidance responses or orientational disability, with special reference to concentrations in the River Rhine – Study Institute for Inland Water Management and Waste Water Treatment (BDW/RIZA) – University of Utrecht, Section Aquatic Toxicology: 77 pp.

Van den Ende, W. P., 1847. Eenige bijzonderheden betreffende den spiering op den IJssel. Versl. werkzh. Ver. tot bevord. Inl. Ichthyol. 1: 55–55, 69–76 (in Dutch).

Van de Ven, G. P., 1976. Aan de wieg van de Rijkswaterstaat. Wordingsgeschiedenis van het Pannerdens Kanaal. Walburg Pers, Zutphen: 1–438 (in Dutch).

Verhey, C. J., 1949. Het voorkomen van de steur (*Acipenser sturio* L.) in de Nieuwe Merwede tussen 1900 en 1931. De Levende Natuur 52: 152–159 (in Dutch).

Verhey, C. J., 1961. De vissen en de visvangst. In Verhey, C. J. (ed.), De Biesbosch, land van het levende water. W.J. Thieme, Zutphen: 142–146; 249–250 (in Dutch).

Verhey, C. J., 1963. Het voorkomen van de steur (*Acipenser sturio* L.) in Nederland na 1931. De levende Natuur 66: 15–16 (in Dutch).

Vincent, P. B., 1894. Notes sur l'alose. Bull. Pêche. Marit.: 427–441, 465–477, 525–534.

Volz, J. & S. J. de Groot, 1992. Erster Nachweis des Störs (*Acipenser sturio*) im niederiändischen Rhein seit 40 Jahren. Fischökologie 6: 3–6.

Youngson, A. F. & I. S. MacLaren, 1998. Relocation of naturally spawned salmonid ova as a countermeasure to patchiness in adult distribution at spawning. Scot. Fish. Res. Rep. 61: 12 pp.

Youngson, A. F. & E. Verspoor, 1998. Interactions between wild and introduced Atlantic salmon (*Salmo salar*). Can. J. Fish. aquat. Sci. 55 (suppl. 1): 153–160.

Hydrobiologia **478:** 219–233, 2002.
P.H. Nienhuis & R.D. Gulati (eds), Ecological Restoration of Aquatic and Semi-Aquatic Ecosystems in the Netherlands (NW Europe).

The state of the art of aquatic and semi-aquatic ecological restoration projects in the Netherlands*

P.H. Nienhuis[1,3], J.P. Bakker[2], A.P. Grootjans[2], R.D. Gulati[3] & V.N. de Jonge[4]
[1]*Department of Environmental Studies, Faculty of Science, University of Nijmegen, P.O. Box 9010, 6500 GL Nijmegen, The Netherlands*
[2]*Laboratory of Plant Ecology, University of Groningen, P.O. Box 14, 9750 AA Haren, The Netherlands*
[3]*Centre for Limnology, Netherlands Institute of Ecology, Rijksstraatweg 6, 3631 CA Nieuwersluis, The Netherlands*
[4]*Department of Marine Biology, University of Groningen, P.O. Box 14, 9750 AA Haren, The Netherlands*

Abstract

The Netherlands are a small, low-lying delta in W. Europe (42 000 km^2; 50°–54° N; 3°–8° E), mainly consisting of alluvial deposits from the North Sea and from the large rivers Rhine and Meuse. The country was 'created by man'. The conversion of natural aquatic and terrestrial ecosystems into drained agricultural land was a major cultural operation over the past 1000 years. Roughly 55% of the country's surface area is still agricultural land. Some decades ago, The Netherlands' landscape was characterised by an armoured coastline and bridled estuaries, a drastically reduced area of saline and freshwater marshes, fully regulated rivers and streams, and numerous artificial lakes. The aquatic ecosystems beyond the influence of the large rivers, the Pleistocene raised bogs and moor lands, have almost been completely annihilated in the past. Acidification and eutrophication led to the deterioration of the remaining softwater lake vegetation. Last but not least, an artificial drainage system was constructed, leading to an unnatural water table all over the country, high in summer, low in winter. Only very recently, some 25 years ago, the tide has been turned and ecological rehabilitation and restoration of disturbed ecosystems are in full swing now, enhanced by the European Union policy to set aside agricultural land in the Netherlands in favour of the development of 'nature'. The state of the art of aquatic and semi-aquatic ecological restoration projects in the Netherlands is given. Starting from the conceptual basis of restoration ecology, the successes and failures of hundreds of restoration projects are given. Numerous successful projects are mentioned. In general, ecological restoration endeavours are greatly benefiting from progressive experience in the course of the years. Failures mainly occur by insufficient application of physical, chemical or ecological principles. The spontaneous colonisation by plants and animals, following habitat reconstruction, is preferred. But sometimes the re-introduction of keystone species (e.g. eelgrass; salmon; beaver) is necessary in case the potential habitats are isolated or fragmented, or when a seed bank is lacking, thus not allowing viable populations to develop. Re-introduction of traditional management techniques (e.g. mowing without fertilisation; low intensity grazing) is important to rehabilitate the semi-natural and cultural landscapes, so characteristic for the Netherlands. For aquatic ecosystems proper (estuaries, rivers, streams, larger lakes) the rule of thumb is that re-establishment of the abiotic habitat conditions is a pre-requisite for the return of the target species. This implies rehabilitation of former hydrological end geomorphological conditions, and an increase in spatial heterogeneity. The 'bottom-up' technique of lake restoration, viz. reduction in nutrient loadings, and removal of nutrient-rich organic sediment, is the preferred strategy. The 'top-down' approach of curing eutrophicated ecosystems, that is drastic reduction of fish stock, mainly bream, and introduction of carnivorous fish, may be considered as complementary. For semi-aquatic ecosystems (river-fed and rain-fed peat moors, brook valleys, coastal dune slacks) it also counts that the abiotic constraints should be lifted, but here the species-oriented conservation strategy, the enhancement of the recovery of characteristic plant and animal species, is mainly followed. An important pre-requisite for the rehabilitation of the original natural or semi-natural

* NIOO Contribution no. 3054.

exist, which prevent the system from returning to a less-degraded state without the input of management effort. The concept was already formulated earlier by Hobbs & Norton (1996), who illustrate their theory with the example of woodlands, currently degraded by grazing, which may recover and regenerate simply by excluding stock. On the other hand, if the system has exceeded a threshold value, removing the degrading influence will not be sufficient to allow transition back to a state suggestive of the original one. For instance, if the grazed woodland is heavily invaded by weeds and the soil structure is altered, exclusion of grazing will not be adequate to promote woodland recovery, and more severe management measures would be needed.

The best known aquatic examples from the Netherlands are described by Scheffer et al. (1993) who group the shallow lakes on the basis of one of the two stable states in which these lakes may find themselves in: the clear-water stage with dominance of submerged macrophytes, benthic diatoms, zooplankton and carnivorous fish, and the turbid-water state with dominance of phytoplankton, especially Cyanobacteria, and of planktivorous/benthivorous fish. The removal of bream from eutrophicated and turbid freshwater bodies, may rehabilitate the lake to its original clear-water situation with aquatic macrophytes. We also know that the reduction of bream standing crop does not always lead to the target situation. Moreover, this grouping of lakes with extreme under-water light characteristics is based on the turbid lakes, which during the restoration process exhibit intermittent periods of improved light climate and increased littoral vegetation rather than a stable situation over several consecutive years (Gulati & van Donk, 2002).

Another example to illustrate the restoration threshold concept is the removal of the accumulated organic sapropelium layer from softwater bodies, in order to restore the original macrophyte vegetation. But again, this recipe does not always work, because oxidation of the sandy sediment and a process of oligotrophication are pre-requisites for the development of the waterplants (Roelofs et al., 2002). In other words, the restoration threshold concept seems theoretically sound, but in practice many of the Dutch ecological restoration projects have gone along a pathway of trial and error: learn from the failures and try to understand the rationale behind the successes.

Whisenant (1999) has suggested that two main types of restoration thresholds are likely: one that is caused by biotic interactions, and the other caused by abiotic limitations. If the system has degraded mainly due to biotic changes (such as grazing-induced changes in vegetation composition), restoration efforts need to focus on biotic manipulations which remove the degrading factor (e.g. the grazing animals) and adjust the biotic composition (e.g. replant target species). If, on the other hand, the system has degraded due to changes in abiotic features (such as through soil erosion or contamination), restoration efforts need to focus first on removing the degrading factors and repairing the physical and/or chemical environment. In the latter case it would serve no use focusing on biotic manipulations, without first tackling the abiotic problems (Hobbs & Harris, 2001). Similar ideas were put forward by Grootjans et al. (1996) and Bakker & Londo (1998).

In fact, Hobbs' & Harris' (2001) conceptual ideas are not as new and original as they are considered to be. It was already Southwood (1977), who hypothesised that ecological theory predicts that the medium, the physical habitat, rather than the organisms, determine the structure of ecosystems. The habitat provides the template on which evolution processes forge the characteristic life-history strategies. Thus, through the effects of habitat conditions on the fitness of individual organisms in geological time, certain combinations of adaptations for survival and reproduction are assumed to be selected. According to Moss (2000), it is a consequence of Southwood's (1977) strictly ecological theory, that emphasis on individual, in policy terms often charismatic species, may be counter-productive for the conservation of aquatic ecosystems. Aquatic ecosystems provide many of the fundamental biogeochemical services upon which the continuity of life depends, and this provides a strong argument for conservation of the functional values in aquatic systems through whole-system reconstruction rather than following the philosophy of terrestrial species-oriented conservation (Moss, 2000).

An example from the Rhine Action Programme may illustrate this line of thinking. The euphoria about 'the salmon is back in the Rhine' (ICPR, 2000) provokes the wrong impression. No wonder that the salmon is back in the Rhine after stocking the tributaries of the river with hundreds of thousands of young individuals. The salmon returns to the Rhine when the fish is able to migrate from the sea to its upstream spawning areas, finding the proper resting habitats, feeding items etc. on its way. Weirs in the harnessed rivers are still preventing this. Maintenance of migratory salmonid stock may require major moderation of forestry, and operations around the spawning rivers;

moreover, conditions in estuaries or on the open sea will affect the recruitment and survival of the stock. According to De Groot (2002), to place much emphasis on recent observations of an odd salmon is, in fact, misleading. Many of the declines of anadromous species can not be attributed to a single factor, such as the construction of dams, pollution, de-watering of streams or sand and gravel extraction. It is better not to single out one causal factor for the decline, as most factors do not necessarily act singly, but rather in concert. The chain of required habitats forms the template for the return of the salmon.

The Dutch applications

Which ecological concepts have been applied during the restoration processes of Dutch ecosystems? Did the authors of this book refer to the restoration threshold concept? These questions will be answered further in this section. Concerning the restoration of estuarine and coastal ecosystems, both Bakker et al. (2002) and de Jonge & de Jong (2002) stress the restoration of abiotic habitat conditions as a pre-requisite for the return of the desired species. Bakker et al. (2002) define restoration measures as an attempt to bring back the destroyed habitats to an original state. In the context of salt-marsh restoration this implies, changing the present state of agriculturally intensified sites into the former, more natural ones. The effects of restoration management can only be measured after setting the targets for both abiotic and biotic factors. The conditional abiotic targets are to restore hydrological, geomorphological and saline, tidal conditions. Once this 'new' tidal regime has been restored, the biotic targets will be reached with namely, livestock grazing or without human interference, viz. a zonation of characteristic plant and animal communities, integrating the suitability of a-biotic conditions and plant traits, comprising the entire series from pioneer community to mature salt marsh vegetation. Defining targets and monitoring the effects of ecological restoration should be strongly interrelated. According to Bakker et al. (2002), monitoring the ecological developments in the field is a pre-requisite for the evaluation of the effects of ecological restoration measures.

De Jonge & de Jong (2002) go even a step further in their ambitions to restore the estuarine systems. The development of coastal ecosystems is the result of complex processes of large-scale and small-scale functional relations. Consequently, restoration of these systems has to be realised primarily by influencing the acting processes, and this is the only way to a sustainable preservation or recovery of ecological elements. Focusing only at one particular species (e.g. breeding terns) or a specific habitat (e.g. salt marsh) easily ends up in a sort of 'bio-agriculture' where the natural processes have been mainly excluded. Such an approach does not exclude the species or mono-habitat oriented strategy, but one should be conscious of the limitations and the unpredictability of such a biotic repair.

As said above, de Jonge & de Jong (2002) stress the need for a thorough knowledge of the processes responsible for ecosystem functioning in restoration projects. Large interventions in ecosystem processes without the proper knowledge about the long-term consequences might have severe, and often unexpected side-effects. Three examples: a reduction in tidal range of only 10% in the Oosterschelde estuary caused the tidal areas to erode severely, and an increase of salinity by only 2‰ in the Grevelingenmeer resulted in the total disappearance of the eelgrass ecosystem including the accompanying fishes and birds. The deepening of the most shallow parts in the main shipping channel in the Westerschelde estuary by about 2 m resulted in significant geomorphological changes in the sand and mud flats.

According to Nienhuis et al. (2002), Verdonschot & Nijboer (2002) and Gulati & van Donk (2002), river and stream restoration and lake restoration in the Netherlands should preferably start with lifting the abiotic constraints in favour of biotic rehabilitation. The excavation of new side channels in the river basins is a good example of successful habitat restoration, because this was followed by a striking increase of aquatic biodiversity (Schropp & Bakker, 1998). Besides lifting the restoration threshold of the abiotic limitations, the Dutch river managers have also focussed on the introduction of keystone species (beaver, stork, salmon), reared specially for such re-introductions.

Holistic ecological theories to promote river catchment management are available in the literature (Townsend, 1996). A common aim of rehabilitation projects is to make the river more natural, and 'naturalness' has become an important element of the international enhancement ethos (de Waal et al., 1998). However, what is considered 'natural' nowadays, might not have had the same qualification centuries ago. Is a floodplain overgrown with hardwood and softwood floodplain forests more 'natural' than floodplain pastures, bordered by hawthorn hedges and studded with cows? This means that each rehabilitation process needs a reference image, i.e. a picture of the former state or condition of the river. The second and possibly the most important step in any rehabilitation scheme is to define a target image and associated project objectives. Such an target image sketches an optimal solution for the river and its floodplains, considering the present conditions, opportunities and constraints.

Although the theoretical concepts for river rehabilitation are predominantly catchment oriented, rehabilitation projects rely heavily on local and regional planning policies. The planning and implementation of rehabilitation schemes is mainly done at the level of the regional water boards, and participation of municipalities and local interest groups is important. For practical reasons, many habitat restoration projects in river basins in the Netherlands start from the 'stepping stone' concept (Cals et al., 1998). This concept is based on the assumption that local management projects should be set up along the longitudinal axis of the river at regular distances, in such a way that a connected chain of nature reserves comprising favourable conditions for the (potential) plant and animal species will be developed. The use and application of ecological theory and empirical knowledge are increasingly playing a role in the rehabilitation schemes, mainly focusing on nature conservation and restoration strategies, i.e. restoring habitat diversity and hence species diversity. Older plans for river rehabilitation were mainly based on physical (hydrological and geo-morphological) criteria, the more recent ones, in contrast, fully involve ecological criteria (Nienhuis et al., 2002).

Lake restoration has proven to be the most successful in 'whole lake experiments' e.g. by the reduction of the external phosphorus loading by more than 50%, and by flushing the lake with water high in calcium and bicarbonate to reduce internal P-loading from the sediments. Additional measures, called biomanipulation or 'active biological management', i.e. drastic reduction of the fish stocks, mainly bream, has led in a number of cases to rehabilitation of the clear water state of the lake. A drawback, however, is that without additional measures such as the introduction of piscivorous fish (e.g. pike), the bream removal should be repeated once every 5–10 years. Again, a combination of abiotic habitat oriented and biotic species oriented measures appeared to be most successful (Gulati & van Donk, 2002).

Grootjans et al. (2002a,b) and Lamers et al. (2002) focus in the restoration projects, respectively dealing with the biotic communities in brook valleys and coastal dune slacks, and river-fed peat moors, predominantly on the vegetation and individual plant species. Herein they follow the terrestrial species-oriented conservation tradition, without ignoring the abiotic demands necessary for a successful restoration (cf. Moss, 2000). According to Grootjans et al. (2002a,b) in the Netherlands success of a restoration project is almost always judged by whether or not a restored site provides good growing conditions for rare or endangered plant or animal species. Thus, a restoration project is considered a success if many 'Red List species' establish themselves at the restoration site and can maintain large populations for at least several decades. Successful restoration is analogous to constructing a field museum for preserving the living part of a lost cultural heritage.

Lamers et al. (2002) stress that fen restoration aims at the recovery of original plant communities and their fauna, often semi-natural, as known from the times before undesirable anthropogenic disturbances occurred, such as desiccation or eutrophication. Restoration should, as Grootjans & van Diggelen (1995) stated, 'be aimed at restoring the fen system', and not at 'restoring fen species'. Restoration ecologists do not use a pre-historic concept for most fen types, but rather a cultural-historical concept, involving as a benchmark the semi-natural landscape of the 19th century or the beginning of the 20th century. Already in 1945, Westhoff (1945) initiated the idea that not only the so-called natural, but also the semi-natural systems in the Netherlands are well worth conserving. The concept implies the re-establishment of traditional management measures.

Ecological restoration: successes and failures

Recent experiences

Strikingly, ecological restoration in the Netherlands is a very young branch of science, which is not more than 25 years old, and in many cases the endeavours are much more recent. Arbitrarily the mid 1970s can be pointed out as turning point regarding the prevention and abatement concerning water pollution and eutrophication. The reduction of point source emissions appeared to be the most effective measure, and toxic heavy metals and persistent organic micro-pollutants where the first chemicals to be banned. Freshwater ecosystems benefited markedly from drastic reductions of the phosphorus emissions. Nitrogen sources, the limiting nutrients in marine ecosystems, appeared much more difficult to combat. The present situation is much better than that 25 years ago, although emissions from diffuse sources and from the polluted sediments of many surface waters are difficult to overcome in restoration projects.

Estuaries, rivers and smaller streams

Ecological restoration in the coastal zone and in the estuaries has been reviewed by de Jonge & de Jong (2002) and by Bakker et al. (2002). Remarkably, nearly all coastal restoration measures in the Netherlands are related to the recently realised engineering works or are their side effects. The Delta Works in the SW Netherlands, executed between 1957 and 1986, were designed to close off the large Rhine-Meuse estuaries from the North Sea for safety reasons, viz. to avoid recurrence of the flooding disaster in 1953. The execution of the Delta Works deprived the region of its estuarine transition zones: the brackish water habitats almost completely disappeared (except in the Westerschelde estuary). It is now clear, some 15 years after the completion of the Delta Works, that ecologically speaking the closure of Haringvliet, Veerse Meer, Grevelingenmeer, and the partial closure of the Oosterschelde have resulted in ecosystems that for one reason or the other do not fulfil the expectations. The negative developments that occurred were generally unexpected, affirming the difficulties of predicting developments after such large scale interventions. Various major schemes are now underway to restore the brackish-water gradients. Smaller, purely technical rehabilitation projects, such as fish passages, can not be regarded as a serious restoration attempt of a former habitat because these actions were restricted to a few specific elements in the foodchain.

Continuous sea level rise, due to global warming, forced the Dutch coastal managers to re-fix and armour the dune ridges along the Dutch coastline. Only very recently have the natural 'walking dunes' been re-introduced, although the stability of these systems can be questioned because of the limited spatial scale of the projects.

As stated by Bakker et al. (2002), salt marshes are threatened ecosystems, having until recently decreased in surface area continuously owing to civil engineering schemes. In order to increase the salt-marsh area, two options are at present in operation, viz. de-embankment of the summer polders while maintaining and reinforcing the main levee, the seawall between land and sea, and an increase in the effects of saline seepage landwards of the seawall by top soil removal. Both the options include the 'natural' restoration of salt-marsh communities (target communities) in former, intensively used agricultural sites. The general policy in the European Union is to set aside the low quality agricultural land, implying that the restoration of salt marshes and halophytic plant communities has recently become a realistic option in the Netherlands. Restoration of salt marshes or halophytic plant communities can not rely on a viable soil seed bank, which might reflect the vegetation prior to embankment. This is caused by the fact that the majority of salt-marsh plant species have a transient or short-term persistent seed bank. Hence, the success of restoration mainly depends on the dispersal of propagules from elsewhere. The development of a salt marsh at the former arable fields takes more time than at the former pastures. Livestock grazing is necessary to maintain a species-rich salt-marsh plant community and the coexistent fauna at the small scale. At the scale of e.g. the entire Wadden Sea area, the combination of both grazed and ungrazed salt marshes contribute to biodiversity.

Nienhuis et al. (2002) reviewed the ecological rehabilitation of Dutch large rivers, particularly that of the Lower Rhine catchment. In the past 25 years, the water quality and – to a lesser extent – the sediment quality, have been improved considerably, hence leading to improvement of biotic diversity. The rehabilitation of the lost or disturbed ecosystems started some 15 years ago. The established boundary conditions for restoration projects are protection against flooding and transport by cargo ships. The use and application of ecological criteria are increasingly playing a role in the rehabilitation schemes. Ecological rehabilitation focuses mainly on nature conservation and restoration strategies, i.e. exploiting the hydrodynamic and morphodynamic potentials of the flowing rivers, and introducing a semi-natural grazing regime by large herbivores. Particularly the creation of new secondary channels contributes to the restoration of riverine habitat diversity and heterogeneity, and hence to species diversity.

The assessment of a number of rehabilitation projects along the Lower Rhine has led to the conclusion that no general patterns have emerged concerning 'success' or 'failure' of specific restoration measures. The assessment based on the 14 scrutinized projects out of some 30 projects in progress, is positive. Recent projects benefited from the experience gained during the earlier projects (viz. in Duursche Waarden the connection of isolated freshwater pools with the main river channel in 1989). All projects started with the dominance of cultivated, over-fertilised and levelled grasslands or maize fields. The actual values of nature in all areas within a few years improved by increasing the spatial heterogeneity by re-introduction of larger hydro-dynamical and morphodynamic forces. This was achieved by excavation of sediment and dredging of silted-up river branches and excavation of secondary channels, in combination with extensive grazing by cattle on the fallow land. These measures enlarged the actual values of nature in all areas within a few years. The water quality in surface water bodies in the floodplains is rather negative, owing to retarded seepage of fertilisers via groundwater. The highly dynamic river Waal offers the best possibilities for the restoration of natural sandy levees and back swamps. It will be very difficult to rehabilitate the hardwood floodplain forests under the present management conditions. A further recovery of the river biota depends on continued repression of pollution, an increase in the morphological dynamics, and the development of the original natural habitats (Nienhuis et al., 2002).

According to de Groot (2002), there are chances for the recovery of some of the anadromous fish species (a.o. the twaite shad population) in the SW Netherlands, when the estuarine gradients will be restored. The return of the sturgeon is unlikely. The salmon, once fished in large numbers, is now the subject of restocking programmes in Germany. Encountering a stray specimen of salmon can be attributed only partly to these programmes. Restocking programmes need to be considerably improved before noticeable success is to be expected.

Postma et al. (1996) made an inventory of the present and past status of riverine landscapes in the Netherlands, classified as ecotopes. At present roughly 75% of the surface area of the river floodplains of the Rhine is cultivated land, either dominating grassland (pastures), fields or built-up area. The current government policy in the Netherlands is to decrease the acreage of cultivated land, to set aside or to abandon agricultural land and to stimulate the development of natural grassland and floodplain forests. The reference image dating back some centuries, shows dominance of hard- and softwood floodplain forests (roughly 60–70%), and this is in sharp contrast with the present situation, where only less than 5% of the floodplain forests is left. The target image shows an intermediate position, a strong stimulus for natural grasslands, brushwood and floodplain forests, at the expense of classical agricultural land. The aim of the government is the doubling of the area of natural landscape along the large rivers to 30 000 ha in 2010.

The Netherlands, rich in water, still has numerous deteriorated streams and brooks, and the techniques of stream restoration are being quickly developed. In 1991, 70 projects were in progress, in 1993 the number increased to 170, and in 1998 to 206. Both hydrological and morphological measures, as well as biological measures are being taken. The catchment oriented measures involve an increase of the water retention capacity of the stream, or the revival of lost inundation areas. The morphological measures deal with the construction of fish ladders, pools, wooded banks, the creation of new meanders and adaptations of stream profile width and depth. The introduction of species – fish; aquatic plants – that occurred in the stream in the past, is considered as an additional measure (Verdonschot & Nijboer, 2002).

Lakes, fens and bogs

Lamers et al. (2002) reviewed the restoration of Dutch mineral-rich peatlands with either surface peat accumulation or submerged peat accumulation. Although some fens are brackish (8.5–85 mmol Cl l^{-1}; mainly in the western part of the Netherlands), the majority are freshwater systems. The focus is mainly on aquatic and semi-terrestrial fens. Lowland fens differ from ombrotrophic mires (bogs) in that the biogeochemical processes in their top layer are strongly influenced by the influx of mineral-rich groundwater or surface water. The assumption that this implies meso- or eutrophic conditions is incorrect. The mineralisation rates and nutrient availability are not necessarily lower in the raised-bog plant communities compared with the fen plant communities. The uptake of nutrients, rather than their availability, appears to form the principal constraint for plant growth in bogs. Fen restoration aims at the recovery and conservation of characteristic, often semi-natural fen systems (flora and fauna), by restoring the hydrology, hydrochemistry and the sediment considered optimal for the fens under restoration.

In contrast to the larger lake types (cf. Gulati & van Donk, 2002), the reduction of turbidity in the small fen lakes by external or internal measures appeared to be often unsuccessful. Such lakes remained turbid or the decrease in turbidity was short lived after taking the measures, and the submerged plant communities did not develop. In the hypertrophic lakes, turbidity is the only possible stable situation. Biomanipulation will only work after PO_4^{3-} concentrations are reduced to a level in which two alternative stable states, the clear water and the other turbid, are possible (Lamers et al., 2002).

Softwater pools and lakes are found scattered over the higher, Pleistocene parts of the Netherlands. These waters possess a highly characteristic vegetation adapted to limited carbon availability. The vegetation of shallow softwater lakes in the Netherlands is strongly endangered, due to atmospheric sulphur- and nitrogen deposition, local eutrophication, habitat fragmentation, afforestation and drainage. Removal of the nutrient-rich, anoxic, organic sediments is a pre-requisite to the restoration of these lakes. In the acidified or acid-sensitive lakes, additional measures against acidification are required. A controlled supply of calcareous, nutrient-poor water is preferable to direct liming. Most softwater macrophytes produce long-lived seeds and regeneration of soft water macrophytes via the seed bank can recur even after 20–40 years of absence in the vegetation. Removal of the recently accumulated matter from eutrophicated softwater lakes strongly stimulates the germination of viable seeds from the seed bank, and the lakes can be completely recolonised by softwater macrophytes within a foreseeable period. Oxidation of the sediment and oligotrophication of both water layer and sediment and the presence of a broad riparian zone are important pre-requisites (Roelofs et al., 2002).

Semi-aquatic brook valleys and dune slacks

Grootjans et al. (2002a) reviewed the restoration of semi-aquatic and terrestrial brook valleys, habitats closely connected to the aquatic peat moor and fen systems, discussed by Lamers et al. (2002). Until recently, restoration measures in brook valley meadows in the Netherlands consisted of re-introducing traditional management techniques such as mowing without fertilisation and low-intensity grazing. In the Netherlands, additional measures, such as rewetting and sod cutting, are now carried out on a large scale to combat negative influences of drainage and acidifying influences by atmospheric deposition. An analysis of successful and unsuccessful projects shows that restoration of brook valley meadows is most successful if traditional management techniques are applied in recently abandoned fields that had not been drained or fertilised. Successful projects have been executed at sites that have been least affected by intensive agriculture and drainage. Thus, the restoration projects should be initiated preferably in areas which have not been affected by drainage and still have some relics of meadow species in the vegetation. The availability of reference communities is, indeed, an acute problem in target areas where intensive fertilisation has taken place for a long time, soil degradation has occurred, and where practically all target species have disappeared in both the existing vegetation and the seed bank.

According to Grootjans et al. (2002b), several environmental conditions influence the recovery and consequent stability of young dune slack ecosystems. These conditions are most favourable when the soil is nutrient-poor and fairly well buffered (pH>6), prolonged flooding with base-rich surface waters occurs, and when the soil remains moist during dry summers, preferably provided by an additional flow of calcareous groundwater from adjoining dune areas. Less favourable conditions occur when a rapid acidification of the top layers occurs during succession, resulting in a fast accumulation of organic matter, or when the dune slack is infiltrated with nutrient-rich (river)water, enhancing eutrophication, hence counteracting the natural nutrient-poor situation. An important pre-requisite for restoration success of the dune slack vegetation is, of course, the presence of a seed bank. The better the development of the seed bank is, the greater guaranty that Red List species (the criterion for success!) will show up during the early stages of succession.

Ecological restoration, predictability and society

Ecological restoration and predictability

What can we learn from the aforementioned successes and failures? Firstly, the abiotic conditions should be restored to establish the target communities. Secondly, the target species must be present or able to arrive at the target sites. It is important to realise the constraints in ecological restoration projects (Bakker & Berendse, 1999). The experimental introduction of target species may show that after initial emergence, these species disappear again (van Duren et al., 1998). Abiotic constraints may comprise, e.g. a too low buffer capacity in seepage water, or acidification or eutrophication from atmospheric deposition. Successes on the re-establishment of softwater macrophytes suggest that a number of species is still present in the long-term persistent soil seed bank, indicating former plant communities. Unfortunately, many other species only have a transient or short-term persistent seed bank (Thompson et al., 1997). This means that ecological restoration needs dispersal of propagules from elsewhere, which can be difficult to accomplish in the present fragmented landscape. Case studies reveal that dispersal, even over a distance of a few kilometres, may be a constraint (Verhagen et al., 2002). In fact, data on seed dispersal are mainly lacking; a database on dispersal of plant species, sustained by experimental evidence, would give insight in potential distribution patterns.

Do we have enough knowledge to predict the outcome of ecological restoration projects? The sequence of prediction, monitoring and evaluation is essential in ecological restoration projects. When a prediction comes through, apparently the researcher and the manager have convincing ecological knowledge of the ecosystem involved. In fact, it is more interesting when a prediction does not hold. In those cases, research is needed to understand the mechanisms and learn the constraints in the ecosystem, at the benefit of future predictions of both manager and researcher. It is important to distinguish between targets dealing with the rehabilitation of structural components such as plant communities or the restoration of processes such as the reduction of nutrient availability. Top soil removal or the introduction of grazers may never be a goal in itself.

There is an ongoing discussion among ecologists in the Netherlands about the predictability of ecological restoration measures. Theoretical modellers (Huisman, 2000a,b) disagree with the overall predictability of ecological systems. The determining factor is not the type of ecosystem but the number of interactions between the different species. The more complex the system the greater is the chance for 'chaotic competition', leading to large fluctuations in the structure of the biotic community. The multi-species phytoplankton communities are often used as examples to illustrate this 'chaos' hypothesis. Theoretical ecologists argue that the creditability of ecology would be at stake if too definite predictions are made, and the calculation of probabilities that certain species will maintain themselves in specific habitats would be more reliable than the statement that certain species will re-appear after specific management measurements. From their focus nature development is as un-predictable as the weather: the general sequence of the developments is, of course, predictable: spring, summer, autumn and winter may be foreseen, but the day to day distribution or intensity of rain, sunshine, fog, etc. can not be forecasted. The same is true for ecological restoration projects: supposing that the proper environmental steering variables have been restored, large-scaled succession patterns can be predicted, but small-scaled distribution patterns, both in space and in time, and the answer to the question which species composition finally would prevail, are largely not predictable.

Lamers & Roelofs (2000) contest the assumption that the species composition is not predictable, and that instead the main lines of succession and development of a (plant) community can be forecasted. The major challenge is to assess the proper steering variables, that determine the major processes in the ecosystem. Our inability to explain the ecological processes in the past has often been attributed to insufficient knowledge of the proper steering variables. 'Chaotic competition' is invariably used as a pretext to accept failures in the ecological restoration projects. This is an inevitable consequence of insufficient application of ecological theory, starting with the ill-founded hypotheses, and mainly based on trial and error. Such an approach may easily lead to a 'laissez faire' policy, and hence to non-repairable damage of ecosystems. The way an ecosystem evolves in response to restoration measures, is largely predictable, provided the main environmental variables are known in enough detail. This will, however, not provide an exact blueprint of the expected species composition and biodiversity. The successful restoration of oligotrophic fens and species-rich fen meadows are mentioned as examples. The management aim was to get back these ecosystems, without expecting a resurgence in the abundance or distributions of the species that had disappeared.

The above discussion is also an ongoing phenomenon in the international literature on ecosystem restoration (Ehrenfeld, 2000; Zedler, 2002). For example, according to Zedler (2000), the restoration of aquatic ecosystems is more complex than implied by early concepts of ecosystem degradation and restoration. Degradation involves many paths of change in species abundances and ecosystem functions, and restoration is at least as complex. Furthermore, models developed for one wetland type appear not to be easily applicable to other types (Ehrenfeld, 2000). Ecological theory has much to offer to the practitioner, but the predictions remain vague. Predictability should improve if generalities are sought where the restoration context and specific restoration actions are held constant. To date, there are too few studies to draw generalizations from within the context of one specific ecosystem type, let alone between the different types and landscape settings. According to Zedler (2000), there is a great need for more habitat-specific advice, such as that which recently became available for riverine wetlands and tidal wetlands (Middleton, 1999; Zedler, 2002).

In this context of predictability in rehabilitation endeavours a study of de Melo et al. (1992) on the rehabilitation of aquatic ecosystems is illustrative. Lake restoration theory (Shapiro & Wright, 1984) is based on the prediction that an increased piscivore abundance will result in a decrease in planktivore abundance, and an increase in zooplankton abundance, as well as an increased zooplankton grazing pressure leading to reductions in phytoplankton abundance and improved water clarity. De Melo et al. (1992) contested this irrefutably accepted trophic cascade 'law' in the generalist literature, which is used worldwide as a lake management tool (cf. Hosper, 1997 for restoration practice in the Netherlands). De Melo et al. (1992) examined 50 papers published between 1961 and 1989, documenting 44 independent food-web biomanipulations in ponds and lakes. The trophic cascade of these 44 case studies revealed data for 118 response cells (relations between trophic levels). Fifty-two represented complete agreement with the predictions of the top-down models. Twenty-one represented complete disagreement and 45 were undecided. On first examination, these results might seem to augur well for the top-down hypothesis; however, a more detailed analysis casts doubts on this conclusion. The undecided or equivocal results showed that the confounding effect was most apparent at the zooplankton–phytoplankton link in the food web. This relationship is dominantly determined by factors, such as the climatic conditions, macrophyte presence and the palatability of the phytoplankton algae, and not by planktivore-mediated alterations. Far from being 'clearly confirmed' (cf. Carpenter, 1988), biomanipulation is clearly at the stage of 'paradigmal crisis' sensu Kuhn (1962). According to de Melo et al. (1992), far too many questions remain unanswered to presently advocate biomanipulation as a justifiable management strategy for lake rehabilitation.

Gulati & van Donk (1989) who reviewed a series of papers on biomanipulation in fresh-water ecosystems and estuaries in the Netherlands, reached similar conclusions as de Melo et al. (1992). The 'top-down' remedial factors to alleviate eutrophication of lake ecosystems can perhaps be considered as complementary to the 'bottom-up' technique of lake restoration, viz. reduction in allochthonous nutrient loading, especially of phosphorus. Thus, this complementary approach of lake restoration may have surprises as well as successes in store for us.

Ecological restoration and society

Ecologists from policy-oriented state institutes are forced to express themselves carefully about the successes and failures of restoration projects, in accordance with the current national policy. 'Failures' of ecological restoration projects are preferably not published: dirty linen should not be washed in public, isn't it? Failures might be interpreted as shortcomings of the national policy, and this will cause damage to the Ministry.

Ecologists from universities and independent research institutes may allow themselves a more critical and independent approach, but their 'truth' is not seldom coloured with emotion, because of the links with organisations for nature conservation. Ecological restoration projects rather frequently suffer from conflicting economic arguments which are weighed against environmental arguments. Ecologists from both sides, conservation ecologists *versus* policy oriented ecologists, are easily opposing each other, particularly when a major political problem is to be debated. The point is that both parties should uphold their scientific integrity.

A well-known and endlessly debated example is the effect of large-scale bottom trawl fishing on marine bivalves, cockles, in the Wadden Sea, an international nature reserve of great size and standing. Large sucking devices, based on the vacuum cleaner principle, plough the bottom of the Wadden Sea, and destroy benthic habitats, such as natural mussel banks and seagrass beds, hence lowering habitat and species diversity. The massive harvest of cockles jeopardises the food quantity of benthos-eating waders and ducks. Ecological restoration of seagrass beds is also negatively influenced by the cockle fishing industry (de Jonge et al., 1996; van Katwijk, 2000). Of course these fishing activities are not allowed unrestrained: closed areas have been demarcated where cockle fishing is forbidden. Interpretation of the adverse effects of cockle harvesting is hampered by natural events such as heavy storms and severe winter frost.

Long and emotional discussions have been published in 'Bionieuws' (1998), the bi-weekly magazine of professional Dutch biologists, concentrating on the pitfalls of 'type I and II statistical failures' that may be made in discussions on conservation biology. The controversy between the shell-fishermen and their department of civil servants, and the nature conservationists and their home front is a reality. The 0-hypothesis states that cockle fisheries has no negative ecological consequences for the Wadden Sea. The type I failure, i.e. to wrongly conclude that there is a negative effect, should statistically be kept as small as possible. The burden of proof is on the side of the ecologists. The type II failure, i.e. to wrongly conclude that there is no effect, while there is indeed an effect, is even more dangerous. The likely chance that a negative effect is wrongly diagnosed is considerable, because the spatial and temporal variability in the system is large. Unsatisfactory research results and a too small number of samples lead frequently to massive type II failures. In simple terms: following a type I failure, the innocent goes to jail; following a type II failure, the villain does not go to jail. Thus, statistically and scientifically both type I and type II failures should be kept as small as possible.

Nature conservationists argue that the burden of proof should be put on the side of the licencees. The Dutch parliament allowed shellfisheries to exploit the cockle resources in the Wadden Sea, without turning the burden of proof onto the shell-fishermen. Cockle fishers argue that the 0-hypothesis has been rejected some time ago. Shell-fish fisheries has always some effect, but the question is whether this effect negatively influences the predatory bird population. The shell fishermen may not disturb the Wadden Sea to such an extent that the Habitat Directive of the European Union is abused. Cockle fishers argue that they are not responsible for the decline in numbers of migratory birds and cockles; this phenomenon is said to be mainly ruled by natural factors such as storms and severe winters.

This discussion makes clear that conservation biology and restoration ecology are more than 'hard' science only. The distinction between scientific and political arguments, consultation of and communication with stakeholders, public acceptance, and the willingness to pay for measures, either by public or private parties, are finally decisive for the progress of the project. Van Diggelen et al. (2002) edited a series of congress papers on ecological restoration, and they reached roughly the same conclusions.

References

Allen, E. B., W. W. Covington & D. A. Falk, 1997. Developing the conceptual basis for restoration ecology. Restorat. Ecol. 5: 275–276.

Bakker, J. P. & F. Berendse, 1999. Constraints in the restoration of ecological diversity in grassland and heathland communities. Trends Ecol. Evol. 14: 63–68.

Bakker, J. P., P. Esselink, K. S. Dijkema, W. E. van Duin & D. J. de Jong, 2002. Restoration of salt marshes in the Netherlands. Hydrobiologia 478/Dev. Hydrobiol. 166: 29–51.

Bakker, J. P. & G. Londo, 1998. Grazing for conservation management in historical perspective. In Wallis de Vries, F. M., J. P. Bakker & S. E. van Wieren (eds), Grazing and Conservation Management. Kluwer Academic Publishers, Dordrecht.: 23–54.

Bionieuws, 1998. Type I en II: over ecologische effectstudies, beslistheorie en een symmetrische bewijsvoering. Bionieuws 4,5,6 &7 (in Dutch).

Blom, C. W. P. M., 1999. Adaptations to flooding stress: from plant community to molecule. Plant Biol. 1: 261–273.

Cals, M. J. R., R. Postma, A. D. Buijse & E. C. C. Marteijn, 1998. Habitat restoration along the river Rhine in the Netherlands: putting ideas into practice. Aquatic Conservation: Mar. Freshwat. Ecosystems 8: 61–70.

Carpenter, S. R., 1988. Transmission of variance through lake foodwebs. In Carpenter, S. R. (ed.), Complex Interactions in Lake Communities. Springer: 119–138.

De Groot, S. J., 2002. A review of the past and present status of anadromous fish species in the Netherlands: is restocking the Rhine feasible? Hydrobiologia 478/Dev. Hydrobiol. 166: 205–218.

De Jonge, V. N. & D. J. de Jong & M. M. van Katwijk, 2000. Policy plans and management measures to restore eelgrass (*Zostera marina* L.) in the Dutch Wadden Sea. Helgol. mar. Res. 54: 151–158.

De Jonge, V. N. & D. J. de Jong, 2002. Ecological restoration in coastal areas in the Netherlands: concepts, dilemmas and some examples. Hydrobiologia 478/Dev. Hydrobiol. 166: 7–28.

De Melo, R., R. France & D. J. McQueen, 1992. Biomanipulation: Hit or myth? Limnol. Oceanogr. 37: 192–207.

De Waal, L. C., A. R. G. Large & P. M. Wade (eds), 1998. Rehabilitation of Rivers. Principles and Implementation. Landscape Ecology Series, Wiley & Sons, Chichester.

Ehrenfeld, J. G., 2000. Defining the limits of restoration: the need for realistic goals. Restor. Ecol. 8: 2–9.

Grootjans, A. P. & R. van Diggelen, 1995. Assessing the restoration prospects of degraded fens. In Wheeler, B. C., S. C. Shaw, W. J. Fojt & R. A. Robertson (eds), Restoration of Temperate Wetlands. Wiley & Sons, Chichester: 73–90.

Grootjans, A. P., L. F. M. Fresco, C. C. de Leeuw & P. C. Schipper, 1996. Degeneration of species-rich Calthion palustris hay meadows; some considerations on the community concept. J. Veg. Sci. 7: 185–194.

Grootjans, A. P., J. P. Bakker, A. J. M. Jansen & R. H. Kemmers, 2002a. Restoration of brook valley meadows in the Netherlands. Hydrobiologia 478/Dev. Hydrobiol. 166: 149–170.

Grootjans, A. P., H. W. T. Geelen, A. J. M. Jansen & E. J. Lammerts, 2002b. Restoration of coastal dune slacks in the Netherlands. Hydrobiologia 478/Dev. Hydrobiol. 166: 181–203.

Gulati, R. D. & E. Van Donk, 1989. Biomanipulation in the Netherlands: applications in fresh-water ecosystems and estuarine waters-an introduction. Hydrobiol. Bull. 23: 1–42.

Gulati, A. D. & E. van Donk, 2002. Lakes in the Netherlands, their origin, eutrophication and restoration: state-of-the-art review. Hydrobiologia 148/Dev. Hydrobiol. 166: 73–106.

Hobbs, R. J. & D. A. Norton, 1996. Towards a conceptual framework for restoration ecology. Restorat. Ecol. 4: 93–110.

Hobbs, R. J. & J. A. Harris, 2001. Restoration Ecology: Repairing the Earth's Ecosystems in the New Millennium. Restorat. Ecol. 9: 239–246.

Hosper, S. H., 1997. Clearing lakes, an ecosystem approach to the restoration and management of shallow lakes in the Netherlands. Ph.D. Thesis, University of Wageningen.

Huisman, J., 2000A. Natuurontwikkeling is even onvoorspelbaar als het weer. Bionieuws 17 (in Dutch).

Huisman, J., 2000B. Ecologen moeten hun hand niet overspelen. Bionieuws 19 (in Dutch).

ICPR, 2000. Rhine 2020 Programme for Sustainable Development and Action Plan 2005. International Commission for the Protection of the Rhine, Koblenz.

Kuhn, T. S., 1962. The Structure of Scientific Revolutions. Univ. Chicago.

Lamers, L. P. M. & J. G. M. Roelofs, 2000. Natuurontwikkeling is wel goed te voorspellen. Bionieuws 18 (in Dutch).

Lamers, L. P. M., A. J. P. Smolders & J. G. M. Roelofs, 2002. The restoration of fens in the Netherlands. Hydrobiologia 478/Dev. Hydrobiol. 166: 107–130.

Londo, G., 1997. Natuurontwikkeling, Backhuys Publishers, Leiden (in Dutch).

Middleton, B., 1999. Wetland Restoration, Flood Pulsing, and Disturbance Dynamics. John Wiley & Sons.

Moss, B., 2000. Biodiversity in fresh waters – an issue of species preservation or species functioning. Envir. Cons. 27: 1–4.

Nienhuis, P. H., A. D. Buijse, R. S. E. W. Leuven, A. J. M. Smits, R. J. W. de Nooij & E. M. Samborska, 2002. Ecological rehabilitation of the lowland basin of the river Rhine (NW Europe). Hydrobiologia 478/Dev. Hydrobiol. 166: 53–72.

Palmer, M. A., R. F. Ambrose & N. L. Poff, 1997. Ecological theory and community restoration ecology. Restorat. Ecol. 5: 291–300.

Postma, R., M. J. J. Kerkhofs, G. B. M. Pedroli & J. G. M. Rademakers, 1996. Een stroom natuur. Natuurstreefbeelden voor Rijn en Maas. Watersysteemverkenningen 1996: 1–102, RIZA-nota 95.060 (in Dutch).

RIVM, 1999. Milieucompendium 1999. Het milieu in cijfers. Rijksinstituut voor Volksgezondheid en Milieu (RIVM) en Centraal Bureau voor de Statistiek, Bilthoven, Heerlen (in Dutch).

Roelofs, J. G. M., E. Brouwer & R. Bobbink, 2002. Restoration of aquatic macrophyte vegetation in acidified and eutrophicated soft water wetlands in the Netherlands. Hydrobiologia 478/Dev. Hydrobiol. 166: 171–180.

Scheffer, M., S. H. Hosper, M. L. Meijer, B. Moss & E. Jeppesen, 1993. Alternative equilibria in shallow lakes. Trends Ecol. Evol. 8: 275–279.

Schropp, M. H. I. & C. Bakker, 1998. Secondary channels as a basis for the ecological rehabilitation of Dutch rivers. Aquatic Conservation: Mar. Freshw. Ecosyst. 8: 53–59.

Shapiro, J. & D. I. Wright, 1984. Lake restoration by biomanipulations. Round Lake, Minesota – the first 2 years. Freshwat. Biol. 14: 371–373.

Southwood, T. R. E., 1977. Habitat, the templet for ecological strategies. J. anim. Ecol. 45: 337–365.

Townsend, C. R., 1996. Concepts in river ecology: pattern and process in the catchment hierarchy. Ergebn. Limnol. 113, Large Rivers 10: 3–21.

Thompson, K., J. P. Bakker & R. M. Bekker, 1997. Soil Seed Banks in North West Europe; Methodology, Density and Longevity. Cambridge Univ. Press, Cambridge.

Van Diggelen, R., A. P. Grootjans & J. A. Harris, 2001. Ecological restoration: State of the Art or State of the Science? Restorat. Ecol. 9: 115–118.

Van Duren, I. C., R. J. Strykstra, A. P. Grootjans, G. N. J. ter Heerdt & D. M. Pegtel, 1998. A multidisciplinary evaluation of restoration measures in degraded Cirsio-Molinietum fen meadows. Appl. Veg. Sci. 1: 115–130.

Van Katwijk, M. M., 2000. Possibilities for restoration of Zostera marina beds in the Dutch Wadden Sea. PhD Thesis University of Nijmegen.

Van Staalduinen, C. J., H. A. van Adrichem Boogaert, M. J. M. Bless, C. J. W. Doppert, H. M. Harsveldt, H. M. van Montfrans, E. Oele, R. E. Wermuth & W. H. Zagwijn, 1979. The Geology of the Netherlands, Meded. Rijks Geol. Dienst 31–32.

Verhagen, R., J. Klooker, J. P. Bakker & R. van Diggelen, 2001. Restoration success of low-production plant communities on former agricultural soils after top-soil removal. Appl. Vegetation Sci. 4: 75–82.

Verdonschot, P. F. M. & R. C. Nijboer, 2002. Towards a decision support system for stream restoration in the Netherlands. Hydrobiologia 478/Dev. Hydrobiol. 166: 131–148.

Westhoff, V., 1945. Biologische problemen der natuurbescherming. In Verslagen van de Natuurbeschermingsdag van de NJN te Drachten: 18–30 (in Dutch).

Whisenant, S. G., 1999. Repairing damaged wildlands: a process-oriented, landscape-scale approach. Cambridge University Press, Cambridge.

Zedler, J. B., 2000. Progress in wetland restoration ecology. TREE 15: 402–407.

Zedler, J. B. (ed.), 2001. Handbook for Restoring Tidal Wetlands. Marine Science Series, CRC Press.